Rethinking the Needham Question

Raymond W. K. Lau

Rethinking the Needham Question

A Non-Eurocentric Framework Transcending Dialogism

palgrave
macmillan

Raymond W. K. Lau
Contemporary China Research Centre
Hong Kong Shue Yan University
North Point, Hong Kong

ISBN 978-981-97-9471-3 ISBN 978-981-97-9472-0 (eBook)
https://doi.org/10.1007/978-981-97-9472-0

Cover Credit: Tuul & Bruno Morandi

This Palgrave Macmillan imprint is published by the registered company Springer Nature Singapore Pte Ltd.
The registered company address is: 152 Beach Road, #21-01/04 Gateway East, Singapore 189721, Singapore

Dedicated to my deceased parents, and to all the scholars who in the past seven decades have contributed to the building of a non-Eurocentric case.

Preface

Up until fifteen months or so ago, it would never have occurred to me that I might be writing a book on the Needham Question (NQ). I was trained as a sociologist with a special interest in theoretical issues, but in the latter 2000s, I became interested in learning about ancient Chinese thoughts.[1] A substantial proportion of the literature concerned discusses ancient Chinese thoughts in comparison to Greece and Europe, hence, I was led into an investigation into ancient Greek thoughts as well. As is to be mentioned later on, it was my work on comparing ancient China with Greece that subsequently led me to the NQ.

As someone not trained in the fields of ancient Chinese thoughts, classical studies, or science history, is it not presumptuous of me to believe that I might have something new to say in these areas? To answer this question, I will have to beg the reader's indulgence and forbearance so I can explain the background that ultimately led to the writing of the present work.

On the occurrence of paradigmatic changes in the development of science, Kuhn (1996, p. 90) observes that 'Almost always the men who achieve these fundamental inventions of a new paradigm have either been very young or very new to the field whose paradigm they change'. Although I was familiar with Kuhn's work, these words of Kuhn's were never on my mind when I began in the latter 2000s what I took for granted

[1] Educated in colonial Hong Kong, under the colonial government's policy of sidelining the teaching of Chinese history and culture in primary and secondary schools, like the vast majority of our and subsequent generations, my knowledge in these areas was then minimal.

to be a purely learning journey—and it goes without saying that not every outsider entering a new field would be able to make new contributions to it. However, almost right from the beginning, I was taken by surprise by the prevalence, in both the Chinese-language and the English-language literature, of essentialist analyses of various types[2] in the comparison between China and Greece/Europe. In my theoretical training as a sociologist, I had come to regard essentialism with great scepticism, hence I was convinced that these analyses must have been mistaken. This was the first time that it occurred to me that I might not simply be learning in a wholly passive way.

Then, when I came to read the philosophers Laozi, Zhuangzi, and Xunzi of the pre-Qin period (i.e. before −221 when the Qin Dynasty was founded) in the original texts, I found that my knowledge and experience as an outsider coming to a new field were of tremendous help in the interpretation of several key issues of their philosophies. Let me illustrate with a case concerning Zhuangzi (ca. -369 to −286), selected from several similar cases because it is especially dramatic.

As a sociologist, I was much attracted by the concept of *habitus* put forward by the French sociologist Pierre Bourdieu. This concept is derived from the phenomenology of the French philosopher Maurice Merleau-Ponty, one of whose major concerns was with the dispositional in human actions. The dispositional refers to that which is non-cognitive, and, as illustrations, Merleau-Ponty referred to various skills-based activities. As a student of Chinese calligraphy and practitioner of several skills-based sports, I could quite readily come to grips with Merleau-Ponty's notion. All this was before I became interested in ancient Chinese thoughts. Zhuangzi and his disciples were fond of making use of parables, and of these parables there are a dozen or so concerned with consummate craftsmanship. Zhuangzi experts have advanced several different interpretations of these parables and their role in Zhuangzi's overall philosophy, such as one, advanced by several leading sinologists, that sees Zhuangzi as referring to the daemonic, an interpretation that I found, to put it bluntly, particularly absurd. With my previous knowledge of the notion of the dispositional, I had little problem in understanding that Zhuangzi was saying that the relationship between the consummate calligrapher's hand (to use calligraphy as illustration) and the brush, the scroll, and the ink is

[2] See Chap. 2 for a detailed analysis of essentialism and its various types.

dispositional and non-cognitive in nature. What, then, is the role of the notion of the dispositional in Zhuangzi's philosophy?

Before I came to read Zhuangzi, I had been a scuba diver for many years. One of the greatest joys that I derived from diving was the following. Beneath the water surface, there was absolute quietness, the only sound that existed was one's own breathing, there was no need for words, and because of which, one was no different from the fish and corals around. This gave me a sublimely serene feeling of being one with nature. In one well-known statement, Zhuangzi states: 'Heaven and earth[3] coexist with me, and the myriad things are one with me. Since [all] are one, is there any need for words?'. The phrase 'one with me' has been given various interpretations, such as the one which argues that 'one' refers to the primal chaos. However, Zhuangzi's meaning was crystal clear to me. As explained, the theme of the craft parables concerns the notion of the dispositional, which refers to that which is non-cognitive. Despite possessing one of the most perceptive minds that humanity ever had, Zhuangzi ironically longed to rise above the cognitive and to relate to nature without the need for words and, hence, thinking, just as in the case of the scuba diver or the consummate craftsman. More than two millennia before anyone suspected the existence of environmental issues, Zhuangzi was already laying down the most thoroughgoing environmental philosophy that the world has yet to catch up with.[4]

By the late 2010s, I felt myself to be reasonably well-grounded to put pen to paper on a comparison between intellectual developments in Greece and ancient China. This was published in 2020 under the title of *Intellectual Developments in Greece and China: Contingency, Institutionalization and Path Dependency* (Lau 2020), in which comparisons are made in the areas of natural philosophy, logical and abstract thinking, approach to the study of nature, mathematics, astronomy, and so on. The subtitle refers to a specially constructed theoretical-methodological framework, to be discussed presently, that underpins my comparison.

In the above, I have explained how my own background as an outsider to the field of ancient Chinese thoughts enabled me to bring some new

[3] The phrase 'heaven [meaning the physical sky] and earth' means the natural world.

[4] My interpretations of the parables and their role in Zhuangzi's philosophy, and of his statement 'Heaven and earth…' have since been published in two peer-reviewed journal articles, firstly, in a Chinese-language version (Lau 2019), and then, in an English version in *Asian Philosophy* (Lau 2021).

ideas to that field that have proven to be of value. Did anything similar occur in my comparison between Greece and China? Over the long process of my empirical research that lasted more than a decade, I came to realize that various observations that I was able to make could best be understood or made sense of by means of certain theories, concepts, and insights from a number of disciplines, including historical sociology, physics, the philosophy of science, evolutionary theory, neo-institutionalist theory, and so on. It will be noted that all of these disciplines, with the partial exception of the philosophy of science, are unrelated to both classical studies and the study of ancient Chinese thoughts. I put the just-mentioned theories, concepts, and insights plus some of my own observations together to construct a single, integrated theoretical-methodological framework that I believed to be useful for analysing intellectual developments, whether of a particular civilization or in a comparative context. This framework enabled the drawing of some unexpected conclusions. The following is one illustration.

Greece's atomic theory, the crown jewel of Greek natural philosophy, is often taken to be proof of the Greeks' purportedly 'unique' ability to engage in abstract and logical thinking. With reference to several components of my above-mentioned theoretical-methodological framework, I show, however, that that is a myth; instead, the theory was a contingent unforeseen outcome of a long non-deterministic process. This process began with a contingent synthesis of mystical thinking and the peculiarity of the Greek copula on the part of Parmenides, for whom the empirical was illusory, while the non-empirical was understood in largely mystical terms. During this process, a couple of conceptual errors were committed by Parmenides and Zeno (a very elementary one in the case of Parmenides), errors that Democritus took up lock, stock, and barrel, and which were indispensable to the formulation of his theory. In my framework, I call something that, on some particular criterion (such as rationality), is originally negative (positive) but which subsequently turns out to serve positive (negative) purposes, an irony of history, which, as something unforeseen, is contingent in nature. Given the semi-mystical origins of the above process, and the indispensability of Democritus' sharing of Parmenides' and Zeno's conceptual errors, the coming up of the atomic theory precisely constituted such an irony of history, as that theory was eventually to constitute the conceptual basis upon which Descartes engaged in his rational mathematization of nature. Contingency constitutes a core, systematically analysable component of my framework, while unforeseen consequence and irony of history constitute key concepts that enable the making sense of many historical developments, including intellectual developments, in a non-essentialist way, as in the case of Greece's atomic theory.[5]

After completing my book, I suggested in the 'Concluding Remarks' that my theoretical-methodological framework could also be useful for analysing the NQ, but, for various reasons, I did not then intend to go on to research into the NQ. However, my book subsequently came to the attention of some members of the Needham Research Institute and the science historian Arun Bala, a leading advocate of the non-Eurocentric dialogical perspective (or dialogism for short) on the NQ, and discussions with them followed. Of particular importance was my rewarding dialogue with Bala that continued throughout 2021 and 2022. Sometime during our dialogue, we explored the possibility of convening an international conference on the NQ. Eventually, a conference titled 'Needham's Dialogical Vision: Understanding Science as a Multi-Civilizational Outcome' was held at the Hong Kong University of Science and Technology (HKUST) on 23–24 March 2023.[6]

For that conference, I prepared a paper titled 'Revisiting the Needham Question: Beyond transmission in critiquing Eurocentrism', which provided a first, preliminary formulation of the argument of the present book. At that time, I already sensed the need to transcend the dialogical perspective, because it suffers from certain fundamental limitations (see shortly

[5] In my book, besides the atomic theory, I provide numerous other cases of such unforeseen consequences and ironies of history in relation to both Greece and China. Incidentally, in relation to the atomic theory, it is worth pointing out that contrary to the common claim, as mentioned, that the Greeks were 'uniquely' capable of engaging in abstract thinking, in Greek natural philosophy, the various *archê* substances proposed by the Milesians (Thales, Anaximander, and Anaximenes), Empedocles, Anaxagoras, and Aristotle (water for Thales, and so on) are all based upon empirical, not abstract, concepts (see note 14 of Chap. 5); Plato's notion of divine Forms constitutes an abstract concept, but it is largely a religious notion; only the atomic theory is based upon an abstract concept, but, as just explained, it was a contingent unforeseen outcome with semi-mystical origins. If one compares and contrasts this with how ancient Chinese thinkers conceptualized the ultimate constituents of empirical things (see Sects. 5.3 and 5.4 of Chap. 5), one would realize that the received wisdom in this regard is exactly the reverse of reality.

[6] The conference was organized by The Joseph Needham Foundation for Science & Civilization (Hong Kong) and the Needham Research Institute (Cambridge), jointly with HKUST, the Faculty of Asian and Middle Eastern Studies of the University of Cambridge, and the Jao Tsung-I Petite Ecole of the University of Hong Kong.

below). However, I wanted to state my position in a more restrained manner, hence, I spoke of 'transmissionism' (meaning focusing solely on the indispensability of knowledge transmissions from non-European civilizations into Latin Europe in the making of modern science) instead of dialogism as such, vaguely leaving room for the possibility of some form of the dialogical position that does not fall into the trap of transmissionism. However, after the conference and further exchanges with Bala, I came to the conclusion that there is no such room as just mentioned that exists, because dialogism's limitations are theoretical and methodological in nature, hence the necessity to point this out in clear terms and call unambiguously for dialogism's transcension.[7] It was then that I made the decision to write the present work.

The same question as mentioned in the second paragraph above needs to be raised again: As a non-science historian, can I claim to have any new contributions to make to the study of the NQ? At the risk of being ridiculed by expert science historians, I believe that, indeed, I can, and that, again, is due precisely to knowledges derived from other disciplinary areas that, as an outsider to the history of science, I am familiar with. To refer to just one example that also concerns the theoretical-methodological framework that I had constructed in my comparison between Greece and China, as a result of my investigation into the NQ, this framework has been refined and further developed. Almost all the components of this framework are derived from non-science history disciplines. Among these components, there are two that I want to highlight in the present context to illustrate the point that I am presently making. These two components are scholarly tradition, analysed with reference to neo-institutionalist theory in organizational sociology and path-dependency theory which was initially formulated in institutionalist economics, and scholarly agency, which is my own application of the sociological concept of agency to the scholarly context. Both scholarly tradition and scholarly agency constitute explanatory variables that, as analysed from the perspectives of the just-mentioned theories and concepts, can be made subject to non-essentialist accounts. To date, that these variables need not, at all, be interpreted essentialistically has not been understood by either Eurocentric or non-Eurocentric (dialogical) scholars in the debate on the NQ. Instead, both camps alike believe, mistakenly, that arguments that make reference to

[7] By transcension, I mean going beyond, not supplanting or abandoning. This will be explained in detail in the Introduction.

scholarly tradition and what I refer to as scholarly agency are necessarily essentialist. This is why scholars of the latter camp invariably shy away from these two variables, which is one important factor responsible for dialogism's intrinsic limitations as a perspective.

So much for the background to, and my justification for, the writing of the present book as a non-science historian. It must be stressed that while I do believe that my outsider background is instrumental to formulating the argument of the present work, no claim is made that anything like paradigmatic change is intended or involved. I can recall that the first time I submitted a paper to a Chinese-language China Studies journal on ancient Chinese philosophy, I encountered stiff resistance to the extent that even my writing style[8] was faulted by one of the referees who could obviously tell that I was an outsider from my writing style alone. A veteran neo-Confucian (see note 8) scholar who had read my manuscript beforehand and liked it told me afterwards that that journal, while being top in the field, was a bastion of extreme conservatism. I have much greater confidence in the openness of the contemporary field of science history, though whether or not my efforts are of any value is, of course, in the hands of the readers.

I would like to end by expressing my deepest gratitude to Arun Bala for his extensive exchanges with me throughout the process of my investigation into the NQ up to the present day. The argument of the present work was first formulated in a provisional way in a paper that I delivered at the previously mentioned 'Needham's Dialogical Vision: Understanding Science as a Multi-Civilizational Outcome' conference held in Hong Kong in March 2023; and an outline of the full development of my argument was presented at the 'Circulating Knowledge: 20 Years On. Research, Translation, Pedagogy' conference held in Halifax in August 2024. I have much benefited from exchanges with, among others, Keekok Lee, John Steele, Roger Hart, Prasanjit Duara, Mei Jianjun, Christian Daniels, Bill M. Mak, Tung-yi Kho, Hyunhee Park, Gordon McQuat, Robin Yates, Fa-ti Fan, and Anna Winterbottom, either during the preparation for or at those two conferences. I would also like to thank the anonymous referees for their very constructive and thought-provoking comments. Marion

[8]Trained in the British abstract theoretical academic tradition, I acquired its writing style, and though I was writing in Chinese, the way I wrote under the influence of that style was at odds with that practised by scholars nurtured in the century-old modern neo-Confucian tradition to which that journal belonged.

Duval and others at Palgrave Macmillan have been wonderful partners to work with. Last but not least, a special word of thanks is due to Dr. Peter Lee, a close friend and associate of Joseph Needham before the latter passed away, and incumbent chair of The Joseph Needham Foundation for Science & Civilization (Hong Kong), for his support over the past several years.

North Point, Hong Kong Raymond W. K. Lau

Contents

CHAPTER 1

Introduction

One manuscript reviewer of the present work has noted that whereas it is primarily addressed to scholars directly engaged with the Needham Question (NQ), it is of relevance to a much wider circle of scholars who are concerned with the issues of Eurocentrism and non-Eurocentrism in one way or another. We think that there are two reasons why this is so. Firstly, the present work's substantive historical analysis in relation to the NQ is underpinned by a specially designed theoretical-methodological framework for analysing intellectual development (extending in various ways to historical development generally) on a non-essentialist basis. The relevance of such a framework extends well beyond the NQ. Secondly, and relatedly, in the literature on the NQ, there is a lot of confusion, among both Eurocentric and non-Eurocentric scholars alike, concerning what essentialism and its critique precisely involve and entail in historical analysis, as a result, we often find Eurocentric scholars assuming that certain types of explanation are 'by nature' supportive of their position, while non-Eurocentric scholars, who share the same confusion, reject those types of explanation on principle thereby losing much potential, non-essentialist explanatory capability. There is, however, no reason why any particular type of explanation is necessarily essentialist, it is how any specific argument belonging to that type is conceived and argued that determines whether or not it constitutes an essentialist argument. Again, in clarifying such confusion, the present work's relevance extends beyond the

R. W. K. Lau, *Rethinking the Needham Question*,
https://doi.org/10.1007/978-981-97-9472-0_1

NQ. Hence, whereas scholars directly engaged with the NQ constitute our primary target, the present work is not restricted to them.

Returning to the substance of the present work, it is not an attempt to provide a full analysis of the rise of modern science, the subject matter of the NQ. Rather, its objective is *methodological*. Chapters 2 and 3 are methodological chapters, Chap. 2 defines the problem and its various aspects for our investigation, Chap. 3 presents our theoretical-methodological framework for analysing intellectual development including the rise of modern science on a non-essentialist basis, and the scope of the following chapters is confined to illustrating the points made in the second chapter, and applying the theoretical-methodological framework laid down in the third.

The NQ is a dual question, incorporating a positive one ('Why had modern science arisen in Europe?', hereinafter positive NQ) and a negative one ('Why had it not arisen elsewhere?', hereinafter negative NQ). The late Angus Graham, the pre-eminent English-language sinologist of modern times, perceptively pointed out that both the negative question and the dual nature of the NQ are conceptually problematic (Graham 1973). In Chap. 2, we explain what this conceptual flaw is, and show that Needham committed it because he remained very much under the influence of Eurocentrism. This becomes especially clear from our examination of the Eurocentric position in its original form with reference to Max Weber, in which it can be seen that Needham's raising of the NQ in dual form was in complete accord with Weber's conceptualization. We define Eurocentrism and essentialism precisely, and analyse essentialism into three varieties, namely, bio-psychological, social, and cultural. With bio-psychological essentialism no longer defended today, at least not openly, Eurocentric arguments take either the social-essentialist or cultural-essentialist forms. In critiquing essentialism, there is often confusion with regards to social and, particularly, cultural explanations. Such explanations need not necessarily be essentialist in nature, if and when the factors concerned can be accounted for on a non-essentialist basis, they constitute legitimate explanations. Though remaining under the heavy influence of Eurocentrism, Needham, on the other hand, called for abandoning it by arguing that Europe had benefited substantially from various transmissions of scientific (in addition to technological) knowledges from China. This was a seminal call that justifiably earned Needham his reputation.

Because both the negative NQ and the dual form in which Needham raised the NQ are conceptually flawed, our focus is on the positive

NQ. Nonetheless, given the fact that, like Needham, Eurocentric scholars invariably raise the NQ in dual form, hence, in critiquing their arguments concerning the positive NQ, it is impossible to avoid critiquing at the same time their arguments concerning the negative NQ, since these latter arguments are meant to support and complement their arguments concerning the positive NQ. As we are more familiar with China, in doing so, we restrict ourselves to its case. Hence, in some of the following chapters and in the appendix, detailed discussions on China will feature on various occasions. It should, however, be noted that our critique contained in these discussions is not meant to address, and does not constitute an analysis of, the negative NQ.

Thanks largely, though far from exclusively, to Needham's efforts in carrying out his above-mentioned seminal call, there gradually emerged a non-Eurocentric position in the study of science generally and the rise of modern science in particular known as the dialogical perspective or dialogism for short. Dialogism argues that it was through the cross-fertilizations of knowledges between different civilizations that science, including the rise of modern science, developed.[1] By the 2000s, a relatively strong case for dialogism with regard to the rise of modern science had been established. Under its challenge, Eurocentric scholars have responded in two ways, either singly or in combination. The first type of response tries to minimize the extent to which Europe had benefited from knowledge transmissions from non-European civilizations; the second type resorts to neo-Eurocentric arguments. The first type takes either or both of two forms, one is to adopt a legal definition of evidence concerning knowledge transmission, in accordance with which many knowledge-transmission claims can be dismissed out of hand without further ado; the other is to make the cultural-essentialist claim that there are 'cultural obstacles' to

[1] The term 'dialogue' was first used by Arun Bala in the title of his 2006 book. Bala has informed us in private communication that he, in turn, has borrowed the term from the title of Needham 1969. This term can be misleading in that it may be taken to imply conscious intention on the part of the parties concerned to engage in dialogue. In our understanding, Bala uses the term to simply refer to knowledge transmissions in the course of contacts between different civilizations that led to intellectual cross-fertilizations, without any necessary implication of the existence of a conscious intention to discuss with one another between individual scholars on the transmitting end and individual scholars on the receiving end of such transmissions. Upon our enquiry with him in private communication, Bala has confirmed that our understanding is correct. Despite the term's potential to mislead, we stick to it as it has become part of the lexicon concerning the NQ.

knowledge transmission. It should be noted that in both of these forms, the first type of response remains within the parameters of conventional Eurocentrism.

In neo-Eurocentric arguments, the role of knowledge inputs from non-European civilizations is, to varying degrees, acknowledged, hence the claim is no longer made, as it is in conventional Eurocentric arguments, that science was all along Europe's 'autochthonous' development, but it is argued that Europe was able to make use of non-European inputs to breakthrough into modern science as a result of certain 'unique' European 'qualities'. In speaking of 'qualities', most neo-Eurocentric arguments take the form of cultural essentialism. It is clear that although it is useful to refer to such arguments as neo-Eurocentric, there is actually very little that is 'new' in them; the only 'newness' consists in their application of conventional cultural-essentialist arguments but restricting that application to the period of the breakthrough into modern science.

Faced with the above responses, dialogism has, on the one hand, not directly tackled the responses that remain within the parameters of conventional Eurocentrism; and, on the other hand, it has been unable to provide any effective counter-argument against the neo-Eurocentric challenge. As the critic of Eurocentrism and neo-Eurocentrism in relation to the rise of capitalist modernity John Hobson once remarked to the leading advocate of dialogism Arun Bala, neo-Eurocentrism has 'stolen the thunder' from dialogism,[2] and Bala has admitted as much to us in private communication (see presently below).

Why has dialogism failed to rise to the above challenges? This has everything to do with the very *theoretical and methodological nature* of dialogism itself. Dialogism is solely concerned with showing the extent to which Europe had learnt from non-European civilizations, but it does not engage in any analysis of the phenomenon of intellectual development itself; that is why when asked by us the question of why, based on what Europe had learnt from non-European civilizations, it was able to make the breakthrough into modern science itself, Bala had to admit that this remains a 'black box'. It is for this very reason that dialogism has not been able to respond to the neo-Eurocentric challenge. Furthermore, in relation to conventional Eurocentric arguments, dialogism has never taken the bull by the horns, i.e. it has never offered any direct rebuttal of specific social-essentialist and cultural-essentialist arguments, these arguments are

[2] Private communication to us from Arun Bala.

simply bypassed. This is why not only has dialogism not been able to respond to the neo-Eurocentric challenge, it has also not tackled the above-mentioned Eurocentric responses to the challenge of dialogism that remain within the parameters of conventional Eurocentrism. The reason why dialogism has never tackled specific social-essentialist and cultural-essentialist arguments is that, like Eurocentric scholars, supporters of dialogism also believe, mistakenly, that the factors resorted to by Eurocentric scholars, such as cultural values, etc., can only be accounted for on an essentialist basis.

The above shows that as a perspective, dialogism suffers from *intrinsic* limitations, and hence needs to be *transcended*. By transcend, we mean *to go beyond, not supplanting or abandoning*. What this means is that we accept the merits of dialogism (demonstrating the indispensability of various knowledge transmissions into Europe from non-European sources in the making of modern science) as a valid and important starting point in the construction of a non-Eurocentric analysis, but argue that in order to satisfactorily resolve the positive NQ on a non-essentialist basis and decisively rebut both Eurocentrism and neo-Eurocentrism, it is necessary to go beyond dialogism in theoretical and methodological terms. To achieve this transcension, what is required is to address what dialogism precisely fails to do, namely, to provide an analysis of the phenomenon of intellectual development itself, i.e. to account for intellectual development including the rise of modern science on a non-essentialist basis.[3] To do this, we propose a specially designed theoretical-methodological framework, first formulated in an earlier project of ours (Lau 2020) and further developed for the present work.

Our framework draws upon theories, concepts, and insights from a variety of disciplines including historical sociology, philosophy of science, physics, evolutionary theory, organizational sociology, and so on, which, together with our own views, are synthesized into a particular approach. Our framework comprises the following components: (a) Sensitive dependence of historical development on starting point; (b) contingency in history as a systematically analysable explanatory variable; (c) the synthesis of

[3] One manuscript reviewer of the present work, noting dialogism's failure to respond to the neo-Eurocentric challenge and our ability to fill this gap, has suggested the term 'neo-dialogical' to us as the counter to 'neo-Eurocentric'. We are most grateful to this reviewer for her/his imaginative and tantalizing suggestion and have considered it carefully; however, we have not taken it up since our transcension of dialogism goes well beyond any 'neo-' refinement of it.

externalist and internalist explanations; (d) scholarly tradition as a self-reinforcing informal institution that can be made subject to non-essentialist analysis; (e) the distinction and relationship between the formal aspects and substantive knowledges of a scholarly tradition; (f) historical development as a path-dependent process; (g) process approach versus mechanistic approach; (h) the internal factor of scholarly agency as a non-essentialist explanatory variable.

While it is impossible to provide in this Introduction even a brief explanation of these elements and of how they are combined into our approach, a few points can be highlighted here to provide a preview, as it were. Firstly, with regard to (a) above, given the role of the Greek tradition in the rise of modern science, in explaining our approach in Chap. 3, we briefly touch upon the issue of how and why the Greeks came to establish their particular tradition, an issue that we analyse in detail in our earlier work (Lau 2020). A non-essentialist understanding of this starting point is crucial to analysing the rise of modern science on a non-Eurocentric basis. The role of contingency in historical development (including intellectual development) cannot be stressed enough; most scholars, such as Needham, are ready to accept the effect of a contingent factor or event when it cannot be denied, but are averse to making contingency an explanatory variable as such, because they wrongly believe that it is merely haphazard and cannot be pinned down in a systematic way. On the surface, 'contingent' and 'systematic' do appear to be antithetical to one another, but this antithesis is only apparent, because while the appearance/occurrence of any *particular* contingent factor/event is accidental, it does not follow that *the presence (indeed, prevalence) of contingency in general in historical processes* cannot be conceptualized and hence analysed in a systematic way. In our approach, we precisely show how contingency, as a general and prevalent phenomenon, can be constituted into a systematically analysable variable with reference to the notion of sensitive dependence on starting point and path-dependency theory.

There are two components of our framework which non-Eurocentric scholars including advocates of dialogism shy away from, almost as a matter of principle, namely, scholarly tradition and scholarly agency. This is because, like Eurocentric scholars, they believe that these variables can only be understood on an essentialist basis. This is a huge conceptual mistake, in consequence to which much potential, non-essentialist explanatory power is lost. We show how scholarly tradition can be analysed on a non-essentialist basis with reference to neo-institutionalist theory in organizational sociology and path-dependency theory in historical sociology, and, by cross-referencing Kuhn's notion of scientific paradigm, that recognition of scholarly tradition as a core explanatory variable enables the direct rebuttal of many cultural-essentialist arguments of both the Eurocentric and neo-Eurocentric varieties. It is also shown how and why scholarly agency can be given a non-essentialist interpretation, and, in Chap. 8, we show how it is precisely dialogism's failure to consider this factor, as part and parcel of its failure to analyse the phenomenon of intellectual development itself, that renders it incapable of accounting for Europe's breakthrough into modern science on the basis of what it had learnt from other civilizations.

The Greek tradition played a crucial role in the making of modern science as it had arisen in Europe, but the role of Greece in the analysis of the NQ has caused a lot of confusion among both Eurocentric and non-Eurocentric scholars alike. The nature of this confusion is encapsulated in the notion of the 'West' as purportedly a 'trans-historical' civilization beginning with Greece. In Chap. 4, it is pointed out that Greece and Latin Europe were two different civilizations, both in objective terms, and in the consciousness of the Greeks themselves, as well as that of Latin European scholars up to the late Renaissance. From Greek times up to the mid-nineteenth century, the notion of the 'West' did not exist, even as a collective identity, let alone a 'trans-historical' entity. In consequence to the Ottoman's conquest of Constantinople and the Byzantine Empire and further advance westwards, the notion of 'Europe' became the collective identity of Latin Europe (and only Latin Europe, excluding Byzantine and the Vikings—which is why we add quotes to the word *Europe* here), replacing its previous identity as (Latin) Christendom. As Latin Europe's scientific developments began to overtake that of the Arabic world, in the environment of rapidly rising racism consequent upon the success of the imperial enterprise and the intense hatred of Islam, Latin Europe's previous intellectual debts to the Arabs came to be increasingly denied while Greece was declared to be 'Europe's' 'childhood'. Science came to be seen as the purely 'autochthonous' development of this 'trans-historical' civilization originating in Greece and consummating in 'Europe'. During the eighteenth century, with France becoming the pre-eminent power and French replacing Latin as the official and high-society language, the notion of Europe shed its previous restriction to the Latin world and became understood in a vague geographical sense. Since in this sense Europe included states which differed greatly in terms of development (the main divide being between western/central Europe and eastern Europe/Russia), whereas the above-mentioned notion of a 'trans-historical' entity developing on an 'autochthonous' basis was an elitist notion that tacitly referenced only the western and central European states, a conceptual inconsistency was created. This was overcome by the invention by the French positivist philosopher Auguste Comte in the 1840s of the notion of the 'West' as a 'trans-historical' civilization that included only the French, the Italians, the British, the Iberians and the Germans, with Greece as that civilizations' origins. This notion was the basis upon which the Eurocentric view concerning the rise of modern science in its original form, as best exemplified by Weber, came to be formulated. Thus, the notion of the 'trans-historical' 'West' is a central pillar of Eurocentrism.

Today, even with Latin Europe's debts to Arabic scholarship fully restored in a supposedly post-racism age, the notion of the 'trans-historical' 'West' still enables some Eurocentric scholars to maintain the original version of Eurocentrism almost in pure form. Deconstructing this notion is, hence, crucial to the critique of Eurocentrism; conversely, by showing that Greece and Latin Europe were two different civilizations, the case for dialogism is much strengthened. In our discussions with non-Eurocentric scholars, we find many of them reluctant to do away with the notion of the 'trans-historical' 'West'. Besides inertia, one reason for that is the mistaking of Europe's *adoption* of the Greek tradition (not only in science and philosophy, but also in various cultural fields) for the *organic development of the same entity* over time at different stages. This mistake can easily be seen from the fact that no one would regard the Arabic world that adopted the Greek tradition and Greece to belong to one and the same trans-historical entity; and the reason for this is obvious, the ancient Greeks were white,[4] while the Arabs were not, which also goes to show that for those who reject racism but continue to speak of the 'West' as a 'trans-historical' entity, complexion might well still somehow carry extra-physical implications without themselves being aware of it.

Given the role of the Greek tradition, a non-essentialist understanding of how the Greeks came to establish their particular tradition is most crucial. This is especially because this issue has always been given essentialist interpretations. In Chap. 3, this issue has been addressed in a preliminary way, Chap. 5 pursues it further from another angle. In the literature, there

[4] White, black, and so on are valid categories of population groups without carrying racist implications (Malik 2008).

is a thesis concerning Greece in comparison to China which is of special significance because, firstly, it is the most sustained and systematic social-essentialist analysis concerning intellectual development available and its social essentialism takes the form of sociological reductionism, hence it provides an ideal case for the critique of social essentialism and sociological reductionism; secondly, it is exactly identical in form to Weber's original formulation of the Eurocentric position; and, thirdly, it also exactly matches the dual form in which Needham raised the NQ, which, as previously remarked, is conceptually flawed. This thesis is what we refer to as the 'agora debate' thesis advanced by the classical scholar G.E.R. Lloyd and the sinologist Nathan Sivin in a collaboration that lasted three decades until the latter's passing away in 2022. A single social factor, namely, the presence or absence of agora debate, is argued to be what accounts for a series of scholarly achievements attained by, and supposedly only by, the Greeks (the search for ultimate foundations, element theories in physics, the 'invention' of the concept of nature, and so on), which the Chinese are said to lack. We show that each and every one of Lloyd's and Sivin's claims is empirically false, which goes to show the theoretical bankruptcy of their social essentialist and sociological reductionist views. In the process of our critique, we also examine the general issues of how to critique social essentialism depending on the different circumstances of each particular argument, and of the flaws of sociological reductionism.

In making claims of knowledge transmissions from China to Europe, Needham often depended solely on chronological priority for support; this is clearly insufficient, and has justifiably been critiqued. In order to minimize the extent to which Europe benefited from knowledge transmissions from non-European civilizations, some Eurocentric scholars, however, go to the other extreme, by demanding evidence of transmission in accordance with the legal criterion and principle of the onus of proof and the presumption of innocence, as in the criminal law court. On the criterion of onus of proof, or more fully, proof beyond reasonable doubt, many knowledge transmission claims can be dismissed out of hand without further ado, and, in accordance with the principle of the presumption of innocence until proven otherwise, the presumption of independent invention when knowledge transmissions cannot be proven beyond reasonable doubt is said to constitute 'common sense'. However, it is common knowledge that in historical investigation, direct evidence often does not exist or has not survived or has not survived in sufficient amounts to satisfy the criterion of proof beyond reasonable doubt. Hence, leading scholars in various historical disciplines, ranging from archaeology, classical studies to historical linguistics, adopt a different criterion, namely, the balance of probabilities or plausibility on the basis of indirect evidence (which, incidentally, also often does not exist or has not survived), and, especially, circumstantial evidence. On this criterion, there is no need to either presume independent invention or, for the matter, transmission or diffusion, since that is wholly decided by the balance of what is more probable or plausible. Chapter 6 examines these issues, in which various cases of knowledge transmission claims and their rejection on the above-mentioned legal basis by Eurocentric scholars are examined. This examination, however, is not in order to adjudicate between those who make and those who reject the claims concerned, but to illustrate methodologically the general principles involved concerning evidence in historical investigation, especially in relation to the debate between Eurocentrism and dialogism with regard to the rise of modern science.

Given the general acceptance of the Arabic world's taking up of the Greek tradition and Hindu mathematics, and of Latin Europe's intellectual debts to Arabic scholarship, it may come as a surprise to find some Eurocentric scholars still arguing that there are allegedly 'cultural obstacles' to knowledge transmission across different cultures. But such arguments do exist up to the present day, and, what is even more surprising is that, on at least one important occasion, none other than Needham had made one such argument, even to the extent of talking about 'ethnic obstacles'. In formulating our theoretical-methodological framework in Chap. 3, we discuss how a non-essentialist understanding of scholarly tradition enables, inter alia, a clear understanding of, on the one hand, the general non-culture-specific nature of scholarly tradition, and, on the other, why there could sometimes occur, *for non-cultural reasons*, obstacles to knowledge transmission across different cultures or societies. In Chap. 7, we employ this framework to examine certain cultural-essentialist claims of 'cultural barriers' to knowledge transmission. This is done by means of a detailed examination of the reception of Euclid's *Elements* in China over the centuries, and of the introduction of the Tycho Brahe model and the telescope to China in the late 16th to early 17th centuries. By coincidence, the introduction of the telescope to China is used by the historical sociologist of science Toby Huff to advance a cultural-essentialist neo-Eurocentric 'intellectual curiosity' argument. We show how Huff's own analysis actually cancels out his own argument, and how his argument betrays an extraordinarily simple-minded conception of how scholarship develops, according to which scholarship develops like blank calculators, each and every one of which is expected to automatically produce the same output given identical inputs.

To explain how and why Latin Europe was able to make the breakthrough into modern science, one must first define what that breakthrough consists of. For our limited purposes (see above second paragraph), we do so with reference to four elements: (1) The breakthroughs achieved by Kepler and Galileo; (2) the methodical and systematic use of the experimental method to test hypotheses about nature; (3) the rise of the mechanical philosophy; and (4) the mathematization of nature. These four elements more or less parallel Cohen's thesis that the breakthrough into modern science consists of what he calls three 'revolutionary transformations', namely, the rise of realist mathematical science (roughly corresponding to our first element); Descartes' kinetic atomism (this coincides with part of our third and fourth elements); and the rise of experimental science (this corresponds to our second element). In Chaps. 8 and 9, we analyse our four elements from the perspective of our theoretical-methodological framework, in the course of which, we contrast our analysis with Cohen's in a critique of the latter.

Due to a contingent coincidence, the Islamic world of the Middle Ages was located between Byzantine and India, in close proximity to both. As a result, when it began to borrow learning from foreign civilizations, it borrowed both the theoretical tradition developed by the Greeks and what we refer to as the computational tradition of Hindu mathematics. In consequence, right from the beginning, there existed a strong trend in Arabic scholarship towards the realist direction in its adoption of the Greek corpus, as exemplified by the optical theory of Ibn al-Haytham and Arabic astronomers' reform of Ptolemy. When Latin Europe learned from Arabic scholarship, this realist trend was also absorbed. In speaking of his first 'revolutionary transformation', Cohen refers to Kepler's and Galileo's work as the transformation of the Alexandrian mathematical science of Ptolemy and others into 'Alexandria-plus', i.e. Alexandrian mathematical science made realist. Cohen's argument completely ignores the fact that late-Renaissance European scholars had absorbed Arabic scholarship's trend towards realism, hence his argument of Alexandria being directly transformed into 'Alexandria-plus' by late-Renaissance Europe is Eurocentric. Such a critique of Cohen's argument can be made within the parameters of the dialogical perspective. However, Kepler and Galileo did go beyond Arabic scholarship, an advance that cannot be explained within those parameters. In Chap. 8, we analyse, on a non-essentialist basis, how a synthesis of contingent external factors in the post-1453 era led, as contingent unforeseen consequences, to a fundamental shift in Latin Europe's scholarly climate, in which crucial changes occurred in the scholarly agency (scholarly values and drives) of late-Renaissance scholars such as Kepler and Galileo. It is from this angle, completely explicable on a non-essentialist basis, that the ability of Kepler and Galileo to go beyond Arabic scholarship can be accounted for, thereby showing the need to transcend the dialogical perspective in critiquing Eurocentrism and neo-Eurocentrism.

In his thesis, Cohen credits Bacon for his third 'revolutionary transformation'. According to him, in the 15th and 16th centuries, there developed a 'specifically European-coloured mode' of 'coercive empiricism', which Bacon 'consolidated' into the experimental method of the seventeenth century. In Chap. 9, we show that Cohen is wrong in both respects. Cohen's 'coercive empiricism' had not arisen from purely internal European dynamics, as claimed by Cohen ('specifically European coloured mode'), but from circumstances arising from the hostile relationships between Latin Christianity, Byzantine, and the Islamic world. The seventeenth century's use of experimentation to test hypotheses about nature had nothing to do with either 'coercive empiricism' or Bacon, nor was there any 'revolutionary transformation' involved at all, because experimentation had existed for a long time in alchemy which was first borrowed from the Arabs. Over several centuries, alchemy had developed experimental processes to prove the corpuscular theory of matter, but alchemy's corpuscular theory (subsequently to be called 'chemical atomism' in the nineteenth century) was experimentally based, and owed nothing to Democritean atomism, which was a physical theory. It was as part and parcel of this tradition that Robert Boyle's mechanical philosophy emerged, which was again experimentally based, and which completely dispensed with the Aristotelian notion of 'substantial form' by explaining, on the basis of the precedent set by his immediate alchemical predecessor Daniel Sennert, qualitative changes in terms of variable accidental relations between component parts.

On the other hand, inspired by Galileo's work on motion, Descartes took up Democritean atomism by giving it a kinetic interpretation. In focusing on the accidental kinetic relations between component parts (Democritean atoms), Descartes' theory, like Boyle's, though sharing no common heritage with Boyle, also embodies the mechanical worldview. To analyse the kinetic relationships between Democritean atoms, Descartes synthesized geometry and algebra so as to provide the mathematical tools required. Since algebra was borrowed from India through Arabic mathematicians, Cohen's description of his second 'revolutionary transformation' as the transformation of Athens (Athenian natural philosophy) directly into 'Athens-plus', i.e. Descartes' kinetic interpretation of Democritean atomism and its mathematization, is Eurocentric because it ignores the role played by algebra in its synthesis with geometry.

After discussing his three 'revolutionary transformations', Cohen then asks why they all occurred at around the same time. There are problems with this very question itself, but what is most interesting is Cohen's answer to his own question. This answer is of the cultural-essentialist variety because it talks about 'unique' European 'qualities' such as 'dynamism', 'curiosity', and so on; at the same time, it is neo-Eurocentric because Cohen accepts, at least verbally, the role played by borrowed knowledges from non-European civilizations, and his argument is that Europe was able to make the breakthrough on the basis of such borrowings due to the above-mentioned 'unique' European 'qualities'. Cohen's answer is nothing new at all, it is, in fact, nothing but the turning of the outdated ideological 'voyages of discovery' narrative into a pseudo-explanation. The reason why Cohen's answer constitutes only a pseudo-explanation is explained by contrasting it to our own analysis of late-Renaissance scholarly agency. In offering his pseudo-explanation, Cohen is able to recruit support from a lengthy comment made by none other than Needham that contrasts Europe with China. Not only is Needham's comment with regard to Europe also derived from the same outdated ideological narrative, we show that each and every element of his comment with regard to China is directly contrary to the simplest historical facts about China. This may surprise many readers who are unfamiliar with China and who take Needham to be the authority on China. Needham's knowledge of China's technological development is certainly formidable, but the same cannot be said of his knowledge of other aspects of China. Further, Needham was certainly aware of at least some of the contrary facts mentioned by us, but his remaining under the ideological grip of the 'voyages of discovery' narrative is such that these historical facts simply escaped him when he made his lengthy cultural-essentialist comment on China that Cohen finds so congenial.

The 'voyages of discovery' narrative pertains to the positive NQ only, since, like Needham, Eurocentric and neo-Eurocentric scholars invariably raise the NQ in dual form, an argument that matches precisely the 'voyages of discovery' narrative in form but which addresses the negative NQ to complement that narrative is required. This is provided by one half of an overall argument that we refer to as the 'maritime versus continental civilizations' (MvC) narrative, which was spawned by the 'voyages of discovery' narrative. According to the MvC narrative, Europe constituted a 'maritime civilization', 'outwardly oriented', 'lured' by and 'curious' about the sea and what lies in its far reaches, and 'fearless' of its hazards, while China, which serves as an illustration of non-Europe, was a 'continental civilization', 'inwardly oriented', 'bound' in 'self-sufficiency' to the land, 'incurious' about what lies beyond this 'self-sufficient' land, and 'shying away' from the hazards of the sea. The 'M' component of the MvC narrative is none other than the 'voyages of discovery' narrative; while Needham's claims about China made in his above-mentioned lengthy comment cited by Cohen for support constitute some of the elements comprising the 'C' component of the MvC narrative. The 'M' component has been subject to adequate critique in Chaps. 8 and 9, in view of the fact that the MvC narrative lies at the root of most cultural-essentialist arguments, which have become especially prominent with the rise of neo-Eurocentrism, a more comprehensive critique of the 'C' component of that narrative is required. We include an appendix especially for that purpose.

In the Conclusion, a chart to graphically represent our approach and overall argument is provided. With reference to an observation of the critic of Eurocentrism Jack Goody that echoes our own view, we finish with a comment on the readiness of Eurocentric scholars, from Weber to Cohen, Huff, and others, to jump to conclusions about non-European civilizations despite their (sometimes self-acknowledged) limited knowledge of these civilizations, as well as propose an explanation as to why such a readiness is rife among such scholars.

Bibliography

Graham, Angus C. 1973. 'China, Europe, and the origins of modern science: Needham's The Grand Titration'. Pp 45-69 in Shigeru Nakayama and Nathan Sivin (eds.), *Chinese Science*. Cambridge, Mass: MIT Press.

Lau, Raymond W.K. 2020. *Intellectual Developments in Greece and China: Contingency, Institutionalization and Path-Dependency*. Newcastle upon Tyne: Cambridge Scholars Publishing.

Malik, Kenan. 2008. *Strange Fruit: Why Both Sides Are Wrong in the Race Debate.* Oxford: Oneworld.
Needham, Joseph. 1969. *Within the Four Seas: The Dialogue of East and West.* London: George Allen and Unwin.

CHAPTER 2

The Needham Question, Dialogism Versus Eurocentrism and Neo-Eurocentrism, and the Need to Transcend Dialogism

First raised in the 1940s, the Needham Question (hereinafter NQ), named after Joseph Needham, is a double question, incorporating a positive question ('Why had modern science[1] arisen in Europe?', hereinafter the positive NQ) and a negative one ('Why had it not arisen in non-European civilizations such as China or India?', hereinafter the negative NQ). Needham sometimes posed the NQ in a slightly different way: Why had China been more advanced than Europe in science and technology up until around 1500, but was then overtaken by Europe? In relation to this way of putting the matter, that China had, indeed, been more advanced in technology is generally accepted even by Needham's critics; however, these critics would dispute Needham's claim with regard to science. In view of this, in order not to get sidetracked by this dispute, it will be more straightforward to address the NQ in its above dual form for the purpose of the present work.

[1] The notion of the scientific revolution occurring in the seventeenth century has been subject to dispute. On this, we adopt Lindberg's (2007, pp. 357–67) view that there is substantial continuity in important respects between pre-1600 and post-1600 developments, but it is still true that there is a qualitative breakthrough occurring around 1600 to justify speaking of post-1600 science as modern science. However, since whether or not this breakthrough constitutes a revolution depends considerably on how the notion of revolution is defined, we refrain from using the term 'Scientific Revolution' itself except in quotes from other authors. For convenience, we shall often speak of the breakthrough of ca. 1600.

R. W. K. Lau, *Rethinking the Needham Question*,
https://doi.org/10.1007/978-981-97-9472-0_2

2.1 The AWD Paradigm, the DML View, Essentialism, and Eurocentrism

Needham was not the first person to have pondered about the NQ. Before Needham, various scholars had already provided different answers to the positive NQ (Cohen 1994). However, for our purposes, it is unnecessary to consider these different accounts. As to the double NQ, also before Needham, the German sociologist Max Weber, while not specifically referring to modern science, had, in fact, provided an answer to it. Indeed, the accounts of some of the scholars, both before and after Needham, who have examined the positive NQ, such as Reijer Hooykaas, Robert Merton, and Joseph Ben-David, are, to varying degrees, inspired by Weber's Protestant-ethic thesis. Weber provides a good starting point for us, though not with reference to that thesis.

Weber is famous for his notion of rationalism (or rationality or rationalization), but his conceptualization of this notion underwent a significant change in the mid-1910s. Initially, in *The Protestant Ethic and the Spirit of Capitalism* of 1904, rationalization was understood as a unilinear process of increasing disenchantment (i.e. demystification) supposed to be specific to European civilization. This position lasted up to around 1911–13 (Whimster and Lash 1987, p. 16). In 1915, however, he re-defined rationality 'in the sense of logical or teleological consistency', in accordance with which non-European religions, same as Protestantism, were also said to be rational though in different forms from the latter (Collins 1985, pp. 70–1). In 1920, Weber published the *Collected Essays in the Sociology of Religion*, in its Introduction,[2] he advanced what we refer to as his 'forms of rationality' thesis, arguing that different civilizations had developed different (sets of) 'forms of rationality'. In accordance with this thesis, he claimed that in a whole range of cultural fields in addition to religion (science, mathematics, historiography, art, architecture, accounting, business organization, religion, and so on), the forms of rationality of the 'West'[3] were distinctive of and unique to it:

> Only in the West does science exist at a state of development which we recognize to-day as valid...in Babylonia and elsewhere astronomy lacked—...—

[2] This Introduction is reprinted in Weber 1992 (pp. xxviii–xlii).

[3] The reason why we put the term *West* and its cognates in quotes (except in quotes from other scholars) will be explained in Chap. 4.

the mathematical foundation which it first received from the Greeks. The Indian geometry had no rational proof…Indian political thought was lacking in a systematic method…A similar statement is true of art…architecture…higher education…book-keeping…. (Weber 1992, pp. xxviii ff)

Science 'recognize[d] to-day as valid' is, of course, none other than modern science. Weber's scope is much broader than Needham's, covering, besides science, political thought, art, architecture, bookkeeping and so on, but his 'forms of rationality' thesis clearly applies to science specifically besides being applicable more generally. Having stated the above, Weber went on to say: 'It is hence our first concern to work out and to explain genetically the special peculiarity of Occidental rationalism', concerning which he said:

When we find again and again that, even in departments of life apparently mutually independent, certain types of rationalization have developed in the Occident, and only there, it would be natural to suspect that the most important reason lay in differences in heredity. The author admits that he is inclined to think the importance of biological heredity very great. (pp. xxxix, xlii)

Weber's admission that he was inclined towards the racist view is now downplayed by contemporary Weberians, but in the context of the early twentieth century, this should not come as too much of a surprise.[4] Weber then admitted that science had not yet been able to prove 'biological heredity', hence, in the meantime, the sociologist would have to make do with seeking an explanation of the 'peculiarity of Occidental rationalism' 'from the peculiarities of the social structure of the Occident' (p. xxxvii) in comparison with that of other civilizations, i.e. with seeking a social explanation.

[4] It seems possible that Weber's re-definition of his concept of rationality and advancing of the 'forms of rationality' thesis were made under the influence of the French philosopher-turned-sociologist Lucien Levy-Bruhl. In 1910, the latter published a work titled in literal translation 'mental functioning in inferior societies' (the English translation is titled *How Natives Think* [Levy-Bruhl 1985]), in which he argued that Europeans and non-Europeans possessed different 'mentalities' or ways of thinking. Weber's 'forms of rationality' clearly echo Levy-Bruhl's 'mentalities'. Incidentally, Levy-Bruhl's thesis was severely critiqued by his mentor, the French sociologist Emile Durkheim, who, like Weber, was one of the founders of sociology as a discipline.

Thus, for Weber, as part and parcel of his 'forms of rationality' thesis, modern science emerged in Europe but not in other civilizations (the double NQ) because of their different 'forms of rationality', which derive ultimately, Weber believed, from differences in 'biological heredity', and certainly from differences in these civilizations' social structures. Weber's answer is relevant to the consideration of the NQ right up to the present day, hence, it bears examination in some detail.

In Weber's conception, different civilizations are like different impervious water-tight compartments, with each *independently* and *autonomously* developing its own particular 'forms of rationality' due to *purely internal* factors. If inter-civilizational cultural exchanges are acknowledged within this conception, such exchanges are assumed to have no significant effect on each civilization's own independent and autonomous development of its particular 'forms of rationality'. We call such a conception, with reference to the 'West', the *autochthonous 'Western' development (AWD*[5]*) paradigm*. To refer to modern science, the AWD paradigm believes that modern science arose out of the autochthonous development of the 'West' as a result of the working of purely endogenous or pristinely or uniquely[6] 'Western' factors. These factors fall into several categories, what Weber called 'biological heredity', social factors (Weber's 'the peculiarities of the social structure'), and, adding to Weber's schema, cultural factors. We refer to the beliefs in 'biological heredity' *bio-psychological essentialism*, in social factors *social essentialism*, and in cultural factors *cultural essentialism*. In the above quotes, Weber was a believer in the former two, though he was certainly also a believer in the last as well. With racist theory now abandoned by most scholars, contemporary AWD paradigm supporters mainly rely on social and/or cultural essentialism.

The above beliefs are called essentialist because they speak of a purported 'essence', whether of a 'racial', social, or cultural kind. Thus, whereas the term 'essentialism' can be used in several different though inter-related senses, in the present work, except indicated otherwise, *essentialism* refers to the view that a particular society's development is due to factors *unique to that society which have arisen in that society through purely endogenous factors or dynamics*—it is these factors or dynamics that

[5] With reference to note 3 above, since putting a letter in an acronym in quotes is odd and inconvenient, we use AWD instead of A'W'D.

[6] The terms 'endogenous', 'internal', 'pristine', 'unique' and their cognates are used synonymously in the present work.

constitute that society's 'essence'. If this definition of essentialism seems clear enough, what essentialism does and does not imply or entail is often less than well understood, hence it is necessary to consider this with some care.

Firstly, in the historical explanation of a particular civilization's intellectual development, factors purely internal to that civilization do, of course, figure, and taking them into account does not necessarily constitute essentialism. To explain, let us take a simple example which is to be discussed in Chap. 3, namely, the unique and peculiar feature of ancient Greek that it had only one existential verb which was also used as the copula.[7] This feature had an important effect on the trajectory of Greece's intellectual development. As something purely internal to Greece, it can only be regarded as a contingent fact. This is so even were one to take the once popular view, which is of a racist nature, that language is a purported embodiment or manifestation of some alleged bio-psychological qualities, since, to mention just one obvious point, as just observed (see note 7), the same feature is absent in other Indo-European languages. Further, in essentialist explanations, the purely internal factors posited are viewed, if only implicitly, in ***normative*** or ***value-laden*** terms as something positive. For instance, as we discuss in later chapters, contemporary essentialists speak of Europe's purportedly unique 'dynamism', 'curiosity' and so on as being responsible for the rise of modern science; in such an argument, while its protagonists do not say so or may even verbally deny it, there is a clear tacit implication that these are positive, value-laden qualities. In contrast, a purely contingent fact such as the above can carry no value-laden, either positive or negative, implications. In sum, in the case of the Greek copula, it is clear that taking into consideration a purely internal factor carries no essentialist implications. Thus, critics of essentialism need to be aware that critiquing it in no way entails any ***categorical*** refusal to consider purely internal factors. We discuss this point further shortly.

Similarly, it is entirely possible to engage in cultural explanation or social explanation without falling into cultural essentialism or social essentialism respectively. Thus, it is completely legitimate to argue that a particular cultural value of a certain civilization's intelligentsia played a contributory role in that civilization's intellectual development, if, for

[7] In other Indo-European languages, the existential verb also served as the copula, but, to take English as an example, in addition to the English verb-to-be, there is an additional existential verb, namely, to exist.

instance, that value had arisen as a result of that civilization's interactions with other civilizations, then such an argument would not constitute cultural essentialism. Similarly for social explanation. For instance, we have argued elsewhere (Lau 2020) that China's early development of a sophisticated social division of labour led to what we call a divorce between mathematics and philosophy, with mathematics becoming the special vocation of hereditary lineages while philosophers were, without exception, untrained in mathematics beyond elementary level. This divorce had significant effects on China's intellectual development in comparison with Greece. In this case, a social institution (mathematics becoming the special vocation of hereditary lineages) is offered as an explanatory variable, but it cannot be considered as constituting any 'unique' 'essence' that marks China off from other civilizations, since it had simply arisen under the above-mentioned circumstances that cannot imaginably be given any essentialist reading.

In the above Greek copula illustration, it is shown how consideration of a purely internal factor need not carry essentialist connotations because this feature of ancient Greek is simply a contingent fact. Let us pursue this illustration a little further. Whereas this feature of ancient Greek did play an important role, it was only one among a number of equally important factors. In essentialist explanations, however, the purely internal factors posited are regarded as the factors *that make the difference* in accounting for the *explanandum*, while any other factor that might be involved is regarded as subsidiary. Further, these factors are supposed to be what marks one civilization off from other civilizations, in other words, they are supposed to constitute the former civilization's endogenously engendered '*essence*' in contradistinction to other civilizations.

In sum, one needs to be very careful as to what essentialism precisely means, implies and entails; critics of essentialism should understand that critiquing essentialism does not rule out the consideration of either purely endogenous factors, social explanation, cultural explanation, or some combination of these. As the present work shows as we proceed, there can be non-essentialist consideration of purely internal factors, and non-essentialist social and cultural explanations.

To continue with Weber, because in his conception, each civilization's development is autochthonous, the 'forms of rationality' that are 'unique' to the 'West' must, therefore, be lacking in other civilizations. We call such a view, with reference to the non-'West', the ***development-marked-by-lacks (DML) view***, i.e. the view that various value-laden developments found in the 'West' are lacking in the non-'West'. It should be noted that under the AWD paradigm's assumption that different civilizations possess different 'essences' unique to each of them, the DML view and that paradigm logically imply and are complementary to one another. When Weber advanced his thesis in the early twentieth century, European knowledge of the cultural developments of non-European civilizations was minimal, hence, it comes as no surprise that many of Weber's DML claims have since been shown to be mistaken, though at conferences that we attended in the mid-2010s, we still came across contemporary Weberian sociologists taking them to be true. To give just a few examples with reference to China, about which we are more familiar, with regard to Weber's claim that only Greek astronomy was mathematically based (see first quote from him above), as is now widely known among sinologists, ancient China initially had both a geometrical model and an algebraic model of astronomy; with regard to his claim that double-entry bookkeeping was unique to Europe, China had also developed an indigenous double-entry bookkeeping system, though it is not identical to the Italian system (Li Xiaolin 1981, 1988; for a brief account in English, see Lin 1992); with regard to his claim that proof was unique to Greece (see above first quote from him), not only did mathematical proof exist from very early on in ancient China, it is now well-known among science historians that mathematical proof, in different forms, existed in various other ancient cultures as well (Chemla 2018).

A term closely related to essentialism is *Eurocentrism*. Like the former, this term can be used in a number of different though inter-related senses, but in the present work, except indicated otherwise, it is understood in the sense of the belief that the development of the 'West',[8] such as the rise of modern science, is due to allegedly 'pristine' 'Western' factors. Hence, Eurocentrism is inherently essentialist, and it hardly needs to be pointed out that both the AWD paradigm and the DML view are Eurocentric.

[8] Europe in the term Eurocentrism refers not to Europe in the geographical sense, but is understood in the civilizational sense and refers to the 'West' (see Chap. 4).

2.2 The Janus-Faced Needham 1: Needham Remaining Under the Influence of Eurocentrism

To return to Needham, it is worth noting how he came to raise the NQ. Needham was originally a biochemist, in 1937, he received three research students from China to study under his supervision. One of them, named Lu Gwei-djen, later recounted: 'As he [Needham] afterwards wrote, the more he got to know us, the more exactly like himself in scientific grasp and intellectual penetration he found us to be; and this led his inquisitive mind to wonder why therefore had modern science originated only in the western world?' (quoted in Winchester 2009, p. 37). In other words, Needham had previously subscribed to the bio-psychological essentialist view concerning the rise of modern science—incidentally, this indirectly shows how common such a view was in Euro-American scholarly circles up to at least the 1930s and 1940s. It was his realization that this view was mistaken that led Needham to raise the question named after him.

Concerning the dual nature of the NQ, the late Angus Graham (1973, p. 51), the pre-eminent sinologist of modern times, once commented that the negative NQ is conceptually problematic. His comment is worth quoting in full:

> One does not ask why an event did not happen unless there was reason to expect it, and nothing in the conditions even of Europe in the sixteenth century justifies thinking of the Scientific Revolution as an event due at a certain point of maturation,[9] as though civilization were an organism *with stages which it passes through unless its development is arrested* (our emphasis). In the absence of grounds for expectation I explain why a house did catch fire (because someone left a cigarette burning), do not go through all the other houses in turn explaining why they did not catch fire (no one was smoking, the wiring was sound, there were no bombs, no lightning). The difference follows from the fact that like effects may have unlike causes; if the event does happen we can select from the possible causes, if it does not we may not be able to enumerate all the unrealized possibilities.

[9] This observation that conditions even of Europe in the sixteenth century do not justify any expectation that the breakthrough of ca. 1600 was going to occur is consistent with the view of most contemporary historians of modern science.

Graham correctly pointed out that one does not ask why an event did not happen unless there are reasons to expect it to happen, and then remarked that one way in which such an expectation can be entertained is to assume civilization to be like an organism that passes through various stages of growth. The further assumption implied in Graham's comment is that all civilizations constitute organisms of the *same* kind and, hence, are expected to pass through the same growing stages. Such an assumption embodies none other than the unilinear view of history, a product of the Enlightenment that Needham obviously shared. This further implies that Europe's development constitutes the *norm* or *standard* that all other civilizations are expected to conform to sooner or later. Needless to say, such a view is Eurocentric, in a sense different from though inter-related to the one used in the present work, i.e. in the sense of using Europe as the yardstick to measure other civilizations. Graham then observed, in terms of our terminology, that under such an assumption, if a certain non-European civilization has not conformed to the norm or standard, then it can only be because its development has been arrested, i.e. there must have been some factors *inhibiting* it from conforming. In relation to this, Graham (1973, p. 52) remarked further:

> When we ask the negative question, we assume that there are necessary conditions for a Scientific Revolution which were *present in Europe but absent in China* (our emphasis).
>
> But the questions [the positive and negative NQs] remain two sides of the same coin only if we think of the event as having a single cause which is both necessary and sufficient...

We may replace 'single cause' in the second passage with 'a single cause or a set of causes' without altering Graham's meaning; moreover, it will not be assuming too much to believe that by 'present in Europe', as necessary or necessary and sufficient condition(s), Graham implied 'present' *due to internal European dynamics.*[10] Thus, in Graham's view, Needham's very raising of the negative NQ and posing of the NQ in dual form betray the following line of thinking: Every civilization is expected to outgrow

[10]We underline 'due to internal European dynamics' because there is nothing wrong, as our previous discussion on what the critique of essentialism does and does not imply or entail shows, with a 'present in Europe but absent in China' argument per se; it becomes problematic only when it is interpreted in essentialist terms to answer the double NQ.

pre-modern science to the development of modern science; Europe was able to live up to this expectation due to factors internal to Europe; China failed to live up to this expectation not only because the just-mentioned factors found in Europe were absent in China, there were factors internal to China which inhibited it from living up to this expectation. In short, though Graham did not speak in such terms, Needham remained under the heavy influence of Eurocentrism, both in the sense used in the present work, and in the sense of using Europe as the standard to measure other civilizations. This is a remarkable insight for Graham was able to conceptually discern it from Needham's very raising of the negative NQ and the dual nature of the NQ.[11] Given the conceptual problems with the negative NQ and the dual nature of the NQ, *in our own attempt to answer the NQ in later chapters, our focus will be on the positive NQ.* Notwithstanding that, however, since, like Needham, Eurocentric scholars invariably raise the NQ in dual form, hence, in critiquing their arguments concerning the positive NQ, it is impossible to avoid critiquing at the same time their arguments concerning the negative NQ, since these latter arguments are meant to complement their arguments concerning the positive NQ. As we are more familiar with China, in doing so, we restrict ourselves to its case. Thus, in some of the following chapters and in the appendix, detailed discussions on China will feature on various occasions. It should, however, be noted that our critique contained in these discussions is not meant to address, and does not constitute an analysis of, the negative NQ. Meanwhile, let us first return to Needham.

Confirmation of Graham's just-mentioned remarkable insight can be found none other than in Needham himself. Firstly, concerning the point about 'inhibitory factors', in his book proposal dated 15 May 1948 to Cambridge University Press to publish what became *Science and Civilization in China* (hereinafter SCC), Needham stated the following under the heading of 'Statement of the Problem':

> Why did their [the Chinese's] science always remain empirical, and restricted to theories of primitive or medieval type? What were the *inhibiting factors in*

[11] Over thirty years after Graham, Goody (2006, pp. 128, 130) observed: 'when he [Needham] discusses the advances of western sciences in recent centuries, he falls back upon accepted notions of the uniqueness of the Renaissance and the rise of the bourgeoisie, of modernization, of capitalism…'; 'Needham's view not only of the Renaissance but of the development of capitalism is not only Eurocentric but follows Weber'. Unlike Graham, Goody arrived at his observations on the basis of substantive arguments made by Needham.

> *their civilization which prevented* the rise of modern science in Asia? (Winchester 2009, p. 171, our emphasis)

Disregarding whether or not, or to what extent, Needham's substantive comments on China's science are valid, not only did Needham speak of 'inhibiting factors...which prevented', he spoke of such factors as being endogenous to China ('in their civilization'). Secondly, over his career, Needham had hypothesized various 'present in Europe but absent in China' factors which, he believed, were responsible for the rise of modern science in Europe. In our view, these arguments of his are all faulty, but for our present purposes, there is no need to go over them one by one, a few brief illustrations will be sufficient. With his Marxist sympathy, the first suggestion that he made, initially in the 1940s and which he retained all his life, is that modern science had arisen in Europe due to the rise of the bourgeoisie to power, which occurred in, and *only* in, Europe. In 1931, the Soviet physicist Boris Hessen proposed that the rise of modern science was in response to the needs of incipient capitalism which threw up distinct problems in several areas of technology such as navigation and ballistics (on the Hessen thesis, see Cohen 1994, pp. 328–33). Under the influence of the Hessen thesis, Needham argued as follows:

> the rise of the merchant class to power...was the indispensable accompaniment and sine qua non of the rise of modern science. (Needham 1969, p. 150)

> Under bourgeois dominance modern science came to birth. I have described how propitious was the mercantile atmosphere for careful and accurate measuring, recording and testing—first this factor and then that, to see which was the more profitable. In that way the elimination of variables became a method, and scientific method was born. (Needham et alia. 2001: 230)

We shall briefly address the Hessen thesis in Chap. 8, meanwhile, the just-quoted statements provide a good illustration of Needham's proneness to make sweeping claims that are clearly at odds with simple historical facts.[12] With regard to 'the rise of the merchant class to power' and 'under bourgeois dominance', the English Civil War did not result in a parliamentary

[12] Ironically, this is in striking contrast to his attention to detail in his empirical investigations into various aspects of China's technological developments.

victory until 1651, half a century after modern science, in Needham's own reckoning, had begun, while Galileo lived in pre-bourgeois Italy. With regard to the scientific method of the elimination of variables, Europe's experimental method of the seventeenth century developed from a long tradition in alchemy, originally diffused from the Arabs (see Chap. 9), not European merchants.

Needham (1969) also argued that the belief, found *only* among the Abrahamic religions, in the Creator-God as divine legislator may have been necessary for the genesis of the notion of law of nature. In this relation it is worth mentioning that whereas Needham's knowledge of Chinese technology was formidable, his knowledge and understanding of, and judgment on, many other aspects of China either leaves much room for desire, or are flatly mistaken. His claim that China did not have the notion of law of nature is a case in point. By the Spring and Autumn period (−711 to −476), Chinese thinkers had arrived at the notion that nature (*tian*天) operates according to a cyclical pattern or mechanism, which they called 'the way of operation of nature' (*tian dao* 天道), in which *dao* means way of operation. Initially, 'way of operation' was probably understood in the sense of regular pattern only, but textual evidence shows that by the Warring States period (−475 to −221) at the latest, it was also understood in the sense of mechanism. In the chapter 'The operating of *tian*' ('*Tian yun*'天運) of the *Zhuangzi* (《莊子》), dating from the -4th to early -3rd centuries, it is stated: 'Who is in charge of it [the operating]? Who is sustaining it? … Or is there a mechanism (*jijian* 機緘) that is responsible for it such that it operates necessarily? Is it that it goes on operating and cannot stop?' (Chen 2012a, p. 295). Here is a clear statement that nature operates necessarily in accordance with its own naturalistic physical mechanisms. Such a notion is as close to the modern notion of law of nature as one can find in the ancient world. As Graham observed in one of the above quotes from him: 'like effects may have unlike causes', even if it was true (which we regard as doubtful) that in Europe, the notion of law of nature had something to do with the belief in the divine legislator, it does not follow, as Needham implicitly assumed, that the same or an equivalent notion could not have arisen where such a belief did not exist. In believing that the notion of law of nature was *unique* to Europe and attributing it to an *endogenous* European factor, Needham was wrong in both respects.

Finally, Needham claimed that the 'West' (by which he included Greece) had a 'predilection for atomism'. The essentialist undertone, or even overtone, of this claim is clear. However, it is, again, contrary to basic historical facts. Atomism was a minority position throughout antiquity; 'Aristotle's natural philosophy [embodying the holistic-organic worldview] was completely dominant in Western Europe between 1200 and 1450. It had no rivals', and it was not finally abandoned until the late seventeenth century (Grant 1996, pp. 162, 203); the introduction of atomism, alongside hermeticism, Neoplatonism and so on, into Latin Europe[13] in the mid-fifteenth century by Byzantine scholars fleeing Constantinople as the Ottomans expanded westwards, did not immediately lead to atomism's rise. 'At the beginning of the seventeenth century...atomism was not widely held in Europe. Indeed, precisely the opposite was accepted by most of the learned community as an article of faith. Material change was generally explained...by the imposition and removal of immaterial forms [an Aristotelian notion]' (Newman 2006, p. 4). Certainly, in the early seventeenth century, atomism began to be taken up, most influentially by Rene Descartes, but by then Galileo had already been making his breakthroughs.[14] Even after that, Webster (1980, p. 2) observes: 'The self-evident impact of Neoplatonism in England after 1660 should discourage any attempt to describe science at the time of the Royal Society in terms of the unquestioned dominance of the "mechanical philosophy"'.[15]

[13] In the present work, Latin Europe refers to the part of Europe (in the geographical sense) in which Latin was both the official and scholarly lingua franca; in the case of scholarly writing, this period lasted until the late seventeenth century, hence covering the entire period with which we are concerned.

[14] As a matter of fact, the rise of the mechanical philosophy in the 1650s was due to alchemical experimentation, instead of the physical theory of Democritean atomism, which was the theory taken up by Descartes. There will be a detailed discussion of this in Chap. 9.

[15] Conversely, Needham claimed that 'the *philosophia perennis* of China was an organic materialism...The mechanical view of the world simply did not develop in Chinese thought' (Needham 1969, p. 20). Together with his claim of the 'Western' 'predilection for atomism', his whole argument sounds decidedly like, and is analogous to, the Weberian 'European versus Chinese "forms of rationality"' argument. Further, it is also clear that, in Needham's argument, 'organic materialism' constitutes, in his scheme of things, an 'inhibitory factor'. But just as his claim concerning the alleged 'Western' 'predilection for atomism' is wrong, so is his claim concerning China. Although this is not the suitable context to go into details, let it be mentioned in passing that whereas the holistic-organic worldview (Needham's 'organic materialism') was indeed dominant, China had developed the mechanical worldview from the -fourth century onwards, a worldview that was upheld by leading thinkers throughout the centuries.

To conclude, despite his rejection of bio-psychological essentialism, Needham remained under the heavy influence of essentialist thinking. Be that as it may, however, as Needham's empirical knowledge of the development of Chinese technological and scientific thoughts deepened, a contrary line of thinking came to be formed by him.

2.3 The Janus-Faced Needham 2: Needham's Non-Eurocentric Rivers-and-Sea Metaphor

In this line of thinking, he argued that over the centuries, there was a steady flow of knowledge (both technological and scientific) from China to Europe, which much enriched the development of science in Europe. Over the years, Needham made numerous such China-to-Europe knowledge transmission claims, the substantiation of which varies in strength from one case to another. In the present context, our interest is not in these individual claims as such but in the point that they embody a ***non-essentialist*** line of thinking on Needham's part, i.e. in making these claims, he no longer believed that each civilization's intellectual development was derived from purely internal dynamics, but involved cross-fertilizations with the intellectual developments of other civilizations.[16]

Needham's conviction that non-European civilizations such as China also had a rich tradition of scientific (in addition to technological) development led him to advance the following well-known metaphor, which he repeatedly referred to over the decades: 'the river of Chinese science flowed, like all other such rivers, into the sea of modern science' (Needham 1981, p. 9); 'Modern science is indeed composed of contributions from all the peoples of the Old World, and each contribution has flowed continuously into it, whether from Greek and Roman antiquity, or from the Islamic world or from the cultures of China and India' (Needham et alia. 2001, p. 25). Hereinafter, we refer to this as the rivers-and-sea metaphor. There are various problems with this metaphor, but we shall not go into them here; the important point about it is its non-Eurocentric standpoint.

[16] Needham did not only speak of China-to-Europe knowledge transmissions, he also mentioned China learning from other civilizations. For instance, in SCC vol. 3 (Needham and Wang 1959, p. 246) he speculated, though wrongly (see Chen 1996, pp. 171–90), that Chinese astronomy originally developed from Babylonian diffusion.

To sum up with regard to Needham's contributions to the study of the history of science, Harbsmeier (1998, p. 420), today's premier sinologist concerning logical thinking in ancient China, has remarked that 'thanks to Joseph Needham more than anyone else, China is now emphatically on the world map of the history of science and intellectual history'. This is no exaggeration, despite the fact that Needham's various specific explanations concerning the rise of modern science (some of which have been discussed above) are generally not taken seriously by science historians.[17] Further, Needham's non-essentialist argument that Europe had benefited substantially from knowledges, in both technology and science, transmitted from China in its development towards modern science has proved to be a great inspiration to many scholars, inducing them to investigate the role of cross-fertilizations between different civilizations in the development of science in general, as well as concerning the rise of modern science specifically. By any standards, Needham's above two achievements are monumental. Nevertheless, the remaining essentialist strain in his thinking must not be overlooked. As we shall see on a couple of occasions in later chapters, Floris Cohen, a leading contemporary Eurocentric scholar on the rise of modern science, has resorted to none other than the authority of Needham's essentialist observations to find support for two central essentialist arguments of his, one querying the possibility of inter-civilizational cross-fertilization of scientific ideas,[18] the other concerning the driving factors behind the breakthrough of ca. 1600 itself.

2.4 Needham's Non-Eurocentric Legacy and the Emergence of the Dialogical Perspective

Following Needham's lead, other scholars also began to recover and restore the development of science in other non-European civilizations besides China, thereby showing that science was a widespread phenomenon. Thus, in 1968, Seyyed Hossein Nasr published *Science and*

[17] In his preface to Part II of SCC VII (the series' last volume), the historian of science Mark Elvin remarks: 'What is hard to come to terms with, almost half a century after the appearance of the first volume in 1954, is the limited assimilation of Needham's work into the bloodstream of the history of science in general; that is, outside the half-occluded universe of East Asian specialists and a handful of experts sensitive to the decisive contributions of comparisons' (Elvin 2001, p. xxv).

[18] This may come as a complete surprise, since did Needham not precisely stress the role of such cross-fertilization? We must admit that we were also completely astounded by this.

Civilization in Islam (Nasr 2003); shortly afterwards, Bose et al. (1971) published *A Concise History of Science in India*. In volume 3 of SCC, published in 1959, Needham and his collaborator Wang Ling discussed Chinese geography and cartography from a comparative perspective, this subsequently led to a flourish of interest in investigating into the cartographic traditions of different civilizations. Of these efforts, the multi-volume *History of Cartography* edited by John B. Harley and David Woodward (1987, 1992, 1994, 1998) that began publishing in 1987, and is still ongoing, is the most significant.

Besides works that recover and restore non-European scientific traditions, some scholars also began to advance analyses that challenge Eurocentric accounts of the history of intellectual developments, including scientific developments. One significant work of this kind concerns Greece. As we discuss in Chap. 4, the ancient Greeks openly acknowledged their intellectual debts to Egypt and west Asia, but for almost two centuries from the late eighteenth century onwards, these debts progressively came to be denied in European academia. Efforts made by Michael Astour and Cyrus Gordon in the 1960s to restore the role of west Asia were dismissed in classical studies circles. In 1987, Martin Bernal, who was not a classical scholar by training, published *Black Athena: The Afro-Asiatic Roots of Classical Civilization, Volume 1 The Fabrication of Ancient Greece*; and in 1991 *Black Athena: The Afro-Asiatic Roots of Classical Civilization, Volume 2 The Archaeological and Documentary Evidence*, in which he attempted not only to restore the role of west Asia, but that of Egypt as well. Thus, while Needham argued, in his non-essentialist moments, that Latin Europe's intellectual development was not autochthonous, Bernal argued that neither was Greece's.

Bernal's two books proved to be hugely controversial. He was severely critiqued by various scholars from the disciplines of classical studies and science history, and in 1996, a collection of such critiques was published as *Black Athena Revisited* (BAR), edited by Mary Lefkowitz and Guy Rogers. Bernal learnt of the plan to publish BAR and approached its editors to be given the opportunity to respond to his critics in the same collection, a request that the editors refused (Bernal 2001, p. ix). Bernal's responses were eventually collected in *Black Athena Writes Back: Martin Bernal Responds to his Critics*, published in 2001. We are not in a position to adjudicate between Bernal and his critics with regard to many aspects of the debate between them. However, much of that debate concerns issues that are irrelevant to us (such as Bernal's claims that parts of Greece were once colonized by Phoenicia and Egypt, and claims concerning Egyptian roots for some ancient Greek words that lacked Indo-European etymologies). What can be said, with regard to what is relevant to us, with certainty is the following. Firstly, it is a fact that there had previously been a general denial, whether or not justifiably on scholarly grounds, of Greece's cultural and intellectual debts to west Asia and Egypt, and efforts made in the 1960s to restore Greece's debts to west Asia were, again, whether or not justifiably on scholarly grounds, dismissed. Secondly, several years after Volume 1 of *Black Athena* was published, the heavyweight classical scholar Walter Burkett (1992) published *The Orientalizing Revolution: Near Eastern Influence on Greek Culture in the Early Archaic Age*—note that it was during the Archaic Age that Greek philosophy was founded in Ionia on the west Anatolian coast by the Milesians, Heraclitus, Xenophanes, and the early Pythagoras; and ten years after Volume 1 of *Black Athena*, another major classical scholar Martin West (1997) published *The East Face of Helicon: West Asiatic Elements in Greek Poetry and Myth.* These works have put Greece's cultural debts to west Asia beyond dispute. Thirdly, whether or not the previous general denial had been made on good scholarly grounds can be seen from Burkett's (1992, p. 34) following remark:

> It is well-known that a large part of the Greek vocabulary lacks any adequate Indo-European etymology[19]; but it has become a fashion to prefer connections with a putative Aegean substratum [i.e. the indigenous tongue before Indo-European-speaking Hellenes arrived in Greece] or with Anatolia parallels [Anatolians speak Indo-Hittite, closer to Indo-European than Semitic, spoken by the Phoenicians], which involves dealing with largely unknown spheres, instead of pursuing connections to well-known Semitic languages...Anti-Semitism was manifest in this case.

That is to say, *precisely as Bernal had argued*, for the *extra-scholarly* reason mentioned by Burkett, scholars had previously disregarded obvious leads suggested by scholarly considerations and instead preferred to deal 'with largely unknown spheres'. Fourthly, to keep things simple by confining ourselves to the case of west Asia in the present context, all by itself, the restoration of Greece's cultural debt to west Asia shows that Greece's cultural (including intellectual) development was not autochthonous, which

[19] This is one of Bernal's main arguments for the cultural influence of west Asia and Egypt on Greece.

is the most fundamental point that Bernal wanted to make that concerns us.[20] Although Bernal's subject matter is confined to Greece, his fundamental argument concerning Greece is crucial for the critique of the Eurocentric view concerning the positive NQ in general, as we shall see in Chap. 4.

From the late 1950s to the 1980s, a most significant development concerning our present discussion occurred. This began with the accidental discovery at Oxford's Bodleian Library of a model of the moon of the previously unknown fourteenth-century Damascus astronomer Ibn al-Shāṭir. This model that pre-dated Copernicus' model by a century was identical to the latter. This discovery sparked off a wave of research by scholars including Victor Roberts, Edward Kennedy, Otto Neugebauer, Noel Swerdlow, and George Saliba, as a result of which, by the 1980s, evidence generally accepted as incontrovertible had been found showing that the models of the five visible planets, the sun and the moon that Copernicus put forth in the early sixteenth century were all taken from al-Shāṭir (Saliba 1994, 2007).

Al-Shāṭir's work was itself based upon the achievements of two preceding Islamic astronomers of the thirteenth century, namely, the Persian Naṣīr al-Dīn al-Ṭūsī, who successfully combined two circular motions into a straight motion in a theorem known in English as the Ṭūsī Couple; and Mu'ayyad al-Dīn al-'Urḍī of the Maragha observatory, who established a new theorem known in English as Urḍī's Lemma with reference to Ptolemy's equant model. With the help of these two theorems, al-Shāṭir was able to convert the eccentric and equant models of Ptolemy into the latter's epicycle-on-deferent model, and in conjunction with modifying Aristotle's physical model, the new model was able to overcome the greatest objection to Ptolemy, namely, that his models were merely mathematical devices without physical possibility. It is, hence, unsurprising that of the four theorems used by Copernicus in his above-mentioned models, two are from Greece, while the other two are precisely the Ṭūsī Couple and Urḍī's Lemma.

[20] The highly respected Goody (2006, pp. 60–5), an anthropologist and not classicist by training, has a similar assessment of Bernal: He differs from Bernal on various details, but agrees with his main arguments. Incidentally, we have spent considerable time on the positive contributions of Bernal because we have come across non-Eurocentric scholars who, influenced by the hostility towards Bernal on the part of various classicists, dismissed him offhand without even reading him, despite Bernal's obvious relevance to the non-Eurocentric case.

Astronomy is only one of the many fields, in both science and philosophy, in which Latin Europe owed an immense and indispensable intellectual debt to Arabic scholarship[21] in the run-up to the late sixteenth century. As this story is now well-known, there is no need to repeat it here, though as we proceed, we shall have occasions in later chapters to address the issue. For a long time until at least the 1970s, the role of Arabic scholarship was reduced in Euro-American academia to that of 'preserving' the Greek corpus without any enrichment of it until it was eventually 'transmitted back to the "West"'. This narrative continues to be peddled today in survey histories of science, but has long been dismissed in serious science history scholarship, though the narrative's spirit continues to live on even among serious scholars. We examine this issue and the above narrative in detail in Chap. 4.

On the basis of the above developments, by the 1990s, increasingly ambitious arguments against the Eurocentric narrative with regard to intellectual (including scientific) development were advanced. In 1991, George G. Joseph published the first edition of his acclaimed book titled *The Crest of the Peacock: Non-European Roots of Mathematics*, in which he argued that mathematics developed in a long historical process going back to Egypt and Mesopotamia, involving along the way cross-fertilizations between, besides Egypt and Mesopotamia, Greece, India, the Islamic world, and China at different stages of the process. Europe's mathematical development in the 16th and 17th centuries was built upon inputs drawn from this process, and hence was indebted directly or indirectly to the above-mentioned civilizations. For instance, Europe learned much of its mathematics from Arabic scholarship, the latter, however, had not developed autochthonously, instead, it in turn owed extensive debts to Indian mathematics in addition to Greek mathematics.[22]

[21] The term 'Arabic scholarship' is used because whereas the scholars concerned included Arabs, Persians (such as Ibn Sina [Avicenna]), Jews, and so on, they all wrote in Arabic. However, at least as far as mathematics is concerned, Joseph (2011, p. xvi) points out that there was also a rich source of non-Arabic mathematical texts, especially those written in Pahlavi, Syriac, and Hebrew, hence he prefers the term 'Islamic' instead of 'Arabic'. But there are problems with this term too; for instance, the influential philosopher living in Cordoba Maimonides was a Jew, not a Muslim, who wrote in Arabic. On balance, we follow the usual practice and adopt the term 'Arabic scholarship'.

[22] Incidentally, regarding Arabic scholarship's debt to Greece, for a long time, it was believed that this taking up of the Greek tradition began only under the Abbasid Caliphate (750–1258). However, Saliba (2007, chaps. 1 and 2) shows that this process had already begun under the previous Umayyad Caliphate (661–750).

In similar fashion to Joseph but covering science as a whole instead of mathematics alone, David Hess, a social anthropologist, published a work titled *Science and Technology in a Multicultural World* in 1995, in which drawing upon the work of Needham, the rehabilitation of the intellectual debts owed by Latin Europe to Arabic scholarship, and the extensive cross-fertilizations between various civilizations in mathematical development, he argues for a multicultural perspective on the history of science by tracing the roots of modern science to non-European cultures as well.

By the late 1990s and especially the 2000s, the Eurocentric narrative—Eurocentric in the inter-related senses of Europe being used as the yardstick to measure other civilizations, European development being unique, and European development being driven by purely internal dynamics—came under attack on a broad front. To refer to just a few examples, Brook and Blue (2002)[23] published an edited volume titled *China and Historical Capitalism: Genealogies of Sinological Knowledge*, which aims to show how 'Western'[24] knowledge of other civilizations such as China is constructed by means of categories that are derived from 'Western' historical experience, hence making that knowledge Eurocentric (in the sense of using Europe as the yardstick to measure). The literary historian Jerry Brotton (2000) analyses how from architecture to medicine, literature to astronomy, art to geography, the Renaissance is unthinkable without the infusion of Levantine influence. The anthropologist Jack Goody (2006) argues, among other things, that institutions that are purported to be uniquely European, such as universities, trading towns, humanism, democracy, individualism, and even romantic love, were in fact also found in various other past societies. The mathematics historian Victor Katz lent support to George G. Joseph's argument for the multi-civilizational origins of mathematics by publishing an edited volume titled *The Mathematics of Egypt, Mesopotamia, China, India and Islam: A Sourcebook* in 2007. The political scientist John Hobson (2006) argues that Europe's assimilation of inventions (ideas, technologies, institutions, and so on) from the Muslims, Indians, and Chinese constituted the pre-conditions that enabled capitalist modernity to emerge in Europe. Most directly relevant to our concerns is science historian Arun Bala's (2006) important work titled *The Dialogue of Civilizations in the Birth of Modern Science*. While this work

[23] Brook is a sinologist, while Blue is an inter-disciplinary scholar who was once an associate researcher at the Needham Research Institute in Cambridge, England.

[24] Quotes around the term *west* are ours, not Brook's and Blue's.

has a similar objective to the above-mentioned work of David Hess published in 1995, it is more sophisticated and technically superior; and it does for modern science in general what George G. Joseph's work does for mathematics, namely, to see the rise of modern science as a multi-civilizational outcome, in the sense that it is the cross-fertilizations between various civilizations in scientific ideas at various times in a long historical process that laid the grounds for the breakthrough into modern science.

The perspective that modern science is a multi-civilizational outcome in the just-defined sense is referred to as *the dialogical perspective/approach* or *dialogism* (as opposed to Eurocentrism) for short; in the present work, we use these terms interchangeably. The term 'dialogue' in this sense first came to prominence with the publication of Bala's major work of 2006 mentioned in the above paragraph; in the early 2010s, John Hobson (see shortly below), with reference to Bala's argument, began to speak of 'dialogism'; a recent (March 2023) international conference was titled 'Needham's Dialogical Vision: Understanding Science as a Multi-Civilizational Outcome',[25] and in his contribution to the conference, the above-mentioned leading contemporary Eurocentric scholar Floris Cohen contrasts his 'comparative' approach to the dialogical approach under the title of 'Dialogue and comparison compared'. We have some reservations about the term 'dialogue' because it *may* be mistaken to imply intentionality and purpose (see note 1 of the Introduction), whereas the cross-fertilizations between different civilizations were less often intentional or purposeful than simply the result of spontaneous cultural exchanges. Nonetheless, this term has become part of the lexicon concerned, hence we follow common practice and adopt it.

Hence, by the 2000s, a relatively strong case had been established that Latin Europe owed substantial intellectual debts to non-European civilizations that enabled it to make the breakthrough into modern science around 1600. Although evidence supporting the claims of such debts varied from case to case—in some cases, such as Copernicus' debt to al-Shāṭir, the evidence is incontrovertible; in other cases, disputes exist with regard to the adequacy of the evidence concerned[26]—the general picture is

[25] The conference was organized by the Joseph Needham Foundation for Science & Civilization (Hong Kong) and the Needham Research Institute (Cambridge) jointly with the Hong Kong University of Science and Technology, the Jao Tsung-I Petite Ecole (University of Hong Kong) and the Faculty of Asian and Middle Eastern Studies (University of Cambridge), and held in Hong Kong on 23rd to 24th March 2023.

[26] We examine the issue of evidence in detail in Chap. 6 below.

unmistakable. When Needham first raised the NQ, the AWD paradigm and the DML view were dominant, but under the above circumstances, it seems that they can hardly be maintained in their original form. Yet, we can still find contemporary arguments that adhere to them in a manner almost identical to that of the early twentieth century. To refer to just one instance, in a work that addresses the NQ with reference to mathematical astronomy, Chen (2009, p. 634) argues in conclusion that 'Western science…is the embodiment of the spirit of its [the 'West's'] entire civilization'. For Chen, this 'spirit' constituted the 'West's' 'essence' which China lacked, and it was this endogenous 'essence' that was responsible for the rise of modern science in the 'West' but not China or anywhere else. This is the AWD paradigm in its purest form, and it is all the more remarkable in view of the fact that Chen gives full acknowledgment to the role played by Arabic scholarship. Chen's argument is of great illustrative value for our purposes, which we examine in detail in Chap. 4.

2.5 Conventional and Neo-Eurocentric Responses and the Need to Go Beyond the Dialogical Perspective

Thus, Eurocentric accounts in relatively pure form still survive up to the present day even in serious scholarship, let alone in introductory pedagogical and popular discourses, in which, as Cohen observes (see Chap. 4) such accounts remain the standard. Nonetheless, conceding that the case for the dialogical perspective cannot be dismissed, other Eurocentric scholars have responded in two different ways which are complementary to one another. Firstly, they try to weaken the dialogical perspective by attempting to minimize the intellectual debts that Europe owed to non-European civilizations in the making of modern science, and this is done, on the one hand, by insisting on a legal criterion concerning evidence of transmission, and, on the other, by arguing that there are 'cultural obstacles' to knowledge transmission. This first type of responses remains within the parameters of what we call conventional Eurocentrism, which we examine in detail in Chaps. 6 and 7. The second type of responses, and this is sometimes adopted by the same scholars who also adopt the just-mentioned first type, involves a shift in the theoretical terrain of argument, a shift first noticed by the above-mentioned critic of Eurocentrism John Hobson.

In a conference paper, subsequently published as a book chapter, titled 'Global dialogical history and the challenge of neo-Eurocentrism', Hobson (2012) states:

> *[G]lobal dialogism is in danger of being outflanked by the emergence of what I will call 'neo-Eurocentrism'*, or 'neo-Orientalism'. This approach has yet to be fully developed though I anticipate its development in the coming years...[sociologist Ricardo Duchesne, a self-confessed Eurocentric] claim[s] that many Eurocentric world historians, especially since about 1990, are sensitive to the point that Europeans often borrowed Eastern 'resource portfolios' [by which Hobson refers to ideas, institutions and technologies] and that it was this 'cosmopolitan' attitude that marked the exceptional quality of Europe and the West. Above all, this focus shifts the theoretical terrain away from pure European inventive genius towards the idea of European creative adaptability...the old markers of Eurocentric world history—the European logics of 'immanence' and 'inventive exceptionalism'—are in the process of being replaced with the neo-Eurocentric markers of the West: the logics of 'emergence' and 'adaptive/imitative exceptionalism'. (Hobson 2012, pp. 17, 24, our emphasis)

Hobson then quotes from Duchesne:

> [A] distinctive *trait* shown by Europeans was precisely their willingness to imitate inventions made by foreigners...A major secret of European creativeness is precisely its multicultural inheritance.... (Duchesne 2006, pp. 76, 79, our emphasis)

Hobson also refers to Warraq's *Defending the West*, which claims that what distinguishes the 'West' from Islam is a question of values and drives: Whereas the former is said to value truth and knowledge for their own sake, the latter is said to have no notion of truth outside the stultifying sphere of Islamic revelation; and whereas the former is said to be driven by a deep curiosity about the world, the latter is said to have shown only contempt for what is foreign. Ibn Warraq is the pen name of an anonymous non-academic writer who has made it his mission to defend the 'West' against Islam. But his claims concerning the central roles of *values* and *drives* and his *essentialist interpretations* of them are widely shared by Eurocentric scholars as well; for instance, the historical sociologist of modern science Toby Huff (2010) has published a book titled *Intellectual*

Curiosity and the Scientific Revolution, the argument of which we examine in detail in Chap. 7.

As previously noted, Hobson's concern is with the emergence of capitalist modernity in Europe, rather than the rise of modern science. But there are close parallels between the two cases (Hobson argues that the technical pre-conditions for the emergence of capitalist modernity were almost without exception all learned from non-European civilizations), hence Hobson's comments and observations are highly relevant to us too, though in referring to his argument, it is necessary to adapt it to our concern. There are several points to note from Hobson's remarks. Firstly, according to the self-confessed Eurocentric Duchesne, the sensitivity of Eurocentric scholars to European borrowings from non-European cultures really began in the 1990s, this confirms our above observation that it was the steady building up of the case for dialogism, first begun by Needham, from the 1950s onwards that forced changes in how Eurocentric scholars argue their case. Secondly, Hobson's 'old' Eurocentrism is equivalent to what we call conventional Eurocentrism, and by the 'logic of immanence' of 'old' Eurocentrism, he is referring to none other than the AWD paradigm's claim of autochthonous development arising from purely endogenous dynamics—in an earlier work, he states: 'the west is understood through a logic of immanence: that it can only be accounted for by factors that are strictly *endogenous* to Europe' (Hobson 2006, p. 296, our emphasis). Thirdly, in response to the challenge of dialogism, Eurocentric scholars now argue that *what enabled Europe to make use of what they had borrowed from non-European cultures to make the breakthrough into capitalist modernity and/or modern science are certain 'unique' European 'qualities'* ('creative adaptability', attitude, trait, values, drives, and so on) absent in other civilizations. We follow Hobson to call this shift in approach *neo-Eurocentrism*, defined in the immediately preceding sentence. Thus, in the rest of the present work, the term 'Eurocentrism' and its cognates will continue to be used in the generic sense, but when reference is made specifically to the above new approach, the term 'neo-Eurocentric' will be used. This is why in relation to some scholars, we refer to them sometimes as Eurocentric and sometimes as neo-Eurocentric, depending on the particular context in which reference to them is made. It will be noted that neo-Eurocentric arguments are invariably (though we have also spoken of 'most' so as not to sound too absolute) of the cultural-essentialist variety as they speak of 'unique' European 'qualities'; hence, the two main neo-Eurocentric arguments that we examine in later chapters (Huff's in Chap. 7, and Cohen's in Chaps. 8 and 9) both belong to this genre.

It is clear that although it is useful to contrast neo-Eurocentric arguments with conventional Eurocentric ones, in terms of substance, there is actually very little that is really 'new' in the former; the only 'newness' consists in their application of conventional cultural-essentialist arguments but restricting that application to the period of the breakthrough into modern science. Nonetheless, their appearance is still of great significance. As seen, Hobson remarks that dialogism is 'in danger of being outflanked' by neo-Eurocentrism, with reference to this, it is interesting to note that in 2021, we engaged in an extensive dialogue with Arun Bala, in the course of which we raised the question with him as to why Latin Europe was able to break through into modern science on the basis of what it had learnt from other civilizations. Bala responded by saying that this remains, in his wording, 'a black box'. It is precisely because Hobson sees the existence of this 'black box' in the dialogical position that he makes his 'danger of being outflanked' remark[27]:

> [N]on-Eurocentric dialogism is being outflanked by virtue of the fact that neo-Eurocentrism can offer an *explanation* of the rise of the West that, I would argue, is currently missing within the non-Eurocentric global dialogical approach. (Hobson 2012, p. 25)

In other words, even if all present claims of knowledge transmissions from non-European civilizations to Europe are sustainable and accepted, there remains one critical element that is missing in the dialogical perspective's answer to the positive NQ. In contrast, neo-Eurocentric advocates have, in Hobson's view, been able to offer an explanation to fill this gap in terms of certain purported 'unique' European 'qualities'. As we show in later chapters, in addition to with regard to capitalist modernity (Hobson's concern), in relation to the rise of modern science as well, the same kind of arguments have been advanced by neo-Eurocentric scholars. Actually, contra Hobson, what neo-Eurocentric scholars have offered do not

[27] In admitting the existence of this 'black box', Bala was effectively admitting that dialogism is in danger of being outflanked by neo-Eurocentrism; this tacit admission was repeated on a later occasion, when Bala mentioned to us in private communication that Hobson had said to him that neo-Eurocentrism has 'stolen the thunder' from dialogism.

constitute explanations, but only arguments,[28] but nonetheless, unlike in the case of dialogism, there exists no 'black box' in the neo-Eurocentric position. Hobson has correctly pinpointed a fatal gap in the dialogical perspective.

To sum up, the objective of the present work is to crack open the above-mentioned 'black box' on a non-essentialist basis in answering the positive NQ, and thereby show why neo-Eurocentrism is as mistaken as the original AWD paradigm. As explained in the Introduction, to achieve this, whereas the dialogical perspective serves as a valid and important starting point, it is necessary to transcend it, since, as a perspective, it suffers from intrinsic limitations in theoretical and methodological terms due to its failure to engage in any analysis of the phenomenon of intellectual development itself. By transcend, we mean going beyond, not supplanting or abandoning, since, as just mentioned, we accept the importance of dialogism as a starting point. To provide a non-essentialist analysis of the phenomenon of intellectual development itself, both in its own right and in relation to the rise of modern science, requires the laying down of a theoretical-methodological framework as the basis upon which substantive historical analysis can be conducted. This is the subject matter of the next chapter. But before proceeding to it, there is one more issue that requires some comment.

Some Eurocentric/neo-Eurocentric scholars refer to their approach as the comparative approach, on which a few words may usefully be said. What does comparison amount to in this approach? Clearly, Weber's contrast between purportedly different 'forms of rationality' between different civilizations is certainly of a comparative nature; it is, unfortunately, also how Needham understood comparison:

> [SCC is to be concerned with] the general history of civilization, and especially the *comparative* development of Asia and Europe...Why did their [the

[28] The reason why Hobson believes that they have offered an *explanation* is simply that he actually accepts the 'creative adaptability' (see above first quote from Hobson) *argument*. But he 'believe[s] that the non-Eurocentric [case] can be saved' (*op.cit.*), by offering an explanation of how and why this 'creative adaptability' had come into being. His explanation takes the form of an 'identity-formation' argument, an argument that he has previously advanced (Hobson 2006). Hobson obviously believes that his argument is of a non-Eurocentric nature. We find Hobson's 'identity-formation' argument unconvincing, but that is outside the scope of the present work. In any case, it concerns the rise of capitalist modernity, and is irrelevant to us.

> Chinese's] science always remain empirical and restricted to theories of primitive or medieval type? What were the *inhibiting* factors in their civilisation which *prevented* the rise of modern science in Asia? It is suggested that…the concrete factors which moulded Asiatic civilization differently from that of Europe are…. (Needham's book proposal for SCC of 1948, quoted in Winchester 2009, p. 171, all emphases are ours)

Thus, for Needham, as it was for Weber, the comparative approach is for the purpose of finding out what was 'present in Europe but absent in China [and elsewhere in Asia]' that *enabled* Europe to break into modern science, and what, in its civilization, '*inhibited*' China from making the same breakthrough. This is also what Floris Cohen means by the comparative approach:

> A more fruitful and realistic way out has been to develop standards for a truly *comparative* (emphasis in the original) history. A history, that is, that takes the fact of Western priority in this particular domain of human achievement [i.e. the emergence of modern science in Europe] as a fact…a remarkable fact that cries out for scholarly explanation through finding out how it is that the Scientific Revolution eluded other civilizations. If approached along such lines, the question is meant to throw light, not only upon Western civilization [i.e. the positive NQ] by bringing into sharper relief how it could produce early modern science [i.e. what *enabled* it to produce], but also upon civilizations p or q by focusing on features uniquely its own less conducive overall to science in the post-17th-century sense [i.e. *inhibitive* factors unique to p or q]. (Cohen 1994, p. 404)[29]

The continuity of thinking from Weber, through Needham to Cohen is remarkable. The historian of Greek science G.E.R. Lloyd, who since the 1990s has also taken an interest in ancient Chinese thoughts, also speaks of the comparative approach:

> But if by now that whole edifice of Greek and later European uniqueness can be seen as a sham, that does not mean that we are left with an unclouded vision of what to say about cultural diversity and the different fortunes of different modes of inquiry and methods and aims of understanding. (Lloyd 2022, p. 32)

[29] In a forthcoming book chapter titled 'Dialogue and comparison compared', Cohen (2024) gives no indication that his understanding of comparison has changed over the years.

The above is taken from Lloyd's most recent book titled *Expanding Horizons in the History of Science: The Comparative Approach* of 2022. Thus, Lloyd is opposed to claims of Greek and European uniqueness, at least verbally. Nonetheless, he also argues that different cultures do have different modes of inquiry and methods and aims of understanding, and to find out what these differences are is what the comparative approach is about. On the surface, this seems like a break away from what Weber, Needham, and Cohen understand by the comparative approach. But is this impression borne out in practice by Lloyd's substantive arguments concerned?

As a classical scholar with a formidable reputation, Lloyd, as noted, subsequently became interested in China as well. For three decades since the 1990s, in collaboration with the late sinologist Nathan Sivin, he has argued for what we call the 'agora debate' thesis, in which he reduces a whole list of intellectual developments supposedly 'present only in Greece and lacking in China' to a single social factor (the agora debate), which was present in Greece but absent in China. This thesis will be examined in detail in Chap. 5, but a quick comment here on the above quote from Lloyd in relation to China is in order. 'Different methods of understanding' pertains, for instance, to the claims that the concept of nature is a *unique* Greek 'invention', and that though China did have various notions surrounding such a concept in different contexts, it *lacked* what Lloyd calls 'a single concept of nature'. 'Different aims of understanding' pertains to the claim that not only did China lack the concept of nature, 'before modern times [i.e. before the late 19th century], Chinese did not *need* a word that meant "nature" (the physical or material universe)' (Lloyd and Sivin 2002, p. 200, our emphasis)—the idea is that because the Chinese had 'different aims of understanding', hence they 'did not need' the concept of nature. Thus, verbally, Lloyd does seem to show an awareness that comparison is not in order to make an argument of the only-in-Greece/Europe-lacking-elsewhere type (or, in his own wording, 'Greek and later European uniqueness'), but, unfortunately, his practice has been the exact reverse.

We also make comparisons, but with a *different conceptualization* to show something else altogether. In our comparison between Greece and China (Lau 2020), we aim to explain on a *non-essentialist* basis why intellectual developments in Greece and China, despite sharing many common concerns and possessing various common concepts, modes and aims of inquiry, and so on, *established different scholarly traditions and followed different trajectories*, such as why mathematical proof in the two civilizations took different forms, why the conceptualization of the ultimate constituent of all empirical things differed in approach, and so on. In such a comparison, not only is it shown that there does not exist a long list of what is allegedly possessed only by Greece but lacking in China, differences in development between the two civilizations are not seen from the perspective of what 'inhibited' China from being like Greece or, for that matter, Greece from being like China; i.e. not seen from the perspective of using one civilization as the yardstick to measure other civilizations.

Because Eurocentric scholars precisely adopt the just-mentioned perspective (if only unawares), in comparisons made by them, there is inevitably an implicit assumption that one particular tradition (the Graeco-European) is more 'worthy' than the rest. However, even if the value of modern scientific rationality is accepted as given, in comparing different traditions, to regard the tradition that eventually led to the development of modern science as being more 'worthy' would be teleological. Some scholars are aware of the pitfall of engaging in teleological thinking, and hence re-formulate the issue in terms of a particular tradition (the Graeco-European) being, in retrospect, more conducive to the rise of modern science.[30] That, however, is also mistaken, because, as we show in the present work, the rise of modern science was *an unpredictable or contingent outcome* of a long historical process, in which many contingent factors extraneous to any single scholarly tradition were involved. From this angle, no single scholarly tradition can be said to be more conducive to the rise of modern science, because *to speak, even in retrospect, of something being more conducive to something unpredictable or contingent can make no sense*.[31] Thus, comparison can be understood in a very different sense from that understood by Eurocentric scholars, and in the following chapters, when we make comparisons as and when so required, we do so in the sense explained in the previous paragraph.

[30] This is Cohen's (1994, 2011, 2024) argument. Cohen accepts Needham's contrast between what he called ancient China's 'organic materialism' and Greece's/Europe's 'mechanical atomism' (see Sect. 2.2 above), and argues that whereas both were of great merit, it can be seen in retrospect that the latter was more conducive to modern science.

[31] Some scholars have raised the counter-factual question of whether or not other scientific traditions could have laid the intellectual foundations for the emergence of modern science in a non-Newtonian form, had modern science not arisen in that particular form in Latin Europe. It would seem to us that such a question can, indeed, be legitimately raised, though the laying down of some basic intellectual foundations and the emergence of some other form of modern science on those foundations are two different things altogether, since, as we stress, history is thoroughly stricken with contingency.

Bibliography

Bala, Arun. 2006. *The Dialogue of Civilizations in the Birth of Modern Science.* Palgrave Macmillan.

Bernal, Martin. 2001. *Black Athena Writes Back: Martin Bernal Responds to His Critics.* Durham: Duke University Press.

Bose, D.M., S.N. Sen, and B.V. Subbarayappa. 1971. *A Concise History of Science in India.* New Delhi: Indian National Science Academy.

Brook, Timothy and Gregory Blue. 2002. *China and Historical Capitalism: Genealogies of Sinological Knowledge.* Cambridge University Press.

Brotton, Jerry. 2000. *The Renaissance Bazaar: From the Silk Road to Michelangelo.* Oxford: Oxford University Press.

Burkett, Walter. 1992. *The Orientalizing Revolution: Near Eastern Influence on Greek Culture in the Early Archaic Age.* Cambridge, Mass.: Cambridge University Press.

Chemla, Karine. 2018. 'Abstraction as a value in the historiography of mathematics in ancient Greece and China: a historical approach to comparative history of mathematics'. Pp. 290-325 in G.E.R. Lloyd and Jenny Jingyi Zhao (eds.), *Ancient Greece and China Compared.* Cambridge: Cambridge University Press.

Chen Cheng-I. 1996. *Early Chinese Work in Natural Science.* Hong Kong: University of Hong Kong Press.

Chen Fangzheng. 2009. *Inheritance and Rebellion: Why Modern Science Emerged in the West.* 陳方正，《继承与叛逆:现代科学为何出现于西方》(北京:三联，2009)。

Chen Guying. 2012a. *The* Zhuangzi*: Contemporary Annotations and Interpretation.* 陈鼓应，《庄子今注今译》，修订重校本(香港:中华书局，2012)。

Cohen, H. Floris. 1994. *The Scientific Revolution: A Historiographical Inquiry.* University of Chicago Press.

Cohen, H. Floris. 2011. *How Modern Science Came into the World: Four Civilizations, One 17th-Century Breakthrough.* Amsterdam University Press.

Cohen, H. Floris. 2024. (in press) 'Dialogue and comparison compared'. Chapter 18 of Arun Bala, Raymond W.K. Lau, and Mei Jianjun (eds.) *Multicivilizational Exchanges in the Making of Modern Science: Needham's Dialogical Vision.* Palgrave Macmillan.

Collins, Randall. 1985. *Max Weber.* London: Sage.

Duchesne, Ricardo. 2006. 'Asia First?'. *The Journal of the Historical Society* 6(1), pp 69-91.

Elvin, Mark. 2001. 'Vale atque ave'. Pp xxiv-xliii in Needham, Joseph, with the collaboration of Kenneth Girdwood Robinson and Ray Huang (Huang Jen-yu), and including contributions by them, Introduction by Mark Elvin, edited by Kenneth G. Robinson. 2001. *Science and Civilisation in China, Volume 7, Part II: General Conclusions and Reflections.* Cambridge University Press.

Goody, Jack. 2006. *The Theft of History*. Polity.

Graham, Angus C. 1973. 'China, Europe, and the origins of modern science: Needham's The Grand Titration'. Pp 45-69 in Shigeru Nakayama and Nathan Sivin (eds.), *Chinese Science*. Cambridge, Mass: MIT Press.

Grant, Edward. 1996. *The Foundations of Modern Science in the Middle Ages*. New York: Cambridge University Press.

Harbsmeier, Christoph. 1998. *Science and Civilisation in China, Volume 7 Part I: Language and* Logic. Cambridge: Cambridge University Press.

Harley, J.B. and David Woodward (eds.). 1987. *Cartography in Prehistoric, Ancient, and medieval Europe and the Mediterranean*. University of Chicago Press.

Harley, J.B. and David Woodward (eds.). 1992. *Cartography in the Traditional Islamic and South Asian Societies*. University of Chicago Press.

Harley, J.B. and David Woodward (eds.). 1994. *Cartography in the Traditional East and Southeast Asian Societies*. University of Chicago Press.

Harley, J.B. and David Woodward (eds.). 1998. *Cartography in the Traditional African, American, Artic, Australian and Pacific Societies*. University of Chicago Press.

Hobson, John M. 2006. *The Eastern Origins of Western Civilisation*. 5th printing with corrections. Cambridge: Cambridge University Press.

Hobson, John. 2012. 'Global dialogical history and the challenge of neo-Eurocentrism'. Pp 13-33 in Arun Bala (ed.), *Asia, Europe and the Emergence of Modern Science: Knowledge Crossing Boundaries*. New York: Palgrave Macmillan.

Huff, Toby. 2010. *Intellectual Curiosity and the Scientific Revolution: A Global Perspective*. Cambridge University Press.

Joseph, George G. 2011. *The Crest of the Peacock Non-European Roots of Mathematics*, 3rd edition. Princeton: Princeton University Press.

Lau, Raymond W.K. 2020. *Intellectual Developments in Greece and China: Contingency, Institutionalization and Path-Dependency*. Newcastle upon Tyne: Cambridge Scholars Publishing.

Levy-Bruhl, Lucien. 1985. *How Natives Think*, authorized translation by Lilian A. Clarke. Princeton: Princeton University Press.

Li Xiaolin. 1981. 'Searching for the origins of our country's double-entry book-keeping'. 李孝林, 〈我国复式记账法溯源〉, 中国商务部《财会参考资料》, 1981(26).

Li Xiaolin. 1988. 'Exploring our country's indigenous double-entry book-keeping'. 李孝林, 〈我国固有复式簿记的探讨〉, 《北京商学院学报》1988年2期.

Lin, Z. Jun. 1992. 'Chinese double-entry bookkeeping before the nineteenth century'. *The Accounting Historians Journal* 19(2), pp 103-22.

Lindberg, David. 2007. *The Beginnings of Western Science*, 2nd edition. Chicago: University of Chicago Press.

Lloyd, G.E.R. 2022. *Expanding Horizons in the History of Science. The Comparative Approach.* Cambridge University Press.

Lloyd, G.E.R. and Nathan Sivin. 2002. *The Way and the Word: Science and Medicine in Early China and Greece.* New Haven: Yale University Press.

Needham, Joseph, with the collaboration of Wang Ling. 1959. *Science and Civilization in China, Vol. 3: Mathematics and the Sciences of the Heavens and the Earth.* Cambridge: Cambridge University Press.

Needham, Joseph. 1969. *The Grant Titration.* Taylor and Francis.

Needham, Joseph. 1981. *Science in Traditional China.* Hong Kong: The Chinese University Press.

Newman, William. 2006. *Atoms and Alchemy. Chymistry and the Experimental Origins of the Scientific Revolution.* Chicago and London: University of Chicago Press.

Saliba, George. 1994. *A History of Arabic Astronomy: Planetary Theories during the Golden Age of Islam.* New York and London: New York University Press.

Saliba, George. 2007. *Islamic Science and the Making of the European Renaissance.* Cambridge, Mass: MIT Press.

Seyyed Hossein Nasr. 2003. *Science and Civilization in Islam*, 2nd edition. Islamic Texts Society.

Weber, Max. 1992. 'Author's introduction'. Pp xxviii-xlii in Max Weber *The Protestant Ethic and the Spirit of Capitalism,* trans. by T. Parsons, Introduction by A. Giddens. Routledge.

Webster, Charles. 1980. *From Paracelsus to Newton, Magic and the Making of Modern Science.* Cambridge University Press.

West, Martin. 1997. *The East Face of Helicon: West Asiatic Elements in Greek Poetry and Myth.* Oxford: Clarendon.

Whimster, Sam and S. Lash. 1987. 'Introduction'. Pp. 1-34 in Idem. (eds.), *Max Weber: Rationality and Modernity.* London: Allen & Unwin.

Winchester, Simon. 2009. *The Man Who Loved China.* New York: Harper Perennial.

CHAPTER 3

Analysing Intellectual Development on a Non-essentialist Basis: A Theoretical-Methodological Framework

The following framework was originally developed several years ago in a related project in which we compared intellectual developments in Greece and ancient China (Lau 2020). Our explanation in that work of how we had arrived at our framework is worth quoting:

> Karl Marx once said that the order of exposition is often the reverse of the order of discovery. In our research, we did not begin with a particular theoretical framework and search for data to fit into it, instead, we put theory aside and conducted our research without any pre-conceived theoretical perspective. It was during the course of this research that we gradually came to realize that our findings could best be understood in terms of the insights of certain particular theoretical perspectives. Hence, the following theoretical framework has been formulated, not at the beginning of our research process, but at the end. (Lau 2020, p. 16)

The same process of empirical research inducing the construction of theory was continued during our investigation into the positive NQ. Since the objects of investigation in our previous and present studies, while overlapping with one another, also diverge from each other in various respects, it is hence unsurprising that our framework has been developed further as a result.

Our framework is formulated on the basis of several theoretical and conceptual perspectives, it is, hence, a unique construction synthesizing

R. W. K. Lau, *Rethinking the Needham Question*,
https://doi.org/10.1007/978-981-97-9472-0_3

these perspectives into a unitary whole that caters specifically to the analysis of intellectual development. Our framework comprises the following components: (a) Sensitive dependence of historical development on starting point; (b) contingency in history as a systematically analysable explanatory variable; (c) the integration of externalist and internalist explanations; (d) scholarly tradition as a self-reinforcing informal institution that can be made subject to non-essentialist analysis; (e) the distinction and relationship between the formal aspects and substantive knowledges of a scholarly tradition; (f) historical development as a path-dependent process; (g) process approach versus mechanistic approach; and (h) scholarly agency, one type of internal factors discussed under (c) above, as a non-essentialist explanatory variable. In the following, we first discuss these components individually, though not exactly in the above order, and then combine them into a single, unitary framework.

3.1 Sensitive Dependence of Historical Development on Starting Point

In historical development, *the starting point of a process* is a crucial determining factor. A slightly different starting point could in time lead to very different developmental trajectories and outcomes. This is an insight of chaos theory in physics (Ruelle 1991) that historians can well benefit from. In physics, a dynamical system is called a chaotic system if even a very small change in its initial conditions will after a time lead to a very different end-point for the system. The different trajectory is deterministic, but since the small change in initial conditions cannot in practice be known precisely, hence the system becomes unpredictable beyond a certain period of time. This is the famous butterfly effect, a term coined by Edward Lorenz, whose work in meteorology was a forerunner of chaos theory. The gist of chaos theory is encapsulated in its notion of sensitive dependence on initial conditions, the meaning of which should be clear from the above, which we adapt as sensitive dependence on a starting point in relation to our subject matter. Analyses which focus only on relatively late stages of different developments are ill-equipped to understand why such differences in development have arisen.

In historical explanation, sensitive dependence on a starting point is especially significant when the starting point itself is produced to a significant degree or even largely by contingency. There will be examples illustrating this in appropriate contexts later on in this chapter. As will be seen, such instances are by no means rare, and recognition of this undermines all deterministic understanding of history, and hence essentialist explanations, since in such explanations, intellectual development is the deterministic product of some purported essence(s). This brings us to the issue of contingency in history.

3.2 Contingency in History as a Systematically Analysable Explanatory Variable

To scholars accustomed to thinking in deterministic terms, *contingency* is difficult to accept. That is why until the end of his life, Einstein refused to accept quantum mechanics, though he could not deny its findings. In sociology, whereas sociologists such as Giddens (1984) and Bauman (1991) abstractly and generally lay emphasis on the role of contingency in social development, most sociologists doing empirical historical analysis are averse to theories in which contingency plays a crucial role (Mahoney 2000, p. 509). Unlike social scientists, historians are usually more ready to accept the role of contingency in history, though, with important exceptions such as Pomeranz (2000),[1] thoroughgoing contingent analyses of important historical developments are less often encountered. That scientific discoveries are often made accidentally is well-known to scientists,[2] but most students of comparative intellectual developments seem not to have taken notice of that or given it the attention that it deserves. Needham is a case in point. He believed that there can only be three kinds of alternative explanations for intellectual developments, namely, bio-psychological essentialism, pure chance, and social explanation (Needham et al. 2001, p. 209), he opted for social explanation and rejected 'pure chance' as emphatically as he did bio-psychological essentialism.[3]

[1] In a scholarly tour de force, Pomeranz provides a meticulous analysis of why industrialization took off in certain particular areas in western Europe and not in certain specific areas in China, both of which were up to the eve of the Industrial Revolution on comparable levels of economic and technological development, on the basis of specific configurations of contingent factors.

[2] See, for instance, Chen Cheng-I (1996, p. 216 n 276) on how the connection between magnetism and electricity was made accidentally by Hans Christian Oersted in 1820.

[3] However, given that Needham adopted a multi-causal approach and his study spans many centuries of development, he had unsurprisingly come across factors that are contingent in nature, and, to his credit, given de facto recognition to their role—for instance, in ibid.,

The recognition of and emphasis on the role of contingency is not confined to the work of historians such as Pomeranz. In recent scholarship, in disciplines ranging from evolution (Gould 1989, 2002 and Part III of 2006) to historical sociology (see below), the role of contingency has increasingly been recognized.[4] In the present work, we do not only give recognition to the importance of contingency, we incorporate it as an integral component of a larger theoretical framework.

The concept of contingency cannot simply be taken for granted, but needs to be defined. Firstly, a particular factor or event is contingent if it cannot be attributed to any particular reason or cause. For instance, Chinese is an isolating language while Greek a fusional language with words often beginning with vowels. These can only be taken as contingently given empirical facts. Secondly, a factor or event can sometimes be subject to explanation, i.e. attributable to a particular reason or cause, but it can still be contingent in relation to our concern, namely, intellectual development. To give a simple example without in any way implying mono-causal determinism, the different geographies of Greece and the Central Plains of China (where dynastic China first hailed) are not contingent in geo-physical terms, but in relation to intellectual developments, they are contingently given facts, and any influence, direct or indirect and great or small, that they might have had on these developments can only be regarded as consequences of contingency. Thirdly, in terms of historical explanation, to say that a factor or event is contingent is to say that its appearance cannot be attributed, directly or indirectly, to the operation of any purported master deterministic logic in history. In other words, the factor or event is unpredictable by any master theory, i.e. it cannot be made subject to (ex post facto) deterministic explanation by that theory—in the present chapter, whenever we say 'unpredictable' or 'predictable' by theory, it is meant in this sense.

The above three points concern defining contingency in relation to *cause*, but there can also be contingency in relation to historical *outcomes*, whether or not the factors that cause them to occur involve any element of contingency. Merton's (1936) notion of unanticipated consequences of

p. 230, where he argued that the Black Death played an important role in facilitating the rise of the bourgeoisie in Europe. However, despite giving de facto recognition to chance when its role cannot be denied, Needham retained his rejection of chance as a matter of principle.

[4]Whereas Marxist theory is deterministic, even the Marxist-oriented geographer Blaut (1993) gives stress to contingency in his critique of what he calls 'the colonizer's model of the world'.

purposive social action[5] is a case in point. However, it is not necessary to confine such consequences to Merton's scenario alone, for even where social agents' subjective purposes or intentions do not figure, it is still possible to speak of such consequences in history from an objective historical angle. Imagine an observer standing outside history, as it were, observing event X occurring at time T1 in which no social agents' purpose or intention is involved, X led (either directly or through mediation) to Y at a later time T2, in some cases, the occurrence of Y in consequence to X is within the range of expectations from a causal point of view (such as extreme corruption of a government leading to its downfall), however, when the occurrence of Y in consequence to X is not reasonably within the range of expectations of the observer, then we can speak of unanticipated consequences. Since the term 'unexpected' used by Merton suggests subjective purposive actions on the part of social agents, hereinafter, we speak of *unforeseen consequences in history* instead, as this latter term does not carry that implication. It should be noted that though unforeseen consequences are unforeseen, they can nonetheless be explained ex post facto, but that does not diminish their contingent character.

In the physical sciences, contingency can be formally theorized, however, in the humanities and social sciences, whereas contingency *as an explanatory variable* can be incorporated as an integral component of a larger theoretical framework (see Sect. 3.5 below), the *occurrence* of contingency itself cannot be formally theorized, and its justification lies in its superior explanatory power in comparison to deterministic theories of history. Finally, to stress the role of contingency is not to rule out deterministic cases in which given a particular set of initial circumstances, certain specific historical developments can be expected to occur, though it *does* rule out purported deterministic master logics in history. We return to this point later on in this chapter.

3.3 Integrating Externalist and Internalist Explanations

In the literature on comparative intellectual development, a distinction is often made between *externalist* and *internalist* arguments. The Hessen thesis and the 'agora debate' thesis of G.E.R. Lloyd and Nathan Sivin,

[5] Merton wrote his article in reaction to and critique of the rise of Keynesian economic interventionism.

both mentioned in the previous chapter, are examples of externalist arguments, as is social explanation in general. Social factors are a major category of external factors, but there are other types as well, for instance, the characteristics of a particular language are an external factor, as far as intellectual development is concerned, but our present concern is with social factors. There are two things to note concerning social explanation. Firstly, as discussed in Chap. 2, social explanation can be either essentialist or non-essentialist. Secondly, even with non-essentialist social explanation, it is important to note what it can and cannot do. What social explanation *can* provide is to identify the conditions arising from social factors that impact upon intellectual development, either positively or negatively or in other ways, but in science and mathematics social factors are unable to explain the ideational aspects of that development (i.e. the emergence and development of ideas). A common trap of social explanation is to fall into *sociological reductionism*, i.e. to reduce ideas to social factors. A typical example of sociological reductionism is the following: 'Each [Greek city-state] government was ruled by law and therefore encouraged its citizens to be able to argue and debate. It was perhaps out of this characteristic that there developed the necessity of proof in mathematics' (Katz 1993, p. 43).[6] A simple way to demonstrate its fallacy is to show that the idea of proof also existed in cultures without the feature of Greek city-states mentioned by Katz, and as pointed out in the previous chapter, proof did exist in other mathematical traditions as well. To give another striking example, in the pre-Qin (before −221) period in China, there was a school of thought known as the Mohist School (墨家), founded by Mozi (墨子fl. -fifth century). Mozi argued that all propositions had to be proved, and proposed a general method of determining truth known as the Method of Three Criteria (*Sanbiaofa* 三表法). As a matter of interest, of the three criteria Mozi proposed, two are empirical evidence and practical applicability. In sum, because in science and mathematics social factors are unable to account for the ideational aspects of intellectual development, social explanation, and, pari passu, externalist explanation in general, are, ipso facto and *necessarily*, insufficient.

Internalist arguments are concerned with the internal pre-conditions and conditions of intellectual development. In our view, these pre-conditions and conditions fall into two categories: (1) The intellectual resources from which a certain development draws upon, which can be

[6] Incidentally, Needham also subscribed to this thesis in accounting for the existence of mathematical proof in Greece (Needham et alia. 2001, p. 210).

subdivided into two types, namely, [a] its intellectual ancestry, and [b] other intellectual sources inputting into it; and (2) the agency of the scholars concerned.[7] Intellectual ancestry refers to ***intellectual or scholarly tradition***[8]; and other intellectual sources refer to knowledges transmitted from other civilizations. ***Agency*** is a sociological concept. It refers to the active role played by social agents; while agency can be made subject to explanation such as with reference to factors external to social agents, it is not automatic; otherwise, it would not constitute an active role, while the actions of all social agents would be identical under the same circumstances. With reference to intellectual development, agency refers to the *scholarly* values, drives, and so on of *the scholars concerned, not* to *society-wide* values and so on—the significance of this distinction will be seen in Chap. 9. Supporters of dialogism do not consider scholarly agency at all, probably because they see any talk of scholarly values, drives, and so on as inherently essentialist. Later in this chapter and in later chapters, it will be shown why scholarly agency can be made subject to non-essentialist analysis. That internal factors are indispensable to account for intellectual development is obvious, but they must be complemented by also taking relevant external factors into consideration; for instance, in Chap. 8, we explain late-Renaissance scholarly agency as contingent unforeseen consequences of a combination of several external factors. From our distinction between external and internal factors, and the categorization of internal factors into different types and sub-types, it is clear why the dialogical perspective is inadequate, for it focuses only on 1b above.

3.4 Institutionalization and Scholarly Tradition as a Non-essentialist Explanatory Variable

Scholarly tradition is a potentially misleading issue. Many scholars mistakenly believe that to speak of scholarly tradition is essentialist. Needham is a case in point: '[China's intellectual development] can never be explained by any deficiencies of the Chinese mind or of the Chinese intellectual and philosophical tradition' (Needham et al. 2001, p. 2). In terms of his

[7] The conceptualization of agency as an internal factor is our own view, not found in extant internalist arguments in the literature.

[8] In the present work, we use the two terms interchangeably, for except on one occasion (in the early part of Chap. 8), we are not concerned with non-scholarly intellectual traditions such as musical tradition or art tradition.

above-mentioned trichotomy of alternative explanations for intellectual development, since intellectual and philosophical tradition cannot be associated with chance, and does not constitute a social factor (tradition is internal to scholars instead of external to them), in this remark, Needham could only be associating it with 'the Chinese mind'. In our view, it is because of this that Needham seldom touched upon scholarly tradition. What is true of Needham—interpreting scholarly tradition in essentialist terms—is also true of some of his critics. For instance, in addressing the NQ with reference to mathematical astronomy, Chen (2009) dismisses Needham's rivers-and-sea metaphor and lays great emphasis on the tradition founded by the Greeks, which he regards to be an embodiment of the 'spirit' or 'soul' (*xinling* 心灵) of the Greek people. Not only is the essentialist interpretation of scholarly tradition shared by Needham and his critics such as Chen, it is also shared by many supporters of the dialogical perspective, resulting in a great reluctance or even outright refusal to consider the role of scholarly tradition by them.[9] However, the establishment of any particular scholarly tradition can be made subject to non-essentialist analysis. Given the importance of this issue, it is necessary to examine it in some detail.

Scholarly tradition constitutes what sociologists call an institution. In daily usage, the word institution is often synonymous with the word organization, but in sociology, the two mean different things. Whereas organizations are formal, institutions are informal such as monogamy being the modern institution of marriage. Abstractly, an institution is a particular routinized way of doing things (social action) in a specific context of social practice, and all that underpins that way of doing things. The nature of institutions has been most thoroughly analysed in neo-institutionalist theory in sociology.[10] According to Durkheim (one of sociology's founders) and Parsons (1951), social action is normative, i.e. it is driven by social agents' internalized social values. Similarly, with reference to the organizational context, Selznick (1949) argues that members of an organization act in support of its goals and in accordance with its procedures because

[9] This observation is based upon our communications with advocates of dialogism; testimony to this can also be indirectly observed from the fact that scholarly tradition rarely figures in their arguments.

[10] Neo-institutionalist theory in sociology may refer either to the choice-within-(institutional) constraint model in economic sociology (Brinton and Nee 1998; Nee 1998; Ingram and Clay 2000), or the non-choice-theoretic study of organizational behaviour (Powell and DiMaggio 1991; Scott 2001). The following discussion refers to the latter only.

these have achieved a value-impregnated status among them. On the other hand, March and Simon (1985) point out that in their daily practice, members of organizations do not engage in deliberation but act in accordance with rules and pre-set routines. This view is in line with the general argument of the phenomenological sociologists Berger and Luckmann (1967) that social action is dispositional and premised upon taken-for-granted social knowledges and belief systems. Integrating the insights of the above perspectives, neo-institutionalist theory argues that social action is underpinned by what Scott (2001, Chap. 3) calls the 'three pillars of institutions': The normative (values, goals); the regulative (routines, rules, procedures); and the cultural-cognitive (taken-for-granted shared knowledges and beliefs). From this perspective, social action is non-instrumental, but *normative-dispositional*—'normative' refers to Scott's first pillar, 'dispositional' to the second and third pillars.[11]

A scholarly tradition is an institution in the full sense of the term as conceptualized above. Scholars raised in a particular scholarly tradition do not only acquire its substantive knowledges, they also internalize its goals, values, and taken-for-granted knowledges and beliefs, adopt its practices (routines, rules, procedures), and act in a normative-dispositional way on that basis in their work. The same individual, if raised in a different scholarly tradition, would end up as a very different scholar. Scholars raised in a particular tradition reproduce that tradition by virtue of their daily work. Once a tradition is sufficiently well established, the factors or conditions that originally gave rise to its birth are no longer relevant to scholars raised in it, since from now on, what shape its members normatively and dispositionally in their daily work are no longer these factors or conditions, but the tradition's institutionalized features. For instance, the early Pythagoreans adopted a particular approach to the study of numbers which played a central role in the founding of Greece's mathematical tradition, this approach was the outcome of certain notions held by the early Pythagoreans (see shortly later), the conjoint appearance of which is best understood in terms of a synthesis of contingency and mysticism, a scholar living several centuries later raised in Greece's mathematical tradition need not herself be a Pythagorean or need not even be aware of the

[11] It may be noticed that this is contrary to what rational choice theory argues. Neo-institutionalist theory does not, of course, deny the role of instrumental rational choice, but argues that much of social action is more appropriately conceptualized as normative-dispositional.

just-mentioned early Pythagorean notions, for her to reproduce, entrench, and further develop that tradition. Because a scholarly tradition is a *self-reinforcing institution*, scholars of a particular *entrenched* tradition do not abandon it other than in the most exceptional of circumstances. Thomas Kuhn's concept of scientific *paradigm* brings this point out especially enlighteningly.[12]

Kuhn (1996) distinguishes between three phases in the development of a scientific paradigm, namely, normal science, crisis, and revolution. In the phase of normal science, the branch of science concerned is dominated by a specific paradigm. Very roughly, a paradigm defines what the legitimate problems to be solved are, often this directly determines the very design of apparatuses to enable the solving of those problems; what phenomena are to be observed and what are irrelevant; how observations are to be interpreted; and so on. If findings do not conform to expectations according to the paradigm, scientists do not question the paradigm itself but regard the experiment to have failed, and attempt to repeat the experiment. Kuhn calls repeated non-conforming findings anomalies. Even in the face of anomalies, scientists do not query the paradigm itself, instead they either put the problem away for a while or introduce modifications to the paradigm to rationalize the anomalies. When anomalies pile up to levels unmanageable by means of modifications, the paradigm enters a crisis. The crisis is eventually resolved, if there is available another paradigm, by the revolutionary replacement of the original paradigm by this alternative, but Kuhn stresses that this replacement is not based upon rationalistic comparison between the two paradigms, but driven by extra-rational factors, a point which, however, is of no concern to us. Kuhn's scientific paradigm is one special type of institutionalized scholarly tradition. While we have much to disagree with Kuhn concerning his overall theory of scientific development, his concept of scientific paradigm is important because it shows *the tenacity of the institutionalized grip* that a scholarly tradition

[12] Arun Bala has suggested to us in private communication that in addition to Kuhn's paradigm, the concept of episteme may also be relevant, for which we are grateful to him. Episteme is Foucault's concept who developed it from the philosophies of the French philosophers of science Gaston Bachelard (1884–1962) and Georges Canguilhem (1904–1995). There are indeed affinities between Bachelard's and Canguilhem's philosophies, on the one hand, and Kuhn's notion of scientific paradigm on the other (Gutting 1989). However, we think that Kuhn's notion is more directly relevant to us and is more familiar to most readers, whereas we are not so sure about the comparability between Foucault's concept and our analysis of scholarly tradition.

has over scholars raised in that tradition even as anomalies continue to occur.

To Kuhn's concept, we may add two points. Firstly, crises appear, if they do, only in branches of science in which empirical testing is possible, in other disciplines (mostly non-science ones), no equivalent to Kuhnian crisis ever appears, which implies that scholars abandoning an entrenched non-science scholarly tradition is even harder to occur, and if it does, is likely to take a very long time and only very gradually. Secondly, anomalies are externally generated in the sense that the external physical world refuses, so to speak, to conform to the paradigm's expectations. In other words, this implies, with reference to scholarly tradition in general, that because a scholarly tradition is institutionally self-reinforcing, its members would not query it except under severe *externally generated impact*, if this does ever occur, which is rare in non-science traditions.

There are two inter-related aspects of a scholarly tradition. In the above, we noted that scholars raised in a particular scholarly tradition do not only acquire its substantive knowledges, but also internalize its goals, values, beliefs, orientations, and approaches, as well as adopt its practices. The values, approaches and so on constitute what we refer to as its *formal aspects*, these formal aspects are *embodied in* the tradition's *substantive knowledges* themselves—which is why we call these aspects formal, with the substantive knowledges constituting contents. Form and content are inextricably linked together, instead of existing independently of one another. A simple example will illustrate this: Greece developed what we call a theoretical tradition in mathematics, the tradition's theoretical approach is embodied in the substantive knowledges comprising Euclid's *Elements*.

To illustrate the internalization of the formal aspects of a scholarly tradition by scholars raised in that tradition, let us take an instance from China. China developed what we call a computational tradition in mathematics. In the Song Dynasty (960–1279), extending into early Yuan (the succeeding dynasty), there was a quartet of brilliant mathematicians whose work predated the Europeans in the same mathematical areas by several centuries, one of whom was Li Ye (李冶). The historian of Chinese mathematics Sun Hong'an (2008) examines the logical structure of Li Ye's *Sea Mirror for Circle Measurements* (*Ceyuan haijing* 测圆海镜), pointing out that Li begins by defining the various concepts to be used in his proofs, specifies a number of what are equivalent to today's mathematical symbols, on the basis of the right-angled triangle theorem deduces 692 propositions that fall into seven types which are equivalent to theorems, and finally, on the basis of these propositions, provides proofs of the 170 problems raised in the work. At the end of his examination, Sun (p 224) observes that these 170 problems are all not practical problems, and that, from today's vantage point, Li could well have formulated them as abstract, general problems in the manner of Euclid's *Elements*. Why did not Li do so but instead formulate them in the form of practical problems?, Sun asks and then states: 'This lies in China's mathematical tradition'. Nurtured in China's computational tradition, Li Ye's approach was a matter of *acquired disposition*, which he practised without giving it any thought.

A scholarly tradition is largely non-culture-specific in the sense that it is entirely possible for a culture which has *not yet* developed any strong scholarly tradition to take up a tradition established by another culture and develop it further. The adoption and further development of the Greek tradition by Arabic scholars is a case in point. The Arabist Gutas (1998, p. 192) has also previously pointed this out: '[the translation movement under the Abbasids] demonstrate[s] for the first time in history that scientific and philosophical thought are international, not bound to a specific language or culture'. Gutas says 'the first time', there is also a second time, namely, the taking up Arabic scholarship from the eleventh century onwards by Latin Europe, which was able to occur because the latter also had *not yet* developed any strong scholarly tradition before then. That *scholarly tradition* is non-*culture*-specific under the above condition can easily be missed, with grave consequences. In his analysis of the rise of modern science, Cohen (2011) speaks of a number of what he calls 'cultural transplantations', such as the transplantation from Greece to Arabic scholars. 'Cultural transplantation' is a misnomer resulting from a conceptual confusion between scholarly tradition and culture. What were transplanted were scholarly traditions, not cultures, *Islamic* scholars did not take up *polytheistic* Greek culture in general, but only its scholarly tradition. In relation to the cross-cultural transmission of scholarly tradition, the following remark made by Joseph (2011, p. 13) is also worth commenting upon: 'The relative ineffectiveness of cultural barriers (or "filters") in inhibiting the transmission of mathematical knowledge (in a number of other areas of human knowledge, notably in philosophy and the arts, the barriers are often insurmountable unless filters can be devised to make foreign "products" more palatable)'. One is taken by surprise by this remark, for, besides mathematics (including mathematical astronomy), were Arabic scholars not also able to take up Greek philosophy without any problem and develop it further? The resistance to cross-cultural transmission of scholarly tradition, where it exists, arises not in accordance with what disciplines are involved, but from another factor.

Where there is already a *deeply entrenched* tradition existing in a particular culture, its scholars will be highly resistant to a significantly different tradition for reasons that we have, partly by drawing an analogy with Kuhn's concept of paradigm, explained above. Many scholars misinterpret such resistance in terms of *cultural* conservatism, thereby giving the matter a *cultural-essentialist* twist. To the extent that a civilization's deeply entrenched scholarly tradition is part of that civilization's culture, it does not appear that any misinterpretation is involved. But this is misleading. That the above is a misinterpretation can easily be seen in Kuhn's analysis of the great resistance to paradigm change, which, in Kuhn's analysis, occurs within *the same* culture. Thus, resistance of the same nature on the part of scholars raised in the deeply entrenched scholarly tradition of a particular civilization to a significantly different scholarly tradition is *intellectual* (pertaining to scholarly tradition), not cultural in nature. Hence, although we have yet to discuss how scholarly tradition can be made subject to non-essentialist analysis, it can already be seen that recognizing scholarly tradition as a legitimate explanatory variable *enables a direct rebuttal* of a particular type of cultural-essentialist arguments—that of alleged cultural conservatism.

The mistaking of scholarly tradition-induced resistance for purported cultural conservatism can also be illustrated by a comment made by Joseph. As mentioned, China developed a computational tradition in mathematics, over the centuries, various mathematicians with a theoretical bent of mind pushed hard at the limits of such a tradition but never succeeded in transcending it. The above-mentioned Li Ye was one of them. Noticing this phenomenon, Joseph (2011, p. 261, our emphases) asks: 'So why were mathematicians of the caliber of Liu [Hui] and Zhao [Shuang] [both were 3rd-century Chinese mathematicians], and later the brilliant thirteenth-century quartet [the Song Dynasty quartet mentioned by us previously], unable to free themselves from the constricting influence of the *Jiu Zhang* [a mathematical work compiled during the -1st to 1st centuries that established China's mathematical tradition and formed the basis of all mathematical education since then]? Was the astonishing *continuity and stability of Chinese civilization* responsible, coupled perhaps with a great *reverence for the past*? Most likely we shall never know'. Lacking the concept of scholarly tradition as a self-reinforcing institution, in this comment, Joseph runs the risk of seeking the answer to his own query in cultural-essentialist terms—the italicized words are indistinguishable from the familiar 'cultural conservatism' theme. Incidentally, the purported 'astonishing stability' of China is a common misunderstanding derived from a dominant ideological narrative, an issue that we shall have occasion to address in Chap. 9.

We mentioned earlier that anti-essentialist scholars are reluctant to recognize scholarly tradition as an explanatory variable because they share the common essentialist misinterpretation of it. This misinterpretation constitutes a *conceptual* error. The mistaking of scholarly tradition-induced resistance or obstacles for purported cultural conservatism shows the *theoretical* consequences of this error. For anti-essentialists to throw the baby (recognizing the role of scholarly tradition) out with the bathwater (avoiding essentialism) results in a *great loss of explanatory capabilities* with dangerous *counter-productive* risks—counter-productive because the mistaken attempt to avoid essentialism leads precisely to the risk of falling into (cultural) essentialism.

Turning now to how the founding and development of a scholarly tradition can be analysed in non-essentialist terms, we use Greece and pre-Qin China as illustrations.[13] Concerning Greece, as mentioned above, the Greeks developed a theoretical tradition in general. The foundations of this tradition were laid by Parmenides and the early Pythagoreans. As Kirk et al. (1983, p. 213) observe: 'it was in South Italy [where Parmenides hailed and Pythagoras established his societies], not Ionia [where Greek philosophy first began], that the two most distinctive elements in modern conception of philosophy was born'. With regard to the early Pythagoreans, very briefly, they believed that things are literally numbers (female is literally 2, male 3, hence marriage is 5, and so on) and the divine cosmos is the 'perfect' number 10. In accordance with their beliefs, in order to seek salvation of the soul, it is necessary to understand the cosmos, which means to understand numbers, hence their approach to the study of numbers was not for the purpose of calculation, which they despised as belonging only to the marketplace, but to study numbers as numbers, which means, in modern terminology, to study numbers by adopting a theoretical approach. Hence, a distinction was drawn between their science of numbers (*arithmoi*) and 'logistics' or practical computation, which became the Greek tradition. The term *arithmos* expressed only a natural number, our conception of real number was unknown. In short, it was as a result of

[13] For a detailed account, see Lau (2020, Chaps. 1, 4, 6 and 10).

this combination of mystical beliefs—which, it should be noted, was a *contingent* occurrence—that their approach to the study of numbers was purely theoretical, making no reference to the real world of daily practice. And it was from such beginnings and under conditions that cannot be discussed here, that Greek mathematics subsequently took the theoretical geometrical turn, while computation atrophied.

With regard to Parmenides, as Guthrie (1965) discusses in detail, his philosophy was the outcome of the synthesis of several elements when he addressed the Ionians' speculations on what the constituent of all things is: (a) the exceptional feature of ancient Greek that there was only one existential verb which was also used as the copula; (b) the essentialist notion of names, which was derived from the belief in sympathetic magic, i.e. the belief that there was an essential connection between a name and what we call its referent in modern terminology, on the basis of which it was believed that each word could have only one 'natural' or 'proper' meaning; and (c) Parmenides remaining under the strong influence of mystical thinking—in the proem of Parmenides's philosophic poem in which he advanced his philosophy, he claimed to have been inspired by the Goddess, on which Guthrie (1965, p. 11–2) comments as follows: the 'general character [of the proem points to the] "shamanistic" strain in early Greek religions. One cannot doubt that the prologue describes a genuine experience'. It is clear that none of these three elements is of an essentialist nature. The essentialist notion of names was common among ancient cultures while the continued influence of mystical ideas and thinking on philosophers was also not uniquely Greek. The above-mentioned exceptional feature of ancient Greek is uniquely Greek, but it is simply a contingent fact, not reducible to essentialist causes. Further, the synthesis of (b) and (c) with (a) is itself a contingent occurrence. Hence, the beginnings of the Greek tradition originated (with reference to Parmenides) from a non-essentialist *double contingency*—this is exactly comparable to the fact that it was a certain contingent combination of mystical beliefs that underpinned the early Pythagoreans' approach to the study of numbers.

In Parmenides' philosophy, reality is said to be constituted by one single unitary, continuous, indivisible, unchanging, and complete Being, as to where it exists Parmenides did not say—most probably, he had not even suspected the existence of such a question—but it is definitely not in the empirical world. This is because according to Parmenides, the empirical world and all human experience of it (its movements, changes, and so on) are illusory. Parmenides arrived at this conclusion by means of some 'primitive' and 'crude' quasi-deduction.[14] Yet, on this basis, he claimed that to arrive at the Truth, that all human experience of the empirical world is useless. For reasons that need not concern us here, Parmenides' philosophy had a huge and extraordinary impact on Greek thoughts—for instance, Plato's doctrine of divine Forms was essentially to render Socrates in Parmenidean form. His just-mentioned claim was instrumental to the founding of Greece's theoretical tradition: Reasoning alone, without reference to empirical experience—just as in the early Pythagoreans' study of numbers, no reference was made to the world of real practices—is all that is required to arrive at the Truth.

Turning now to pre-Qin China, since there is no similar analysis that we are aware of in the English-language literature, our discussion will be somewhat more detailed. From ca. -1600 to −1046, China was under the Shang Dynasty, because writing was invented during its reign, the dynasty was culturally much more advanced than neighbouring tribal groups, and hence enjoyed great legitimacy among them. During Shang time, there was a culturally backward tribal group from today's Shaanxi province to the west of Shang known as the Zhou, who volunteered to undertake military service for the Shang in battling groups further west (Lao Siguang 1981a, pp. 58–9). In −1046, the Martial King of Zhou (*Zhou wu wang*周武王), supported by various allied groups, succeeded in overthrowing the Shang to found the Zhou Dynasty that was to reign for eight centuries. Three years later the Martial King died, and the Shang, supported by its allies, rose in rebellion. The Duke of Zhou (*Zhou gong*周公), brother of the Martial King who acted as regent, led a military expedition, and suppressed the rebellion.

Despite successfully putting down the rebellion, the Duke was keenly aware of the fact that in overthrowing the culturally sophisticated Shang, the Zhou, as a culturally backward people, faced great legitimacy problems with both the Shang people and their allies. With this awareness, in speeches made to the conquered Shang people after putting down their rebellion, the Duke explained why it was right for the Zhou to overthrow the Shang. He stated that the right to rule was given by the supreme deity

[14] 'Primitive' and 'crude' are descriptions used by Guthrie in discussing Parmenides' deduction. 'Quasi' refers to his following comment on Parmenides' reasoning: 'the ordering of its contents in the *guise* (our emphasis) of a formal deductive argument…'. See Guthrie (1965, p. 44, 1969, pp. 192–4, 218).

Heaven (*tian* 天), called the Mandate of Heaven (*tianming* 天命), he then explained that the Xia (夏) Dynasty (which preceded the Shang)[15] originally possessed the Mandate of Heaven, but subsequently lost it because the last king of Xia was corrupt and immoral, the Mandate of Heaven was then passed to the Shang, however, the last king of Shang also became corrupt and immoral, hence the Shang had also lost the Mandate, which had now been passed to the Zhou. This legitimation myth invented by the Duke of Zhou proved to be an epoch-making turning point in China's history.[16]

In Zhou's legitimation myth, the supreme deity *tian* was now no longer a whimsical being, as Shang's supreme deity was for them, for its giving or withdrawing of the mandate to rule was now governed by a non-supernatural principle, namely, whether or not the ruler accorded with moral virtuousness. Further, how to define this moral virtuousness was not the prerogative of *tian*, but was left entirely in the hands of humans.[17] Xu Fuguan (1975, pp. 22, 28, 61) calls this the humanitization (*renwen hua* 人文化) of religion. In his words, the supreme deity had now been nominalized (*xushe* 虚设), '*the basis of* humans' *confidence* gradually *passed from the deities to their own selves*', 'in early Zhou [i.e. after the founding of the Zhou Dynasty], offering sacrifices to the ancestors[18] *had changed its meaning from that of a religious nature to that of a moral nature*', '*the development of the spirit of humanitization* during the Spring and Autumn period [-1076 to -771] *thoroughly transformed* traditional religion into *morality*' (all emphases in the original). Lao Siguang (1981a, pp. 68–70) similarly notes that the 'intrinsic special character' of the resultant culture of Zhou was 'the affirmation of the power of human self-conscious efforts', 'the tradition established by this culture de-emphasized the mystical and the other-worldly'.

[15] Because of the lack of pre-Shang written records, there is a dispute as to whether or not the Xia Dynasty really existed. However, recent archaeological discoveries have tilted most historians towards accepting its existence.

[16] The following discussion is based upon the analysis of the venerable neo-Confucian historian of ancient Chinese thoughts Xu Fuguan (1975, Chap. 2), and in agreement with him, that of the leading historian of ancient Chinese philosophy Lao Siguang (1981a). There is also an account of the Duke's invention in Creel (1970, Chap. 5 and Appendix C).

[17] The contrast with what the Bible says concerning the Ten Commandments cannot be more striking.

[18] Both the Shang and the Zhou rulers believed that upon death their ancestors became lesser deities living as 'guests' of the supreme deity.

In accordance with their own legitimate myth, the Zhou needed to prove their moral strength. Thus, the moral-ethical notions of *de* (德 meaning virtue in the present context), *de xing* (德行 virtuous behaviour) and *de xing* (德性 virtuous character), among others, began to appear in early Zhou texts. In order to put virtue into practice, the Zhou established an elaborate system of rites (*li* 礼). Rites refer to rules of social etiquette for different individuals and social strata in different contexts of social practice. For the Zhou, the main objective of making people observe rites was to foster moral consciousness.[19] Thus, for instance, observing sacrificial rites transcended instrumental purposes, but became a moral end in itself. This was why Confucius was later to remark 'worship the deities as though they really existed' (*jishen ru shen zai*祭神如神在), in other words, practising worshipping rites was a moral end in itself even if the deities did not really exist.

This is, of course, not to say that as a result of the humanitization of religion, supernatural beliefs and practices had disappeared under the Zhou. Far from it. Poo (1998, p. 61) has summed up the situation well:

> The idea that a person's moral behavior was the sole reference by which the supernatural beings judged his supplications became popular among intellectuals of the Eastern [Z]hou period[20]…These different understandings of the relationship between human beings and spirits [offering sacrifices versus moral virtuousness], therefore, represent the difference between the religious mentality of average persons and that of the more reflective intellectuals and writers. In the Eastern [Z]hou period…the idea of "*do ut des*" [offering sacrifices in order to be blessed by the deities] was obviously still a widely accepted attitude among the ruling class…

In other words, as far as members of the ruling stratum and grassroots folks were concerned, supernatural beliefs continued to remain entrenched, but, in contrast, the humanitization of religion was deeply influential among intellectuals.

The need for the Duke of Zhou to invent a legitimation myth was a *contingent* event (a culturally advanced dynasty being overthrown by a

[19] As a matter of interest, this coincided with Michel Foucault's influential view that social discipline is achieved not so much through ideology but through social practices.

[20] The Eastern Zhou period refers to the period after −771 until the end of the dynasty. Poo focuses on this period because China's intellectual development, while originating earlier, began to rapidly take off after −771, and Poo's focus is on China's intellectual stratum.

backward tribe), it led to the humanitization of religion as an *unforeseen consequence*, further, this unforeseen consequence itself led to *another unforeseen consequence*, namely, the establishment of what we call a *naturalistic intellectual tradition* that first emerged late in the Western Zhou period (−1046 to −771). We can define naturalistic in this descriptive way: in this tradition, mystical notions are largely absent, and even when they are present, they generally play no role in the formation of concepts, doctrines, and theories. While we can, obviously, not give an account of this naturalistic tradition here, this can be briefly illustrated by the above quote from Confucius, in which he took an agnostic attitude towards the existence of deities, as well as by other similar remarks that he made, such as 'the master does not talk about weird phenomena [i.e. the mysterious], brute force, rebellion and deities' (*zi bu yu guai li luan shen* 子不语怪力乱神), which precisely shows the point that we are making, namely, that mysticism was sidelined by Zhou intellectuals whose focus was on humans. This strikes a sharp contrast with Greece where generally mysticism continued to exert an enormous influence on intellectuals.[21] Elsewhere (Lau 2020), we show how this difference between China (in the era before −221) and Greece had a huge impact on the different natural philosophies developed in the two places (this difference will be shown in Chap. 5), as well as on the difference in their approach to astronomical modelling (this difference will be briefly touched upon in the course of the discussion in Chap. 7). An indication of this can be seen from a point made in the above. Parmenides, who co-founded the Greek theoretical tradition along with the early Pythagoreans, claimed, on the Goddess' inspiration, that the empirical world and all human experience of it are illusory; in China's naturalistic intellectual tradition, such a notion could simply never have arisen.

Thus, in the cases of both Greece and China, a non-essentialist account can be given for the initial founding of a scholarly tradition. In both cases, among other factors, *contingency* played a central role in the starting point of the developmental process. In discussing the reproduction of a well-established scholarly tradition in the above, we mentioned that once

[21] For those who are accustomed to the view that ancient China was imbued with mysticism, let it be noted that mystical thinking made a huge comeback consequent upon the founding of the Qin Dynasty in −221 with the First Emperor engaging in a wholesale destruction of learning, which changed the entire course of China's intellectual development with the exception of astronomy and mathematics.

well-established, the factors or conditions that originally gave rise to the tradition's birth are no longer relevant to scholars raised in it, since from now on, it is the institutionalized features of the tradition that ensure its reproduction. We would like to make the following comment in relation to this with reference, again, to the case of Parmenides. In Parmenides' philosophy, there is a distinction between reality and appearance, the latter of which being illusory. Despite the semi-mystical origins of this notion, over two millennia later, an anti-mysticism scholar such as Marx could inherit it and make it a core basis of his entirely rationalistic political-economic theorization in *Capital*.[22] We call this kind of situation *ironies of history*, which can be in either direction—in contrast to the above case, early intellectual efforts that conform to scientific principles and spirit may subsequently prove counter-productive. The history of intellectual developments worldwide is full of such ironies. Alchemy, as another instance, originated from mystical beliefs and objectives, yet it led to the development of the experimental method. The prevalence of such ironies goes further to show the importance of contingency in history, since ironies are one form of unforeseen consequences.[23]

[22] Reference, for instance, the title of Goldway's (1967) paper: 'Appearance and reality in Marx's *Capital*'.

[23] Incidentally, we arrived at our notion of ironies of history during our investigation into Greece and China, in which we realized that some of our findings could best be made sense of by a crucial argument advanced by the neo-Darwinist palaeontologist Steven Jay Gould. Gould critiques the conventional understanding of evolution as a teleological process driven by a purported inexorable linear tendency and argues that evolution is a contingent process, in which what we call ironies of history often occur. The most well-known example of such ironies is the extinction of dinosaurs, which had been adaptively superior in the then prevailing ecology of the earth until the earth was struck by a massive asteroid 65 million years ago. Under the resultant new ecology, the dinosaurs' features that previously made them adaptively superior now, in an ironic twist, made them unable to adapt leading to their extinction; in contrast, mammals, which had previously been adaptively inferior became adaptively superior under the new ecology. The earth being struck by asteroids is not a contingent event—this happens all the time, but in relation to evolution, to be struck by an asteroid of that size and at that point in time in the evolutionary process is a contingent event. Had the asteroid been smaller, or had it struck at another point in time, the result would have been much different, and *Homo sapiens* might not have appeared at all.

3.5 Contingency, Institutionalization, and Path Dependency

As should have become obvious by now, the role of contingency in historical development cannot be emphasized enough. As noted above, Needham dismissed 'pure chance' as an explanatory variable; we think that this is because, for him, as for many others, contingency cannot be systematically pinned down as an explanatory variable. That, however, is not the case. It is necessary to recognize that whereas the appearance/occurrence of any *particular* contingent factor/event is accidental, it does not follow that the prevalence of *contingency in general in historical processes* cannot be conceptualized and hence analysed in a systematic way. Since the 1990s, there is a theory in historical sociology known as *path-dependency* (PD) theory; although its subject matter (social, economic, or political developmental processes) is different in nature from ours, parallels can still usefully be drawn between the processes that it is concerned with and the processes that we are concerned with.[24] PD theory was founded by economist P.A. David (1985) in a seminal paper concerning the QWERTY-formatted keyboard.[25] According to neo-classical theory, market forces would ensure that the most efficient keyboard design would become the industry standard. David argues how the QWERTY keyboard, despite not being the most efficient, became the industry standard. Details of his argument need not concern us here, except to note that a crucial contention of David's, and of the concept of path dependency in general, is that the outcome of a PD process is unpredictable by (deterministic) theory—neo-classical theory in this case.[26] In other words, PD is in contrast to historical explanation in terms of deterministic theory. Sociologists such as Aminzade (1992), Griffen (1992), and Sewell, Jr. (1996) quickly came to realize the usefulness of PD in historical-sociological analysis.

Mahoney (2000) distinguishes between two kinds of PD processes. In the first kind, called 'self-reinforcing sequence', a contingent event or factor sets into motion a *self-reinforcing process* in which a particular pattern is *institutionalized* and hence reproduced over time. The self-reinforcing mechanisms are deterministic, but because the event or factor that kicks

[24] Path-dependency theory is applied in various branches of the social sciences, here we discuss it with reference to historical sociology only.

[25] So called because this is the sequence from left to right of the first six keys of the top row of the alphabet keys.

[26] Cf. the third point we made in the above in defining contingency.

off the process is contingent, hence the outcome is unpredictable by theory. In this scenario, development is said to be *locked in* along a specific trajectory as a result of the self-reinforcing mechanisms, which pose *obstacles to trajectory change*. These mechanisms include such factors as sunk costs, durable commitments, and learning curves (Aminzade 1992, p. 463). Once there is lock-in, the self-reinforcing mechanisms will ensure the *reproduction* of the institutionalized pattern, and the contingent event or factor that originally kicked off the process subsequently becomes irrelevant to that reproduction (Stinchcombe 1987, pp. 103–4).[27] Due to the operation of these mechanisms, breaking away from the institutionalized trajectory rarely occurs, which is why in the QWERTY keyboard case, the inefficient QWERTY design remained as the industry standard despite challenges from more efficient designs that subsequently appeared. If trajectory change does occur, it can only be due to some huge external impact, and not from internally generated cause.

It is clear that the concept of self-reinforcing PD is eminently applicable to analysing the foundation and development of scholarly traditions in a non-essentialist way. As shown in the cases of Parmenides, the early Pythagoreans, and Zhou's legitimation myth, in the founding of a particular scholarly tradition, contingent factors may play a central role; a scholarly tradition is an informal institution, scholars nurtured in a particular tradition acquire its institutionalized goals, values, etc., and reproduce that tradition by virtue of their daily work, i.e. the institutionalized goals, values, etc. constitute self-reinforcing mechanisms. And, also as said, scholars nurtured in a particular entrenched tradition are locked into its pattern of doing scholarship, and hence highly resistant to change. In the above, this kind of lock-in was illustrated in our discussions of Kuhn and the Chinese mathematician Li Ye.

In Mahoney's second kind of PD process, called 'reactive sequence', a certain event A produces another event B in reaction, which in turn produces event C in reaction, and so on, until the process terminates with, say, event H. For Mahoney, the processes of generation from A to B to C and so on until H are deterministic, hence he stipulates A to be contingent, in order to make the outcome unpredictable by theory. In our view,

[27] This echoes the point we made earlier about what shapes members of a scholarly tradition which is sufficiently well established, normatively and dispositionally in their daily work, are the *institutionalized* features of the tradition, with the factors or conditions originally giving birth to that tradition being no longer relevant.

in order to make the various intermediate processes (A to B, B to C, etc.) all deterministic, one must specify some general mechanism that makes them so, and, unlike in the case where self-reinforcing mechanisms exist, there is no reason to expect that such a general mechanism does exist. Hence, we modify Mahoney's concept of *reactive-sequence PD* in the following way. A does not necessarily lead to B, it may instead lead to X or Y depending on what intervening factors are present, but the occurrence of B in reaction to A is completely explicable ex post facto; similarly for B to C, and so on. Hence, the occurrence of H from A through a number of intermediate reactive steps is explicable, but each step is non-deterministic, and H cannot be predicted by theory on the basis of the initial condition A, for it is *the path itself* that determines the final outcome.[28]

The PD perspective enables us to understand how and why contingency as an explanatory variable can be conceptualized and analysed in a systematic way. In the case of self-reinforcing PD, a particular contingent factor or event gives rise to a predictable process, though because that factor or event itself is contingent, the outcome is neither necessary nor predictable (by theory). On the other hand, contingency can be dismissed as haphazard *only if one begins with the assumption* that historical outcome is predictable on the basis of some deterministic initial conditions, such as some posited 'essence' of a particular society. In such a case, contingency, if given occasional de facto recognition when its effects cannot be denied, only constitutes an *intervening* variable that *only appears occasionally* to interfere with the *basic* deterministic process of development. If, however, it is the path of a historical process itself that determines its outcome which is neither necessary nor predictable, as conceptualized by the notion of reactive-sequence PD, then the *basic* process of development is contingent in nature, and contingency as a general phenomenon becomes a *core, endemic* variable, instead of being only something haphazard that occasionally interferes with necessary and predictable developments.

From the perspective of PD, *timing* is a most crucial factor in influencing the trajectory of a developmental process. A particular factor or event intervening at a particular time may have very different effects on the path

[28] Our modification of Mahoney is similar to the view of Goldstone (1998). In Lau 2020 (Chap. 5), we provide an analysis of how Parmenides' proposition, inspired by the Goddess under conditions mentioned previously, that empirical reality is illusory led through twists and turns in a reactive-sequence PD way to the emergence of the crowning jewel of Greek natural philosophy, namely, atomism. This case does not only illustrate reactive-sequence PD, but also constitutes a supreme irony of history.

taken by that process than had it intervened at another point in time. This can clearly be seen with reference to the cross-cultural adoption of scholarly traditions, discussed in the previous section. In the case of the Abbasids' (and, on a smaller scale, the Umayyads' before them) sponsorship of the adoption of learning from the Greeks and the Hindus, this occurred *before* Islamic society had developed any scholarly tradition, hence, no intellectual obstacles existed to the adoption; likewise, in the case of Latin Europe's learning from Arabic scholarship from the 11th and especially the twelfth century onwards, again, no existing, deeply entrenched and significantly different scholarly tradition posed any obstacles to it. In Chap. 7, we show with reference to Euclid's reception in China, how a different timing had opposite effects on the reception. Another illustration of the crucial importance of timing concerns evolution, as observed in note 23 above.

Related to timing is *the order in which events occur*, which can also exert great influence on the trajectory of development. From this angle, it is easier for scholars raised in a theoretical tradition to incorporate knowledges from an empirical-pragmaticist tradition than vice versa. This is because for scholars from a theoretical tradition, if and when they come into contact with knowledges from the latter kind of tradition that can demonstrate their practical superiority, this immediately constitutes an *external impact* for them.[29] Ptolemy provides an illustration of this.[30] In Greece, the early Pythagoreans were responsible for the notion that because the divine cosmos is 'perfect', celestial motions must be circular since roundness is 'perfect'. The notion of 'perfect' circular celestial motion is flatly contradicted by the observed irregularities of the movements of the visible planets. Whether or not the great mathematician Eudoxus (ca. -390 to −337) was a Pythagorean,[31] he set out to defend this a priori Pythagorean premise, even against Plato who otherwise was a staunch follower of the Pythagoreans in his post-Socratean days, by devising a geometric astronomical model of nested spheres, which constituted a classic case of what Popper critiques as ad hoc modifications. Because Eudoxus' sole concern was with defending the above-mentioned premise,

[29] This did not apply to the cases of Parmenides and the early Pythagoreans, who were *founding* a theoretical tradition, instead of being raised in such a tradition, on the basis of their being enmeshed in a nest of deeply held mystical beliefs.

[30] The rest of this paragraph is based on Lau (2020, Chap. 11).

[31] There is dispute in the literature concerning this issue, which very much depends on how one would define a Pythagorean.

he was not concerned with whether or not his model was physically possible (it was not), or possessed any operational capacity (it did not), or with whether or not it could stand up to empirical scrutiny (it could not, and he made no attempt to assess that). Like it was for Parmenides, for Eudoxus, reasoning alone (in Eudoxus' case, geometrical reasoning), without reference to the physical, the empirical and the practical, is all that is required to arrive at the Truth. It was in this way that Eudoxus founded Greece's astronomical tradition. Ptolemy was steeped in that tradition, but he happened to live several centuries after Eudoxus in Egypt,[32] in a scholarly environment in which there was a strong presence of the Babylonian practical-observational astronomical tradition. Astronomy, if it is to really constitute astronomy (which involves making predictions and serving calendrical purposes), must, in the final analysis, reckon with the practical and the observational. Hence, the Babylonian tradition's practical-observational superiority constituted an external impact on Ptolemy, which is why, although he continued to upkeep the defence of the a priori premise of 'perfect' circular motions (which made his models physically impossible), he incorporated the Babylonian tradition in his work.

By contrast, for scholars raised in an entrenched pragmaticist tradition, unless knowledges from a theoretical tradition can demonstrate that they can produce superior practical benefits, which would constitute an external impact (in Chap. 7, there will be an instance of this in relation to the introduction of the Tycho Brahe model to China), it is difficult for abstract theory, *as abstract theory*, to strike a chord with them. As the case of Li Ye above shows, it is not because they are incapable of thinking theoretically, but a matter of acquired goals, values, dispositions, and so on.

There is often a conflation between PD and sensitive dependence on initial conditions. As seen above, in the latter, the unpredictability (after a certain period) of a chaotic system results entirely from the fact that small changes in initial conditions cannot be known precisely, but the process leading from the slight change in initial conditions to its final outcome itself is completely deterministic; nor is the small change in initial conditions itself necessarily contingent, it is simply unknowable precisely. In

[32] This is another instance of the role of contingency in historical development, namely, the close proximity between Greece, Egypt, and Babylon. By way of contrast, as another contingent fact, China developed intellectually in isolation for more than one and a half millennia before first contact with non-Chinese inputs such as Buddhist logic.

contrast, in self-reinforcing PD, the initial condition is contingent, and in reactive-sequence PD, the process is non-deterministic.

3.6 Process Approach Versus Mechanistic Approach

Let us now combine the above into a unitary framework. Historical outcomes (such as the rise of modern science) are the product of historical processes; in studying historical processes, an analysis of starting point is of supreme importance, a relatively minor difference in starting point may in time lead to hugely different developmental trajectories and outcomes, lacking an understanding of this, explanations that focus simply on the outcome are likely to be inadequate or mistaken. Contingency plays a huge role in history, which may sometimes be largely responsible for the starting point of a particular historical development. In the study of intellectual development, both external and internal factors need to be considered; with regard to external factors, it is necessary to guard against falling into reductionism, it is also necessary to recognize the limits of externalist explanation[33]; internal factors can be categorized into different types, one concerns the agency of scholars (to be discussed in the next section), another concerns inputs from foreign scholarly sources, a third concerns scholarly tradition. The ideational aspects of a scholarly tradition cannot be reduced to external factors, although the process of its founding and development is invariably impacted by the effects of external factors. A scholarly tradition is a self-reinforcing institution in the full neo-institutionalist sociological sense, established from a particular starting point in which contingency may well figure substantially and through a process of institutionalization. Scholarly tradition consists of both formal aspects and substantive knowledges, with the former embodied in the latter; a scholarly tradition is reproduced and further developed through its institutionalized self-reinforcing features in the daily work of scholars raised in that tradition. In intellectual development, scholarly tradition exerts its effects in a deterministic way in the sense that these effects are explicable within determinate parameters; however, intellectual

[33] In general, since externalist explanations do not consider internal factors, these limits are obvious, but other than that, one can also think of the fact that intellectual development has a logic of its own which is independent of external factors, such as a particular idea invoking a specific response, a scenario that is well-captured by the concept of reactive-sequence PD.

development is a path-dependent process not only in the sense of self-reinforcing PD, but also in the sense of reactive-sequence PD, hence, the process as a whole is contingent and unpredictable,[34] with the result that the outcome is not a necessary one.

In formulating our framework, our intention is not to rule out deterministic theories in principle. Goldstone (1998) astutely observes that there is no one correct type of universally applicable theories to explain history, where similar patterns are observed across diverse societies, deterministic theories are likely to apply, where unique and exceptional outcomes are observed in a particular society, PD most likely applies. For instance, Greece's theoretical tradition in mathematics is in contrast to the computational tradition developed by the Chinese, the Egyptians, the Babylonians, and the Indians. Hence, the Greek development constitutes a unique exception, to account for which PD is best equipped. Similarly, the breakthrough into modern science constitutes a unique event, to account for which we believe our non-essentialist and non-deterministic framework is best suited.

As may have been noticed, our framework adopts a *process approach*. Such an approach sees developmental outcomes as being produced by a long series of linked moments in a continuous process, in which some factors (such as scholarly tradition) may exert their effects throughout or through most of the process, while other factors appear at different stages of the process exerting their effects on that process at those particular stages. Most of these factors do not directly produce the final outcome, but nonetheless contribute to the production of the final outcome. The process approach is in contrast to what we call the *mechanistic approach*. Such an approach sees intellectual outputs as being produced separately and individually in the same mechanical way, like billiard balls hitting one another, by the same factor or configuration of factors. Thus, Weber argued that *each and every one* of the 'unique' 'forms of rationality' found in, and allegedly only in, Europe has been produced in the same way, probably by the Europeans' 'biological heredity', and certainly by Europe's 'peculiar social structure'.

To give an indication of what is involved in a process approach, let us illustrate with regard to the point about different factors playing different

[34] On an a priori basis, this is so because, for one thing, of the many variables involved in the process, some of them will invariably be contingent in nature; of course, that this is, indeed, so in any particular case will have to be demonstrated a posteriori.

roles at different stages of the process that eventually lead, in a non-deterministic way, to the final outcome, with reference to a development that directly bears upon the rise of modern science. In the following discussion, the italicizing of certain particular words serves to highlight the elements of contingency, institutionalization, and path dependency in our explanatory framework. Despite Galileo's critique of the scholastics, the *institution* of the theologian-natural philosopher played a significant role. In a later chapter, we discuss their contributions in terms of ideas. For now, we focus on their contribution with reference to an external factor: 'The European university system, with its comparatively huge turnout of people with at least a nodding acquaintance with Greek thought, provided a deeper and also wider soil on which true mathematical talent could grow and on occasion flourish than in Islamic civilization…' (Cohen 2011, p. 206). Taking this to be valid for the sake of argument, it is the institution of the theologian-natural philosopher which enabled the European universities, founded on the basis of cathedral schools that trained the clergy, to contribute in the way mentioned by Cohen. How did this institution come about? To answer this question, it is necessary to go back to the early history of Christianity.

Christianity was born into a world in which secular (Greek) learning was already well established. Further, Christianity grew very slowly, taking four long centuries until it was made Rome's state religion. Thus, during this long period, Church Fathers were confronted with the question of how to deal with secular learning. Some opted to reject it altogether, saying that it was incompatible with theology; but the majority opted to find an accommodation with it by saying that it served to enable better understanding of the Bible. As is well-known, the latter is known as the handmaiden approach, on the basis of which the institution of the theologian-natural philosopher was established. Thus, this institution arose out of a double *contingency*: Christianity being born into a world of established secular learning, and the long time it took for Christianity to triumph. In contrast, under circumstances that are beyond our present scope, Islam spread by means of state conquest, and within a century, it had spread across the entire Mediterranean all the way to Spain, in relation to intellectual development, this constitutes a contingent fact; and all this occurred—as another contingent fact—before the taking up of Greek learning on a massive scale under the Abbasid caliphate. As a result of this double *contingency*, in Islamic civilization, there was strict separation between theology and secular learning. Thus, two different double contingencies produced two different outcomes that had a direct bearing on the long-term intellectual developments in Europe and Islamic civilization that Cohen talks about. Confining ourselves to Europe, on the surface, the above double contingency, which can in no way be interpreted in essentialist terms, seems far removed from the rise of modern science, yet, in a *path-dependent* way, it contributed to the latter in the way mentioned by Cohen through giving rise to the institution of the theologian-natural philosopher. Such an understanding can be achieved only by adopting the process approach.[35]

3.7 Scholarly Agency as a Non-essentialist Explanatory Variable

It was mentioned in the above that the Duke of Zhou invented a legitimation myth, which had profound unforeseen consequences. Why, instead of continuing to rely on coercive rule after suppressing the rebellion by the defeated Shang upon the death of the Martial King of Zhou, did he choose to invent such a myth? That might have been a wise choice, but that does not go any way towards explaining why he chose to be wise. It is difficult, indeed impossible, to subject this kind of social agency—the subjective choices of individual social agents—to explanation. But this is not the kind of agency that we are concerned with. Instead, as already observed in our discussion of internal factors in Sect. 3.3 above, the agency that we are

[35] A caveat should, perhaps, be added. The point of our argument is to show that without the institution of the theologian-natural philosopher, which was the outcome of a double contingency, the European university would not have had the positive effects mentioned by Cohen. It does not constitute any endorsement of some Eurocentric claims made about the European university found in the literature. Firstly, the university is not a European innovation, Goody (2006, pp. 222–32) has shown that universities and other institutional forms of higher learning and research had existed elsewhere predating the European university. Concerning other institutional forms, as Goody points out, while the university form 'was important in the long run but [that] did not mean that institutions of higher learning could not function in other fashions'. Secondly, the claim has often been made that the European university was different in that it enjoyed scholarly autonomy due to its unique corporate status, and that this autonomy was instrumental to the rise of modern science. But operational autonomy need not result in a progressive bent of mind, it may well instead lead towards the opposite direction. As is well known, the universities constituted a bastion of conservatism against the rise of modern science: 'the Aristotelians of the universities...was hostile to the new science emerging in the sixteenth and seventeenth centuries, thus forcing the latter to develop outside the universities, especially in the newly established scientific societies' (Grant 1996, pp. 168–9). And, in contrast to the autonomous universities, these newly established scientific societies all depended on political patronage.

concerned with refers to scholarly values, drives, and so on, which can be made subject to analysis—on a non-essentialist basis.

To begin with a typical argument that anti-essentialists find objectionable, the science historian Ronan (1983, p. 186, our emphases) states:

> [D]espite the fact that in so many fields Chinese science reached at a very early date a level of knowledge equal or superior to that of Europe in 1500,[36] there was no 'scientific revolution' in China: the breakthrough into the era of powerful modern science occurred in Europe not in the East. Why should this have been?...In China, the urgent ***impulse*** to exploration, to new ***discovery for its own sake*** never developed as it did in Renaissance Europe,[37] and there was no ***aspiration*** to break the mould of existing orthodoxies as inspired men such as Galileo. And one of the reasons for that must lie with the prevalence of the efficient but traditional bureaucracy of China, its rules and outlook defined by Confucius many centuries before.

Firstly, let us note en passant that Ronan's argument is self-contradictory, for Confucian culture had been dominant in China since the Han Dynasty (−206 to 220), hence, if 'its rules and outlook' were obstructive in the way described by Ronan, how, then, could it have been possible, in Ronan's own assessment, for Chinese science to have been 'equal or superior to that of Europe' until 1500? This, however, is not our concern here.

Ronan's view constitutes a case of cultural essentialism which is especially popular today—cf. Hobson's discussion of neo-Eurocentrism mentioned in Chap. 2. Discovery for its own sake is a value, both impulse and aspiration constitute drives. In the above, when we discussed internalist and externalist explanations, we mentioned that internal factors fall into two main categories, one of which concerns scholarly agency, and in briefly explaining what agency refers to, we mentioned values, drives and so on of

[36] Ronan shares Needham's view that China led Europe up to 1500 not only in technology but also in science, which is unsurprising, since he has abridged the various volumes of SCC into shorter versions.

[37] The term 'Renaissance' is somewhat confusing. In literature (e.g. Dante, 1265–1321) and painting (e.g. Giotto Bondone, ca. 1276–1337), or art in general, the Renaissance is generally taken to have begun in the late thirteenth century; in historical periodization, however, since the late Middle Ages is generally taken to end around the mid-fifteenth century, the Renaissance refers to the period from the second half of the fifteenth century to ca. 1600. Thus, in the present work, in the context of scholarly development (science and philosophy), by the Renaissance or Renaissance Europe, we refer to the second of the above senses; on the other hand, in the context of art, we use the term in the above first sense.

the scholars concerned. Anti-essentialists find talk of these factors objectionable because it is taken for granted, by both essentialists and anti-essentialists alike, that the latter can only be understood in essentialist terms. For instance, the previously mentioned Marxist-oriented geographer Blaut (1993, p. 208 n 2) has remarked that to accept the role of agency, 'one would have to believe that there is something absolutely fundamental in European culture...that makes Europeans different from other humans. This admits a good part of the Eurocentric claim that Europeans are unique among humans'. Such a view is, however, mistaken. In discussing the formal aspects of a scholarly tradition, we mentioned that these aspects include values, orientations, approach, drives, dispositions, and so on, and we referred to the case of the Chinese mathematician Li Ye to illustrate how practising the computational approach was a matter of acquired disposition for him. Since the establishment of a scholarly tradition can be made subject to non-essentialist analysis, it follows that speaking of approach and acquired disposition can also be done on a non-essentialist basis. The same is true when it comes to scholarly agency in general. In Chap. 8, we will come to the 'impulse' and 'aspiration' of scholars such as Galileo; for now, to illustrate why the 'qualities' mentioned by Ronan can be made subject to non-essentialist analysis, let us focus on values alone.

What Ronan calls 'discovery for its own sake' is part and parcel of 'knowledge is its own reward', which, as is well-known, is a value of the Greek scholarly tradition that was taken up by Latin Europe. Much essentialist capital (sometimes with bio-psychological undertones) has been made of this value of the Greeks, but Guthrie (1969, p. 18) comments on it as follows: 'It is too easily assumed that the Greeks as a whole believed in an ideal of knowledge for its own sake, divorced from practical aims, and despised the useful arts, and there is some justice in recent claims that this generalization results from the academic habit of relying too heavily on Plato and Aristotle as representative of the Greek mind'. Thus, Guthrie warns us against assuming that it is in the 'spirit' or 'soul' of the Greeks generally to pursue knowledge for its own sake. How did Plato and Aristotle come round to such a value?

The first thing to note is that whereas the Greek scholarly tradition consummated in Plato and Aristotle, its foundations were not laid by them, but, as pointed out above, by Parmenides and the early Pythagoreans. As seen, Parmenides certainly did not pursue knowledge for its own sake, rather, his philosophy was inspired by the Goddess mentioned in the proem of his philosophic poem under circumstances already discussed. Pythagoras and the early Pythagoreans believed in the immortality and transmigration of the soul, a belief that had diffused from India to Greece (Kahn 2001, p. 19), that the universe is a 'perfect' number, and that salvation of the soul from the transmigration cycle and its return to the soul of the universe had to be sought from understanding the universe.[38] Their philosophy and approach to the study of numbers were the outcome of seeking salvation under the circumstances of such a combination of mystical ideas[39] and Pythagoras' inheritance of the Milesians' natural-philosophical project. Like Parmenides, the early Pythagoreans were not seeking knowledge for its own sake.

The same applies to the central figure of Socrates: 'Xenophon lays great emphasis on the primarily utilitarian character of Socrates's arguments. In general, this was right'; '[W]hat was his [Socrates'] motive [in insisting on defining terms]? Again our authorities are unanimous, that it was not scientific but practical' (Guthrie 1969, pp. 420–1, 431). Plato himself was far from being motivated by seeking knowledge for its own sake: '[Plato was] under the influence of two related motives. He wished first of all to take up Socrates' task at the point where Socrates had had to leave it, to consolidate his master's teaching and defined it against inevitable questioning...It fitted in with his second motive, which was to defend, and to render worth defending, the idea of the city-state...For it was by accepting and developing Socrates's challenge to the Sophists that Plato thought this wider aim could be most successfully accomplished' (Guthrie 1967, p. 81). Even when Plato had outgrown Socrates' influence on him, 'It was certainly his intention that many of his pupils should leave the academy for politics' (Guthrie 1978, p. 23).

[38] On Pythagoras and the early Pythagoreans with regard to the present discussion, see Lau (2020, Chaps. 10 and 11).

[39] That intellectual developments, even crucial ones, can originally have been inspired by mysticism is something that has long been recognized. Both Kuhn (1957, pp. 128–31) and Yates (1979[1964], pp. 153–4) suggest that Copernicus's heliocentrism was inspired by Hermetic sun worship, which has caused much controversy. Kuhn (1996, pp. 152–3) further notes Kepler's sun worship for his adoption of heliocentrism—this is much less controversial for Kepler regarded the sun as God-the-father. Dobbs (2008[1975]) has suggested that Newton's concept of force may have been inspired by his alchemical pursuits. It must, however, be stressed that to point out extra-rational inspiration is not to say that that inspiration cannot subsequently be developed along lines that conform to rational or scientific criteria.

The regard of knowledge being its own reward had a lot to do with Greece's theoretical scholarly tradition. In what we call a pragmaticist tradition, which prevailed in Babylon, Egypt, India, and China, such an idea, *as a value*, was most unlikely to have arisen, despite the fact, confining ourselves to China with which we are more familiar, that numerous leading intellectuals did *in practice* pursue their studies in a way that can only be described as seeking knowledge for its own sake. For instance, one would be hard-pressed to try to find any practical purposes being served by the spatiotemporal propositions of the mid-Warring-States-period thinker Hui Shi (惠施), or for the extensive investigations into various areas of science engaged in by the premier Song Confucian scholar Zhu Xi (朱熹). As consummators of Greece's theoretical tradition, the foundations of which were, to repeat, laid by Parmenides and the early Pythagoreans, reaping all the benefits (and hazards, for that matter) that it brought, Plato and, especially, Aristotle were in a position to revere knowledge for its own sake. As previously said, in studying historical processes, analysis of starting point is of supreme importance; and once a tradition has been established, its inheritors need not share the motives of its founders to reproduce it. Thus, it matters not one iota to the regarding of knowledge being its own reward to be established as a tradition that this tradition owed its origins to thinkers for whom there did not exist any such motive or value, and, much later, for Latin European intellectuals to pick up that tradition and value.

Not only can the values of a scholarly tradition be made subject to non-essentialist analysis, the same applies to values *impacting on the intellectual climate in general.* Lindberg (2007) speaks of Latin Europe's 'turn to rationality' beginning in the eleventh century. Why such a turn, i.e. why had rational learning become a value among intellectuals? Was it due to some essentialist European quality? The answer is not difficult to find. 'By the end of the eleventh century, Christian Europeans were well aware that the Arabs in Spain and Sicily had attained a far higher level of knowledge and expertise in natural philosophy and the sciences than they had. Although they were undoubtedly aware of this long before the end of the eleventh century...' (Grant 2004, p. 165). It was *also* during the eleventh century that Latin Europe's counter-offensive against Islam began—the Spanish Reconquista began in 1063 and the first Crusade to the Holy Land started in 1095. That Latin Europe was launching a counter-offensive against Islam while feverishly learning from the Arabs at the same time has sometimes been commented upon, but looked upon simply as an irony. But it is far more than simply an irony.

Dante (1265–1321) consigned Muhammad to the eighth circle of the Inferno. Before Dante, Pope Innocent III (1160 to 1216) had described Mohammad as the 'Beast of the Apocalypse', 'Islam was seen as the negation of Christianity; Muhammed as an imposter, an evil sensualist, an Antichrist in alliance with the Devil. The Islamic world was seen as anti-Europe' (Kabbani 1986, Chap. 1, the quote is on p. 5).

> Indeed the return to the classical world was seen as the answer to the threat from Islam to European culture. George Sarton commented that the humanists' love of Greece and Rome was nourished by their *hatred of Islam*. To counter the culture of the Arabic world, the revival of antiquity was carried forth on many fronts. He wrote that the Renaissance was 'partly a reaction against influences (especially those represented by Avicenna and Averroes). The *anti-Arabic* drive was in full swing in Petrarch's [1304-1374][40] time'. (Goody 2004, p. 48, our emphases)

> [The 15th-century mathematician Regiomontanus] treated mathematical science in Islam as just a foreign interlude, with the Arabs temporarily preserving for Europe its Greek heritage. (Cohen 2011, p. 101)

In other words, learning feverishly from the Arabs was in order to overcome Latin Europe's knowledge deficit in relation to their hated enemy that Latin Europe was painfully aware of. That was why: 'The translations [of the first translation movement], however, were overwhelmingly in the realm of science and natural philosophy', not in 'literature, poetry [or] history' (Grant 1996, p. 172). It can, hence, be seen that the 'turn to rationality' had not arisen endogenously through purely internal Latin European dynamics, but was engendered by the above circumstances in Latin Europe's external relations with the Islamic world. This rivalry with and hatred of Islam is a perennial undercurrent in the entire period that we are concerned with, i.e. from the 11th to the 17th centuries, that impacted importantly on the agency of Latin European intellectuals. We return to this point in Chap. 8.

[40] Incidentally, it is worth noting that Dante was a friend of Petrarch's father; and that they all hailed from Tuscany.

Bibliography

Aminzade, Ronald. 1992. 'Historical sociology and time'. *Sociological Methods and Research* 20(4), pp 456-80.

Bauman, Zygmunt. 1991. *Modernity and Ambivalence*. London: Polity.

Berger, Peter, and T. Luckmann. 1967. *The Construction of Social Reality*. New York: Doubleday Anchor.

Blaut, James. 1993. *The Colonizer's Model of the World*. London: Guilford Press.

Chen Cheng-I. 1996. *Early Chinese Work in Natural Science*. Hong Kong: University of Hong Kong Press.

Chen Fangzheng. 2009. *Inheritance and Rebellion: Why Modern Science Emerged in the West*. 陳方正，《继承与叛逆:现代科学为何出现于西方》(北京:三联，2009)。

Cohen, H. Floris. 2011. *How Modern Science Came into the World: Four Civilizations, One 17th-Century Breakthrough*. Amsterdam University Press.

Creel, Herrlee. 1970. *The Origins of Statecraft in China, Volume 1, The Western Zhou Empire*. Chicago: University of Chicago Press.

David, Paul A. 1985. 'Clio and the economics of QWERTY'. *American Economic Review* 75(2), pp 332-7.

Dobbs, B.J. Teeter. 2008[1975]. *The Foundations of Newton's Alchemy*. Cambridge University Press.

Giddens, Anthony. 1984. *The Constitution of Society*. Berkeley: University of California Press.

Goldstone, Jack A. 1998. 'Initial conditions, general laws, path dependence, and explanation in historical sociology'. *American Journal of Sociology* 104(3), pp 829-45.

Goldway, David. 1967. 'Appearance and reality in Marx's *Capital*'. *Science and Society*, 31(4), pp 428-47.

Goody, Jack. 2004. *Islam in Europe*. Polity.

Goody, Jack. 2006. *The Theft of History*. Polity.

Gould, Steven Jay. 1989. *Wonderful Life: The Burgess Shale and the Nature of History*. New York: W.W. Norton.

Gould, Steven Jay. 2002. *The Structure of Evolutionary Theory*. Cambridge, MA: Belknap.

Grant, Edward. 1996. *The Foundations of Modern Science in the Middle Ages*. New York: Cambridge University Press.

Grant, Edward. 2004. *Science and Religion 400B –AD 1550*. Baltimore: John Hopkins University Press.

Griffen, Larry. 1992. 'Temporality, events, and explanation in historical sociology'. *Sociological Methods and Research* 20(4), pp 403-27.

Gutas, Dimitri. *1998. Greek Thought, Arabic Culture. The Graeco-Arabic Translation Movement in Baghdad and Early Abbasid Society (2nd-4th/8th-10th centuries)*. Routledge.

Guthrie, W.K.C. 1965. *A History of Greek Philosophy, Vol. 2, The Presocratic Tradition from Parmenides to Democritus*. Cambridge University Press.

Guthrie, W.K.C. 1967[1950]. *The Greek Philosophers: From Thales to Aristotle*. Routledge.

Guthrie, W.K.C. 1969. *A History of Greek Philosophy, Vol. 3, The Fifth Century Enlightenment*. Cambridge University Press.

Guthrie, W.K.C. 1978. *A History of Greek Philosophy, Vol. 5, The Later Plato and the Academy*. Cambridge University Press.

Gutting, Gary. 1989. *Michel Foucault's Archaeology of Scientific Reason*. Cambridge: Cambridge University Press.

Ingram, Paul, and K. Clay. 2000. 'The choice-within-constraints new institutionalism and implications for sociology'. *Annual Review of Sociology* 26, pp 525-46.

Joseph, George G. 2011. *The Crest of the Peacock Non-European Roots of Mathematics*, 3rd edition. Princeton: Princeton University Press.

Kabbani, Rana. 1986. *Europe's Myth of the Orient*. Indiana University Press.

Kahn, Charles H. 2001. *Pythagoras and the Pythagoreans A Brief History*. Indianapolis: Hackett.

Katz, Victor. 1993. *A History of Mathematics: An Introduction*. New York: HarperCollins.

Kirk, G.S., J.E. Raven, and M. Schofield. 1983. *The Presocratic Philosophers*, 2nd edition. Cambridge: Cambridge University Press.

Kuhn, Thomas. 1957. *The Copernican Revolution*. Harvard University Press.

Kuhn, Thomas. 1996[1962]. *The Structure of Scientific Revolutions*, 3rd edition. Chicago: The University of Chicago Press.

Lao Siguang. 1981a. *New Edition History of Chinese Philosophy*, Volume 1. 勞思光, 《新編中國哲學史(一)》(台北: 三民書局, 1981a).

Lau, Raymond W.K. 2020. *Intellectual Developments in Greece and China: Contingency, Institutionalization and Path-Dependency*. Newcastle upon Tyne: Cambridge Scholars Publishing.

Lindberg, David. 2007. *The Beginnings of Western Science*, 2nd edition. Chicago: University of Chicago Press.

Mahoney, James. 2000. 'Path dependence in historical sociology'. *Theory and Society* 29(4), pp 507-48.

March, James and H. Simon. 1985. *Organizations*. New York: John Wiley.

Merton, Robert. 1936. 'The unanticipated consequences of purposive social action'. *American Journal of Sociology* 1(6), pp 894-904.

Nee, Victor. 1998. 'Sources of the new institutionalism'. Pp. 1-16 in Mary Brinton and Victor Nee (eds.), *The New Institutionalism in Sociology*. New York: Russell Sage Foundation.

Needham, Joseph, with the collaboration of Kenneth Girdwood Robinson and Ray Huang (Huang Jen-yu), and including contributions by them, Introduction by Mark Elvin, edited by Kenneth G. Robinson. 2001. *Science and Civilisation in China, Volume 7, Part II: General Conclusions and Reflections.* Cambridge University Press.

Parsons, Talcott. 1951. *The Social System.* New York: Free Press.

Pomeranz, Kenneth. 2000. *The Great Divergence: China, Europe and the Making of the Modern World Economy.* Princeton: Princeton University Press.

Poo Mu-chou. 1998. *In Search of Personal Welfare A View of Ancient Chinese Religion.* Albany: State University of New York Press.

Powell, Walter, and P. DiMaggio. 1991. 'Introduction'. Pp. 1-38 in Idem. (eds.). *The New Institutionalism in Organizational Analysis.* Chicago: The University of Chicago Press.

Ronan, Colin. 1983. *The Cambridge Illustrated History of the World's Science.* Cambridge: Cambridge University Press.

Ruelle, David. 1991. *Chance and Chaos.* Princeton: Princeton University Press.

Scott, W. Richard. 2001. *Institutions and Organizations*, 2nd edition. Thousand Oaks: Sage.

Selznick, Philip. 1949. *TVA and the Grass Roots.* Berkeley: University of California Press.

Sewell, Jr., W.H. 1996. 'Three temporalities: Toward an eventful sociology'. Pp. 262-4 in Terence J. McDonald (ed.), *The Historic Turn in the Human Sciences.* Ann Arbor: University of Michigan Press.

Stinchcombe, Arthur. 1987. *Constructing Social Theories.* Chicago: University of Chicago Press.

Sun Hong'an. 2008. *Ancient Chinese Mathematical Thoughts.* 孙宏安, 《中国古代数学思想》(大连:大连理工大学出版社, 2008).

Xu Fuguan. 1975. *History of Chinese Doctrines of Human Nature: The Pre-Qin Period.* 徐复观, 《中國人性論史先秦篇》(台北:商務, 1975).

Yates, Frances. 1979[1957]. *Giordano Bruno and the Hermetic Tradition.* University of Chicago Press.

CHAPTER 4

Greece and the Needham Question

The Greek tradition played a crucial role in the making of modern science as it had arisen in Europe, but the role of Greece in the analysis of the NQ has caused a lot of confusion among both Eurocentric and non-Eurocentric scholars alike. The cause of this confusion is the following. In both scholarly and popular discourses, it is taken for granted that there exists a 'trans-historical' entity known as the 'West' which began with Greece and which has continued up to present-day Euro-America, inclusive of Australia and New Zealand. But, in fact, the notion of the 'West' has not always existed, and its existence as a 'trans-historical' entity is a fiction, as Greece and Latin Europe were two different and separate societies and civilizations. The creation of this fiction was closely connected to the construction from the late 18th to the early 20th centuries of the narrative that this 'trans-historical' entity (the 'West') had *autochthonously* produced a particular civilization (which included modern science) called 'Western' civilization. In other words, the notion of the 'West' is part and parcel of the formation of the Eurocentric[1] view and the basis of the AWD paradigm. The anthropologist Wolf (1997 [1982], pp. 5–6, our emphases) has pointed this out way back in the early 1980s:

[1] At the risk of belabouring the obvious, let it be mentioned that Europe in Eurocentrism is meant in the cultural-civilizational sense.

R. W. K. Lau, *Rethinking the Needham Question*,
https://doi.org/10.1007/978-981-97-9472-0_4

> We have been *taught*, inside the classroom and outside of it, that there exists *an entity called the West*, and that one can think of this West as a society and civilization *independent of and in opposition to* other societies and civilizations. Many of us even grew up *believing* that this West has a genealogy, according to which ancient Greece begat Rome, Rome begat Christian Europe, Christian Europe begat the Renaissance, the Renaissance begat the Enlightenment... Such a developmental scheme...a tale of...the unfolding of a timeless *essence*.

Regrettably, four decades after Wolf, non-Eurocentric scholars continue to speak of the 'West', possessing the purported 'trans-historical' genealogy as pointed out by Wolf, as readily as Eurocentric scholars do—in accordance with what they have been taught into believing.[2] The most crucial condition for making the 'West' 'trans-historical' is to make Greece part and parcel of it, and indeed its very origins. Among post-Wolf non-Eurocentric scholars, we are aware of only one instance in which the regarding of Greece as the alleged origins of the 'trans-historical' 'West' is problematized, namely, Hobson (2006, pp. 227–8, our emphases): 'European thinkers suddenly [in the late 18th to early 19th centuries] elevated Ancient Greece to the birthplace of European civilization...the *fabrication* of Ancient Greece was crucial to the *Eurocentric* construct'. Unfortunately, however, Hobson does this in a work that is titled *The Eastern Origins of Western Civilisation*, employing precisely the Eurocentric terminology ('Western') that, in his own words, 'fabricat[es]' Greece to be 'the birthplace of European ['Western'] civilization'.

In this chapter, we deconstruct[3] the notion of the 'trans-historical' 'West' by means of what Foucauldians call a genealogical analysis, that is, by tracing its historical genesis, with the objective of showing that *Greece and Latin Europe were two different societies instead of being succeeding stages of one and the same 'trans-historical' entity, with one (Greece) allegedly developing into the other (Latin Europe), through the interlude of Rome and the early Middle Ages, as an internal and autochthonous phenomenon of that 'trans-historical' entity.* This deconstruction is indispensable for the debunking of the Eurocentric position. The question may be raised as to

[2]We can testify to this from our own communications with various non-Eurocentric scholars.

[3]By deconstruct, we mean debunking the notion of the 'West' which is a construction, no allusion is made to 'deconstruction' in the sense of Derrida, with whose theory we have little in common.

whether or not deconstructing the notion of the 'West' is as important today with the rise of neo-Eurocentrism as it was previously, since unlike conventional Eurocentrism, neo-Eurocentrism no longer speaks of the development of science up to and including the breakthrough into modern science, as an autochthonous development all the way from Greece down to the seventeenth century. The answer to this question is an emphatic 'yes'. Firstly, the rise of neo-Eurocentrism does not mean that conventional Eurocentric arguments are no longer made; far from it, as shown in Sect. 4.7 below, given the role of the Greek tradition in the rise of modern science, without the deconstruction of the notion of the 'West', some Eurocentric scholars, who fully accept that Arabic scholarship took Greek learning to new heights as well as accept that Latin Europe was indebted to Arabic scholarship, i.e., accept the dialogical position with regard to these episodes, can continue to argue that the rise of modern science was an embodiment of the 'spirit' of 'Western' civilization. *Without deconstructing the notion of the 'West', dialogism faces the danger of being reduced to a sideshow*. Secondly, by showing that Greece and Latin Europe do not belong to the same 'trans-historical' entity, but constituted two different societies, the case for dialogism can, on the contrary, be immensely strengthened.

Before going into our analysis, a terminological note is in order: In the present work, except in quotes from other scholars, when the terms *west* (and its cognates) and *east* (and its cognates) are used *without* quotes, they are meant in the locational-directional sense, this applies to proper nouns as well such as Western Roman Empire and Eastern Roman Empire; when the terms *'West'* (and its cognates) and *'East'* (and its cognates) are used *with* quotes, they refer to notions which constitute our present subject matter. Also, except in quotes from other scholars, we use the term *Europe* (and its cognates) in a loose geographical sense—'loose' because geographical demarcation at the margins can be subject to dispute, which need not concern us here; when it is used in a non-geographical sense, such as in the cultural-civilizational sense (for example, see note 1), it will be qualified as such.

One final note: In *daily discourse* today, the 'West' is often used as a collective noun to denote a group of countries, in contradistinction to, for instance, the 'global south'. Of course, even as such a collective noun, the idea that the 'West' as purportedly a 'trans-historical' civilization is never far beneath the surface; nonetheless, on these occasions, this idea remains only in the background, hence, the use of the term *'West'* as such a collective noun may, perhaps, be somewhat 'excusable', as it were. However, in the context of the *scholarly discussion of intellectual development*, the term is inseparable from the Eurocentric position, hence, it needs to be strictly avoided by non-Eurocentric scholars.

4.1 The Notion of the 'West' as a 'Trans-historical' Entity Originating in Greece

Let us begin by showing the taken-for-granted-ness of the notion of the 'trans-historical' 'West' in scholarship. Again, we start with Weber's preface to the *Collected Essays on the Sociology of Religion* (1920–21). As mentioned in Chap. 2, Weber stated that in science, mathematics (he cites ancient *Greek* geometry as illustration), historiography (he cites *Thucydides* as illustration), art, bookkeeping, etc., the 'West' (he used the term 'Occident') has developed 'unique' 'forms of rationality' (Weber 1992, pp. xxviii ff., all emphases in quotes from Weber are ours). 'It is hence,' he stated, 'our first concern to work out and to explain genetically the special peculiarity of *Occidental* rationalism'. He further stated: 'When we find again and again that, even in departments of life apparently mutually independent, certain types of rationalization have developed in the *Occident*, and only there' (pp. xxxix-xlii). He believed that the underlying reason lies in 'biological hereditary', but remarked that before science can prove that, the sociologist will have to be contented with seeking an explanation from 'the peculiarities of the social structure of the *Occident*' (p. xxxvii), in comparison with that of other civilizations.

In the *Cambridge Illustrated History of the World's Science*, the science historian Ronan (1983, p. 124, our emphasis) describes the work of the ancient Greeks as 'mark[ing] the first great concerted effort of *western* man [sic] to understand the workings of nature'. Again, the category of 'western' goes back to the Greeks. This notion of the 'West' also constitutes 'common sense' among non-Euro-American scholars too. For instance, the Chinese philosopher of science Wang Shanbo (1996, p. 11, all emphases are ours) states: '*Europe* [obviously meant in the cultural-civilizational sense]...developed a different context, and that is the tradition of establishing laws of nature...the *Middle Ages* and the *Renaissance* played a special role...the modern scientific tradition was born in this kind of context. Since *ancient Greece*, the spirit of establishing laws of nature developed in *the West*…'. This is all in accordance with what, as Wolf points out, we have been taught to believe in.

Concerning the terms 'West' (and its cognates 'Western', 'Western man [sic]', 'Occident', etc.) and its purported counterpart 'East' (and its cognates), Needham (1981, p. 107) once commented:

> nothing annoys me more than the use of the words 'Eastern' and 'Western', still worse 'Oriental', because the Arabs, and the Indians and the Chinese, differed from each other really even more profoundly than the Europeans differed from some of them.

The careful reader will have noticed that whereas Needham explained why the term 'Eastern' is problematic, he provided no explanation as to why the term 'Western' also 'annoyed' him. Instead, despite complaining about the term 'Western' in the same breath as complaining about the term 'Eastern', he routinely used the term 'West' and its cognates without any hesitation. For instance, in the same work as the above quote from him is taken, within the space of eight pages (to illustrate with just this instance which is chosen randomly), we find him using that term or its cognates a total of four times: 'Bury showed long ago that before the time of Francis Bacon only very scant rudiments of the conception of progress are to be found in *Western* scholarly literature'; 'the idea of…was much more customary in medieval China than anywhere in the pre-Renaissance *West*'; 'the widespread *Western* belief that…'; 'The Hebrews were perhaps the first *Westerners* to give a value to time' (ibid., pp. 116–7, 118, 121, 123, all emphases are ours).

For something to exist trans-historically as the *same* entity, it must possess some *immutable* attribute that does not change over time, otherwise, it cannot be said to have remained the same entity. To speak of the 'West' as a 'trans-historical' entity, what can this something be which is immutable? Can it be culture? A moment's thought will show that it cannot, for culture develops and changes over time. For instance, in terms of religious belief, in what sense can we regard Greece, with its polytheistic religion, as belonging to the same entity as say pre-Reformation Christian Europe (i.e., the part of Europe under the Roman Catholic Church) with its monotheistic religion? Or, with respect to the cultural area of learning, in what sense can Greece be regarded as belonging to the same entity as the Barbarian kingdoms? Thus, in the term 'Western culture or civilization', what is implied is not that the 'West' is 'trans-historical' by virtue of its culture or civilization, rather it is assumed that the culture or civilization that is being spoken of has been produced over time through succeeding stages by one and the same 'trans-historical' entity which is called the 'West'.

If not culture, can it be language, i.e., the Indo-European language? The trouble with this is that both Sanskrit and ancient Persian also belong to the Indo-European language family. Are Sanskrit speakers and ancient Persians 'Westerners' then? As seen, Needham regards, as do many others after the Holocaust, Israelis as 'Westerners', but Hebrew belongs to the Semitic family of languages, which is a branch of the larger Afroasiatic family, not the Indo-European.[4]

Can it, then, be geography that makes the 'West' 'trans-historical'? Disregarding the issue of how to set geographical limits, even if we subscribe to geographical (inclusive of climatic) determinism, the geographies of, for instance, Greece and the Rhine area (from where the Franks hailed) are very different. As is to be seen later, no less than Aristotle asserted that the Greeks were 'superior' to Europeans (he did not count the Greeks as Europeans) precisely because of differences in the climate of the two places.

If neither culture, language, nor geography can serve as the supposedly immutable attribute that makes the 'West' a 'trans-historical' entity, what else can serve to do so? Or is the existence of such an entity a fiction altogether? To answer these questions, it is necessary to begin with the issue of situating Greek civilization in history.

4.2 Situating Greek Civilization in History

The term 'Europe' originated from the ancient Greek myth in which the Phoenician king in Tyre had a daughter named Europa. The supreme divinity Zeus fell in love with Europa, turned himself into a bull and appeared in front of her, she climbed on his back, whereupon Zeus dived into the sea, eventually taking her to Crete. Etymologically, 'Europa' possibly originated from the Semitic word meaning 'land of the sunset' or 'darkness'. Did the myth imply that Greece originated from Asia culturally? In any case, by the time of Herodotus, the Greeks divided the world geographically into Asia, Libya (renamed Africa since Roman times) and

[4] There is a claim that Hebrew was linked at a very early date to the Indo-European family; though the claim is only linked to, not belonging to, it has still not found general acceptance among historical linguists (Fischer 1999, p. 67).

Europe, but the word 'Europe' 'had very little general meaning in the Greek and Hellenistic worlds. The unit...which mattered was the Mediterranean' (Hay 1968, pp. 1–2; Hay 1980, p. 1).

The eastern Mediterranean was a cultural melting pot where Europe, Asia, and Africa merged. Before the -fifth century, the Greeks regarded their neighbours as 'partners, equals and cultural leaders'.[5] In −492 and −480, the Persians twice attacked Greece, many city-states were ready to negotiate with the Persians, but Athens formed a confederation with some other city-states to fight the Persians. The defeat of the Persians greatly raised Athens' status. From then on, Athenians and their followers distinguished Greeks from 'barbarians'. The Greek word *barbarous* originally meant unclear speech, subsequently extended to mean non-Greek speakers. The distinction between Greeks and 'barbarians' originally targeted only the Persians,[6] but it later came to cover all non-Greeks.

To enhance its leadership legitimacy, Athens contrasted its own democracy with 'barbarian' dictatorship, pointing its finger not only at Persia, but also Greek city-states ruled by tyrants and not submitting to Athenian leadership. Further, the Athenians ridiculed Thebes and Argos (both not submitting to Athens) for being founded by 'barbarians' (see Sect. 4.6 below). 'Barbarians' were described as stupid, cowardly, brutal, slavish, and so on. Aristotle completely subscribed to this, in Book 7 of the *Politics*, he advanced a geographical deterministic argument, saying that due to Europe's cold weather, Europeans were courageous but lacked skill (intellect), due to Asia's hot weather, Asians possessed skill but were slavish, whereas the Greeks, due to Greece's suitable climate, not only possessed skill, but were also courageous. Aristotle did not count the Greeks as Europeans.

Aristotle attributed skill to Asians because he well knew that Greek culture and learning neither had anything to do with those whom he called Europeans, nor were they autochthonous, but were rooted in Africa (Egypt) and west Asia. For instance, the alphabet came from Phoenicia; the role of *The Iliad* in the founding of Greek culture is well-known, less well-known is the fact that it is actually based upon the Babylonian myth

[5] This paragraph and the next one are based upon Hall (1989, pp. 4–5, 10–2, 16–7, 59–60, 166–9 and 179), except indicated otherwise.

[6] Concerning the Greeks' labelling of the Persians as 'barbarians', Goody (2006, p. 27) observes: 'In fact the Persians were as "civilized" as the Greeks, especially among the elite. And they were the main way in which knowledge coming from literate Ancient Near Eastern societies was transmitted to the Greeks'.

The Epic of Gilgamesh (West 1997); the wooden horse is probably the single most important theme of all Greek mythology, it came from an Egyptian myth (ibid., pp. 487–8). As is to be discussed in more detail in a later section, long before the Greeks, the Egyptians had developed a substantial mathematical tradition, and many mathematics historians are of the view that statements made by various Greek thinkers themselves that the Greeks first learned geometry from Egypt are reliable. In the *Phaedrus*, Plato had Socrates say: 'He [Thoth, the Egyptian divinity of wisdom] invented numbers, arithmetic and geometry' (Bernal 1987, p. 106; cf. Katz and Lumpkin 1995).

It is true that Greek learning made an epochal breakthrough since ca. mid -fifth century, but neither had its culture and learning developed autochthonously, nor did they have anything to do with the Europeans whom Aristotle referred to. This was once commonly acknowledged but subsequently denied (see later). Today, some scholars unfettered by Eurocentrism are re-stating this, in addition to scholars already cited, Delanty (1995, p. 16) states: 'We often forget that...the world of antiquity was oriental not western'. By 'oriental' and 'western', Delanty probably means Egypt and Babylonia and western Europe respectively. Similarly, Campbell and Sherrard (1965, p. 71) state: 'in the ancient world...Greece was linked spiritually and culturally to the East'. *In sum, whether in culture and learning or in Greek consciousness* (e.g., *Aristotle), non-Greek Europe had nothing to do with and no significance to the Greeks at all. The two were completely different societies and cultures.*

4.3 Collective Identities from Roman Times to the fifteenth Century

Europe also meant little in Roman consciousness, the significance of the term 'Europe' still lay in the Europa myth (Hay 1968, p. 1), and the 'Roman Empire was as much oriental as it was Hellenic' (Delanty 1995, p. 20)—it should be noted that *neither what Delanty refers to as 'oriental' nor 'Hellenic' had anything to do with non-Roman Europe.* After a fifty-year civil war, Diocletian divided the empire into four administrative regions in 285. In 306, Constantine I reunified the empire, moved its capital to Byzantium, renaming it Constantinople. After his death, the empire alternated between unification and division, before it remained divided from 395 until the Western Empire's demise in 480. The terms *west* and *east* were used in the locational-directional sense during Roman times, hence after 285, the two halves of the empire were called the Western Empire and the Eastern Empire (Delanty 1995, p. 23). Christianity was born during Roman times, being persecuted until Constantine I's Edict of Milan of 313. The emperor was himself baptized in 324. At that time, the Nicene creed and the Arian creed differed over the Trinity doctrine. The Nicene creed expanded its influence over time and was made the then-unified empire's state religion in 380. The ecumenical council held in 381 made the Arian creed a heresy.

Barbarian incursions intensified during the civil-war period. 'Barbarian' was the term Rome gave to the peoples living outside the empire's boundary (Noble 2006, p. 10). The incursions involved only movements of military bands, ranging in size from several hundred to tens of thousands, not large-scale population movements (Halsall 2007, pp. 144–5). Barbarians included Huns, Germans, Slavs, and so on, our focus is on the Germans. The term 'German' was not the Germans' own, but a label applied to them by the Romans (Todd 1992, pp. 8–9). German romanticism in the 18th and 19th centuries claimed that the early Germans were a 'pure' people, this has now been universally rejected by historians (Todd 1992, pp. 9–10; Noble 2006, p. 31). The Germans included Franks, Visigoths, Ostrogoths, Lombards, Saxons, Vandals, and so on. None of them, too, constituted any 'pure', discrete people, but were confederations of individuals from various tribes, being formed, disbanded, and re-formed many times (Noble 2006, pp. 16, 29). Their cohesion rested on allegiance to military leaders, who claimed either descent from distant heroes (real or legendary) or connections to distant important battles (again real or legendary). These military confederations constituted the royalty and aristocracy in the areas conquered by them, having little in common culturally and socially with the indigenous population (Geary 2002). Hence, *the various Germanic groups did not have any sense of collective Germanic identity*; though their languages belonged to the same Indo-European branch, whether or not they were mutually intelligible is open to question (Todd 1992, pp. 8, 13).

Rome offered land to the Barbarians in return for service in the Roman army, setting Barbarians against Barbarians. In 470, the Barbarian leader Odoacer deposed Western Emperor Romulus Augustulus, gaining control of the Italian peninsula. Acknowledging Roman legitimacy, he agreed to honour Julius Nepos, who was in control of Dalmatia, as Western Emperor in return for being made viceroy of Italy by the Eastern Emperor. But when Nepos was assassinated in 480, Odoacer took over Dalmatia as well, marking the formal demise of the Western Empire. By 500, various Barbarian kingdoms had been established outside of the Byzantine Empire, ushering in the Middle Ages.

Even while still being persecuted, Christian churches had been set up in various places, with the church in Rome (city) gradually gaining leadership status. The bishop of Rome (city) was subsequently called Pope. After Constantine I moved the then unified empire's capital to Constantinople, its church rose greatly in status. The ecumenical council of 381 made the church of Constantinople second in status only to the church in Rome (city), but this was not accepted by the latter. Since then, conflict over jurisdiction (over other churches) plagued the relationship between the two churches, which subsequently extended to conflicts over rituals, practices, and even doctrine, eventually leading to their mutual ex-communication in 1054.

During the fifth century, the Barbarian kings converted to the Arian creed (declared heretic since 381) one after another, except for the Frankish king Clovis I who converted to Necene Christianity around 500. This is why relations between the Franks and the Vatican were especially close throughout the Middle Ages. Subsequently, the other Barbarian kings also switched to Nicene Christianity. By the late eighth century, the Barbarian kingdoms in central and western Europe including the British Isles had all sworn allegiance to the Vatican; on the other hand, Byzantine followed what later came to be called Orthodox Christianity. After the two churches' mutual ex-communication, the Vatican was also called the 'Western church', while the Orthodox church the 'Eastern church', in which names, the terms *west* and *east* were used in the locational-directional sense. At the same time, states adhering to the Vatican had all adopted Latin as the official language, while, in contrast, Byzantine adopted Greek as the official language in 620. Thus, although both were Christians, the Latin Christian world and Byzantine were two very different cultures and societies, possessing *different identities*. As a matter of fact, the cultural and social divergence between these two regions had already begun during Roman times from the fourth century onwards, as noted by Goody (2006, p. 224):

> Until the late third century, the east and west of the Roman empire had had a common culture, with almost identical mosaics being found a thousand miles apart. Then the west dropped the use of Greek and for many reasons

> the gulf grew wider. There, large tracts of Roman territory passed into 'barbarian' control in the fifth century and Italy by its end became an Ostrogothic kingdom.

Islam was founded in the early seventh century and spread very rapidly across west Asia to north Africa. In 711, Islamic forces entered Spain through Gibraltar, by 788, the whole of Spain came under Islamic rule. As said, the Barbarians at first did not have any sense of collective identity, however, after they had all sworn allegiance to the Vatican in the late eighth century and with Spain completely fallen to the Muslims at precisely the same time (a *contingent* coincidence in terms of our theoretical-methodological framework), there emerged such a sense among them. It was during this period that the term *'Christendom' as a religion-based collective identity* came into being, but *it was restricted to the Latin Christian world* (Hay 1968, pp. 27–30), an indication of the cultural gulf between it and Byzantine. During the eighth century, a book titled *Chronicle of 754* was published in Islamic Spain, in the book, the term *'European' as an identity* appeared for the first time, referring to the Latin Christians led by the Franks who fought against the Muslims at the Battle of Tours in 732 (Hay 1968, pp. 25, 58). Hay explains the relationship between 'Christendom' and 'Europe' in this newly emerged sense of collective identity in this way: 'the word ['Europe'] is devoid of any meaning beyond the strictly geographical…Europe, one must conclude, is a word lacking sentiment, Christendom a word with profound emotional overtones'. Leyser (1992, p. 30) similarly states: 'what gave shape, relevance and duration to the notion of Europe from the early 7th century onwards was Rome, or rather the Roman church and its head, the papacy'. In other words, 'Europe' as a collective identity referred to Latin Christendom and was decidedly subsidiary to it.

To the Latin Christians, since the Vikings were not Christians, they were regarded as 'barbarian' and not considered part of 'Europe'; likewise, Byzantine was also not considered to be part of 'Europe' for obvious reasons (Ullmann 1969, pp. 137–9; Leyser 1992, p. 41)—for this reason, hereinafter, when we speak of 'Europe' in this sense of collective identity of Latin Christendom, we use the term in quotes (except in quotes from other authors), since it excludes many Europeans in the geographical sense. Conversely, the Byzantines regarded the Latin Christians as 'barbarian', 'inferior' not only to themselves, but also to the Muslims (Campbell and Sherrard 1965, pp. 74–5). There was no sense of collective identity between the Latin Christians and the Byzantines. *Latin Christendom's religion-based collective identity, in which the term 'Europe' referred to itself alone, lasted until the fifteenth century* (Hay 1968, Chaps. 4 and 5).[7] As in Roman times, the words *east* and *west* were used in the locational-directional sense throughout this period (Leyser 1992, p. 40), as already seen in the distinction between the 'Western Church' and the 'Eastern Church'.

The Byzantines' view of Latin Christendom as being 'inferior' was, on the criterion of learning, not groundless. Towards the end of ancient Greece's scholarly golden age, there emerged a trend of writing handbooks of the teachings of leading scholars. Rome's knowledge of ancient Greek thoughts was only skin-deep since it depended on these handbooks to write encyclopaedias. Beside encyclopaedias, during Roman times, commentaries on ancient Greek classics were also written, but these were done by Greek-speaking Byzantine scholars or their associates in the Eastern Empire. Knowledge of Greek scholarship in the Barbarian kingdoms was negligible. As Grant (2004, pp. 83–95, 137–45) remarks: 'Until the late twelfth century, knowledge of science and natural philosophy in the Latin West was at a low level. Very little of Greek science had been translated into Latin'.

In sum, there was no collective identity among the Barbarian kingdoms until Islam's conquest of Spain. Their *confrontation with the Muslims*, occurring precisely at the time when they had all converted to Roman Catholicism, made them acquire such an identity for the first time as Latin Christendom.[8] In this identity, 'Europe' excluded both the northern Europeans and the Byzantines. *In terms of learning and culture generally, the 'Europe' of Latin Christendom and ancient Greece constituted two different civilizations.* As is to be mentioned below, the thirteenth-century philosopher Roger Bacon referred to the *pagan, polytheistic* Greeks as '*them*'.

[7] Parenthetically, in the early Middle Ages, some in the Church associated the term 'Europe' with the story of Noah's three sons in the Bible. This, however, is a minor issue which we bypass.

[8] In this particular case, the Saussurian perspective (due to Ferdinand Saussure) that meaning (identity) derives from difference (Christian versus Muslim) applies.

4.4 'Europe' Replaced Christendom as Latin Europe's New Secular Collective Identity

In 1453, the Islamic Ottomans conquered Constantinople and changed its name to Istanbul, marking the end of the Byzantine Empire. Since Roman times, Europe's great demand for spices from India and southeast Asia had been satisfied through the intermediary of Arab merchants; after 1453, Europe became intent on finding a way round the Arabs to grab the spice trade from them. This was the impetus driving the so-called 'voyages of discovery' (Keay 2005). Maritime trade in the Indian Ocean and the South China Sea had previously been conducted, mainly by Arab, Indian, southeast Asian, and Chinese merchants—peacefully on strictly commercial terms. Because the Europeans came to Asian waters with a predatory motive and mentality (to grab the spice trade from the hated enemy, namely, the Islamic Arabs), henceforth, this trade became predicated on violence (Pomeranz and Topik 2006). At the same time as Portugal and Spain began to colonize Asia and the Americas, in 1517, Luther published his *Ninety-Five Theses.* Latin Christianity split down the middle into two.

Latin Christianity's split into two had a huge impact on Christendom as a religion-based collective identity; as a result of which, it gave way to what had hitherto been its subsidiary description. *'Europe' replaced Christendom as Latin Europe's new secular collective identity* (Delanty 1995, Chap. 3). The first book with 'History of Europe' as its title appeared in Florence in 1566 (Dionisotti 1971, p. 13). As previously and until the early nineteenth century, the terms *west* and *east* continued to be used in the locational-directional sense For instance, in 1834, British Foreign Secretary Palmerston spoke of Britain's alliance with France, Spain, and Portugal as 'states of the West', against the Holy Alliance of the 'East' (Russia, Prussia, and Austria); Hegel (1975, p. 197) stated that 'World history travels from east to west: for Europe is the absolute end of history, just as Asia is the beginning'.[9] Up until the turn of the eighteenth century, 'Europe' as a collective identity was still restricted to the Latin realm. However, by the eighteenth century, with France becoming the pre-eminent power and French replacing Latin as Europe's official and high-society language, the notion of Europe had shed its previous restriction to the Latin world, and became a vague, general geographical notion.

[9] Hegel's glorification of Europe, or, more specifically, Prussia as the final stage of the historical self-unfolding of Reason can be ignored for our present purpose.

It was during the eighteenth century that the term *west* also began to be used, but only very rarely, to refer to Europe in this vague geographical sense: 'the term began to be used, in French more than in English, in the eighteenth and increasingly in the early 19th century, but in imprecise and incoherent ways, and always interchangeably with the term "Europe"' (Varouxakis 2019, p. 414). Among the very few Britons who used the term in this sense, Varouxakis mentions the historian and Whig politician Thomas B. Macaulay (1800–1859), and the historian Edward Augustus Freeman (1823–1892). The notion of the 'West' as a collective identity and 'trans-historical' entity was yet to be born.

Meanwhile, with the success of colonization, racism quickly emerged and became widespread in the sixteenth century. The British priest Samuel Purchas (ca. 1577–1626) said: 'The Qualitie of Europe exceeds her Quantitie, in this the least, in that the best of the world' (Hay 1968, p. 120). Thus, the collective identity of 'Europe' (still restricted to the Latin world in Purchas' time) became suffused with the notion of differential 'racial' intelligence between different peoples. As Delanty (1995, pp. 84–5, our emphasis) notes: 'European consciousness…was closely linked to racism…*The core component of secular European identity was race*'. In the eighteenth century, the term 'civilization' came into being (Williams 1976). To the French *philosophes*, it connoted reason and progress, both of which the rest of the world was supposed to lack (Hay 1968, p. 122; Woolf 1989, p. 96). To Enlightenment thinkers who stressed universalism, the collective identity of Europe, shed of its previous restriction to the Latin realm and understood in a vague geographical sense, took precedence over being French, German, etc. (Delanty 1995, p. 71).

4.5 Latin Europe's Scholarly Debt to Arabic Scholarship and Its Subsequent Repudiation Through a Purported 'Return' to Greece

If it was their confrontation with Islam upon the Muslims' conquest of southern Italy and Spain that made the Barbarian kingdoms acquire their collective identity as Christendom, the early groundwork towards the construction of the 'trans-historical' 'West' was also inseparable from *Latin Europe's hatred of Islam*. To begin with, it is necessary to briefly repeat Latin Europe's scholarly debt to Arabic scholarship, though that story has now become well-known among students of science history.

As Islam expanded geographically, from the late eighth century onwards, Arabic scholars began to import ancient Greek learning on a large scale (Lindberg 1978, pp. 55–8), a trend that, as Saliba (2007, Chap. 2) points out, had already begun under the Umayyads. In a whole range of scientific or proto-scientific disciplines such as astronomy, optics, dynamics, medicine, experimental alchemy, mathematics, and so on, Arabic learning took developments, as is now generally acknowledged contrary to previous denials, to a whole new level above that of the Greeks. The Islamic rulers in Spain highly treasured learning, the collection of Cordoba's royal library is reputed to reach 400,000, though this is probably an exaggeration. In the second half of the tenth century, individual Latin scholars came across Arabic learning in Islamic Spain, and realized how backward Latin learning was. During the eleventh century, just as, and precisely as (as we pointed out in the previous chapter), Latin Christendom began its counter-offensive against Islam, individual Latin scholars began to translate Arabic books into Latin. Then in the twelfth century, Latin scholars went to Spain and Sicily, re-conquered from the Muslims, to translate Arabic books (both original Arabic scholarship and Arabic translations of ancient Greek texts) into Latin on a massive scale (Lindberg 1978, pp. 58–79). For a long time since the twelfth century, science was acknowledged by Latin scholars as 'the view of the Saracens' (Goody 2004, p. 61).

In philosophy, Arabic scholars who inherited the Aristotelian tradition, were also dominant. Between 1268 and 1274, Giles of Rome of the University of Paris wrote *Errors of the Philosophers* to condemn secular philosophy for having breached the Church's limits of tolerance. Six philosophers were singled out for critique, other than Aristotle, four were Muslims (in chronological order, al-Kindī, Ibn Sīnā [Avicenna], al-Ghazālī, and Ibn Rushd [Averroes]), the sixth being the Jew Maimonides, writing in Arabic in Cordoba (Grant 2004, pp. 178–9; Goody 2004, p. 57). The preponderance of Arabic learning in Latin European philosophical circles can thus be seen. No less than Roger Bacon (1220–1292) said: 'Philosophy is the special province of the unbelievers: we have it all from them'; by 'unbelievers', Bacon referred to the ancient Greeks and the Muslims (Southern 1962, p. 59).

Given the hatred of Islam, the preponderance of Arabic learning made some Latin European intellectuals very uncomfortable. The first reaction came from literary circles in the form of the rise in Italy of Renaissance humanism (Dante, Petrarch, and others) in the thirteenth century. To repeat a quote from the previous chapter, 'George Sarton commented that the humanists' love of Greece and Rome was nourished by their hatred of Islam…' (Kabbani 1986, p. 5). The reason why the hatred of Islam led to the love of Greece (for simplicity, we put Rome on one side hereinafter) is simple: Because Arabic learning itself had inherited from the Greek tradition, *the love of Greece was a means to repudiate the Islamic contribution*. For convenience, hereinafter, we refer to this *newly arisen love of Greece as a means to repudiate Europe's debts to Islam* a *purported* 'return' to Greece—we realize that the term 'return' can be misleading, because it may be misunderstood as referring to Latin Europe, as a later stage of one and the same trans-historical entity, reclaiming from a previous stage of the same entity its own once-forgotten heritage, but we are unable to come up with a better and equally convenient term to replace it, hence, to prevent misunderstanding of our meaning, we specify that this 'return' is purported only; in the following, whenever the term 'purported "return"' is used, this point needs to be borne in mind. That the purported 'return' to Greece was a means to repudiate the debt to Islam is especially clear when it spread beyond literary circles. To repeat another quote from the previous chapter, the fifteenth-century mathematician Regiomontanus 'treated mathematical science in Islam as just a foreign interlude, with the Arabs temporarily preserving for Europe its Greek heritage' (Cohen 2011, p. 101). As mentioned above, Arabic scholars did not only 'preserve' the Greek corpus, they took it to new heights, as is now generally acknowledged. The simplest example is, of course, the case of mathematical astronomy that has been mentioned in Chap. 2. But that did not stop the reduction of Arabic scholarship to a mere 'preserving' role first begun by Regiomontanus from gathering pace over the centuries until it later became taken for granted:

> Throughout the Middle Ages, the West's debt to Islam was too obvious to need acknowledgement…[but] The early historians of science also minimized Arab achievement. Carra de Vaux [in 1931] remarked that its mathematics 'reminds one of good secondary textbooks', and Dreyer [in 1906] asserted that 'they left astronomy pretty much as they found it'. (Stock 1978, p. 12).[10]

This reduction of the role of Arabic learning persisted well into the second half of the twentieth century. Hall (1962, p. 6) remarked as follows in the 1960s:

[10] In the 1974 edition of de Vaux's book, the statement concerned was eventually deleted.

> Europe took nothing from the east without which modern science could not have been created; on the other hand, what it borrowed was valuable only because it was incorporated into the European intellectual tradition. And this, of course, was founded in Greece.

For Hall, there is, firstly, a 'trans-historical' entity called Europe[11] which began with Greece; secondly, Latin Europe's learning from Arabic scholarship is dispensable, for even without it, Europe could still have created modern science.[12] Hence, Greek learning, produced by the founders of the 'trans-historical' Europe as a culture and civilization was 'preserved' in pure state, as it were, until it was eventually 'transmitted back', in that pure state, to itself. It, then, logically follows that modern science emerged as a result of the purely 'internal' and 'autochthonous' development of this 'trans-historical' entity. In the 1970s, major scholars such as Lloyd (1973, p. 173, our emphasis) could have remarked as follows as a matter of course: 'The history of the contributions of the Arabs and the transmission of scientific knowledge *back into the West*...'. Even today, as Cohen (2012, p. 340) notes, survey histories of science still standardly assign to the Arabs 'the humble role of dutiful babysitters for the Greek corpus'. But we have run ahead of ourselves, so let us retrace the process of the construction of the 'trans-historical' 'West'.

4.6 The Construction of Europe as a 'Trans-historical' Entity 'Autochthonously' Responsible for Its Achievements Due to Its 'Essence' Understood in 'Racial' Terms

It is necessary to begin with a caveat: Much of this section concerns Greece, and Bernal is often referenced. As mentioned in Chap. 2, Bernal's *Black Athena* is highly controversial and unceremoniously dismissed by many of his critics.[13] Much of this controversy (such as Bernal's claims that

[11] Writing in the 1960s, Hall's Europe clearly refers to a cultural-civilizational entity and not to a geographical unit.

[12] This is not Hall's view only, see Bala's (2006, pp. 66–8) critique of the claim that Copernicus' *De Revolutionibus* might 'just as well have been written in the second century CE by a successor of Ptolemy'.

[13] During a discussion that we had with a non-Eurocentric scholar, Bernal was mentioned, whereupon he remarked that *Black Athena* is 'infamous'. He admitted that he had not read Bernal and was relying on book reviews. Lest, given the many hostile reviews of *Black Athena*, such cases are not uncommon even among non-Eurocentric scholars, it would be useful to explain what *exactly* that we are borrowing from Bernal.

parts of Greece were once colonized by Egypt and Phoenicia) is irrelevant to our concerns, while most of our references to him below concern developments in European academia in the eighteenth century and thereafter, and are, hence, not affected by this controversy. What concerns us is Bernal's fundamental argument, with which we agree and which we borrow from him, namely, (a) that Greece's cultural development in general and intellectual development specifically were not autochthonous, but had roots in Egypt and west Asia; and (b) that the previous denials—some of which still persist today—of these roots in European academia were motivated (consciously or not) by the *extra-scholarly* factor of racism (inclusive of anti-Semitism). In Chap. 2, it was pointed out that Bernal's argument, inclusive of both (a) and (b), with regard to west Asia has now been vindicated by authoritative studies restoring and detailing Greece's debts to west Asia published by major classicists. Here, we would like to make a brief general comment on his argument with regard to Egypt. Incidentally, denials, past and present, in European academia of intellectual debts to non-European sources are by no means news, in addition to the case with regard to west Asia, as shown in the previous section, European debts to Arabic mathematical astronomy, to mention only it, have been denied right up to the 1960s and 1970s. In Chap. 6, we show how some contemporary Eurocentric scholars attempt to do the same up to the present.

That Greece and Egypt had cultural contacts with one another is beyond doubt. Thus, for instance, as mentioned in Chap. 2, the theme of the wooden horse came from an Egyptian myth (West 1997, pp. 487–8); the idea of the free-standing statue came from Egypt (Goody 2006, p. 34). Concerning scholarly development, as an illustration for our present purposes, one bone of contention between Bernal and his critics concerns, firstly, the level of development of Egyptian mathematics and, secondly, whether or not geometry was invented by the Egyptians from whom the Greek had originally learnt it. With regard to the former point, Otto Neugebauer, who held Mesopotamian mathematics in high regard, did not have a high opinion of Egyptian mathematics, but various historians of mathematics have since contradicted him.[14] In other words, Bernal's

[14] Bernal (2001, Chaps. 10 and 11) provides an account of this debate. For an updated discussion on the level of development of Egyptian mathematics, see Joseph (2011, Chaps. 3 and 5).

argument on this point is not his own invention, but based upon the views of various mathematics historians.

Concerning the above latter point, as is well-known, it was the Greeks themselves who, one after another, stated that geometry was invented by the Egyptians from whom they had learnt it. C. Froidefond has even published in 1971 a collection of references to this made by Plato and Aristotle (Bernal 2001, p. 436 n 35). For a very long time—to a considerable extent, until today—these statements made by the Greeks themselves have been treated with scepticism by modern science historians. On what grounds? None other than on the assumption that Egyptian mathematics was not up to scratch. If this assumption is questionable, as various mathematics historians have argued (see note 14), then so is such scepticism. As early as in the 1960s, the science historian Giorgio de Santillana (1963, p. 814) has argued that what the Greeks themselves said about learning geometry from Egypt has to be taken seriously; with reference to Eudoxus, Greece's greatest mathematician before Archimedes, who is generally agreed to have gone to Egypt to study, he said of the above scepticism: 'We are asked to admit, then, that the greatest mathematician of Greece learned Egyptian and tried to work on astronomy in Egypt without realizing that he was wasting his time'.

Bernal's severest critic with regard to the role of Egyptian science (including mathematics) in early Greek developments is science historian Robert Palter (1993). In response to Palter's critique of Bernal, mathematics historians Victor Katz and Beatrice Lumpkin (1995, pp. 10–1) have intervened to defend Bernal. The following passage from them is worth quoting:

> Not only is Pythagoras supposed to have studied in Egypt, but so is Thales, the supposed father of Greek geometry. Also Oenopides, Herodotus, Heron of Alexandria, Diodorus Siculus, Strabo, Socrates (through Plato) and Aristotle—all say that geometry was first invented by the Egyptians and then passed on to the Greeks. The question always seems to be, in this regard, what we mean by geometry. If, by geometry, we mean an axiomatic treatment with theorems and proofs in the style of Euclid, then it is clear that this was a Greek invention. But mathematicians have always known that, in general, one does not discover theorems by the axiomatic method. One discovers theorems by experiment, by trial and error, by induction, etc. Only after the discovery is there a search for a rigorous 'proof'.

In other words, one can certainly claim, contrary to repeated statements made by one Greek thinker after another, that the Greeks invented geometry ex nihilo, as it were, instead of having originally learnt it from the Egyptians, by defining geometry in such a way (that only an axiomatic-deductive system constitutes geometry) as to make the claim true by that definition.[15] Otherwise, there is every reason to believe that what the Greeks themselves said is historical fact. In which case, Bernal's argument that Greece's intellectual development, instead of being autochthonous, had part of its roots in Egypt is on solid grounds. This is the first part of Bernal's fundamental argument ([a] in the opening paragraph of this section) with which we agree and borrow from him; and this is not affected by the controversy over Bernal's other arguments.

The second part of Bernal's fundamental argument ([b] in the opening paragraph of this section) is that previous denials of Greece's debts to Egypt and west Asia were motivated by racism. It has already been shown above that from the sixteenth century to, let us say, the mid-twentieth century, racism permeated European society, hence, there is no reason to believe that it did not also deeply influence many scholars, whether or not on a conscious level. For instance, as mentioned in Chap. 2, Weber openly acknowledged that he was inclined to it; while Needham also clearly subscribed to it until his interactions with his Chinese research students convinced him otherwise. As already noted, this part of Bernal's fundamental argument has now been vindicated with regard to west Asia. Are we to suppose that something similar did not also occur with regard to Egypt? Does that sound plausible at all? This is for the reader to judge for herself. But the following is worth drawing her attention to. As is to be explained below, up until at least the fifteenth century, the Egyptians had been regarded as black, then, as the esteem of the Egyptians grew in Europe during the seventeenth century, they became categorized as white; but after that, European academia's view of Egypt underwent a dramatic change during the 18th and 19th centuries, and in the nineteenth century, the Egyptians were re-categorized as 'Oriental'.

[15] Some twentieth-century logicians and mathematicians have argued that Euclid does not actually constitute an axiomatic-deductive system (see the account in Hart 2013, pp. 206–9). For the purpose of the present work, this issue can be bypassed.

Let us begin our account with the issue of 'race'.[16] In order to ground racism 'scientifically', eighteenth-century European scholars resorted without exception to Aristotle's simple-minded climatic determinism. In 1775, J.F. Blumenbach, on the basis of the work of the French naturist Comte de Buffon, advanced the first comprehensive racist doctrine. Based upon the Bible story that Noah's ark landed on the southern slopes of the Caucasus, he argued that that is the birthplace of post-Flood humankind. Those whom he called 'Caucasians' were, in his view, the most beautiful and talented, as they migrated outwards, those heading towards Europe remained beautiful and talented due to the latter's temperate climate. The term 'Aryan' also came into use around this time, and became interchangeable with 'Caucasian' (Bernal 1987, p. 220).

In 1786, Sir William Jones of the British East India Company gave a report to Calcutta's (now Kolkata) Royal Asiatic Society, in which he pointed out that Sanskrit, Old Persian, Greek, Latin, and so on had common roots. Though this was not an entirely new view, philologists today regard Jones as the father of historical linguistics (Fischer 1999, p. 157). According to historical linguistics, on the basis of similarities of grammatical forms and paradigms (but not vocabulary) between certain different languages, it can be determined that they have a common origin, and a proto-language can be reconstructed as this common origin (ibid., p. 61). By the 1820s, linguists had completed reconstructing proto-Indo-European, and postulated Central Asia to be where it originated. Subsequently, this homeland was moved west to the Black Sea-Caspian Sea area (Bernal 1987, pp. 12, 226–30; cf. Keay 2011, Chap 2).

Re-locating Indo-European's homeland to the Black Sea-Caspian Sea area made it overlap with the provenance of post-Flood humankind in Blumenbach's account. Indo-European was thus taken to be the 'superior' language of the 'superior' 'Caucasian race'. In accordance with Blumenbach's Aristotelian climatic determinism, as the Persians and *Aryas*[17] migrated southwards, they 'deteriorated' due to the hot climate there, thus although they spoke 'superior' Indo-European, they still

[16]The following account is highly condensed and schematic, but for our purposes, a more nuanced account, which would require much more space than is called for, is an unaffordable luxury. Hopefully, no serious mis-representation occurs as a result of this. In a chapter titled 'The invention of Antiquity', Goody (2006, pp. 26–67) provides an account that differs somewhat from but is complementary to our account.

[17]The *Aryas* (from whose name the term 'Aryan' is derived) were the Sanskrit-speaking people who migrated south to India, and who subsequently founded Hinduism.

constituted an 'inferior race'. Only 'Caucasians' migrating to Europe retained the 'superiority' of both the 'Caucasian race' and the Indo-European language.

The Greeks openly and repeatedly acknowledged their cultural and scholarly debts to Egypt. We have pointed this out with regard to mathematics above, in relation to culture in general, Plutarch, for instance, stated in *On Isis and Osiris* that Greek religion had roots in Egyptian religion. Up until the eighteenth century, this was generally accepted in European academia. But things subsequently changed in the 18th and 19th centuries. Even before Blumenbach, associating dark skin with inferiority had already become widespread since the late seventeenth century. The ancient Greeks were white, in contrast, Egyptians were not only Africans, many were black.[18] Herodotus referred to the Egyptians as having 'black skins and woolly hair'. In early Christian portrayals, Egyptians were also often black. Until the fifteenth century, the British still saw Egyptians as black. In the seventeenth century, European esteem for the Egyptians grew, and, contrary to previous characterization, Egyptians became seen by many scholars as white. Reverence of the Egyptians continued into the eighteenth century on the part of some scholars. For instance, Sir William Jones remarked in 1784: '*Egypt* appears to have been the grand source of knowledge for the *western*...[part] of the globe'. Under these circumstances, some such as Blumenbach continued to count the Egyptians as white, while in the eighteenth century, Egyptian myths were placed under the 'Western' category in the library of the University of Gottingen (Bernal 1987, pp. 242–6, 476–7, 479, quote from Jones is on p. 479).

[18] Whites, blacks, etc. are valid categories, without carrying racist connotations. To explain this briefly (see Malik 2008), all humans have the same set of genes, there are some genes which can take different forms called alleles, the frequency distribution of alleles can vary statistically between different population groups, with varying degrees of overlap between these distributions. Population groups can range from 'continental groups' to the inhabitants of different cities within the same country. Such differences in the frequency distribution of alleles can result in physical differences such as in complexion, digestion, immunity to certain viruses, and so on. If we can divide the human population in terms of complexion, we can equally divide them in terms of digestion and so on. But no such division carries any extra-physical implications. Thus, it is possible to roughly—roughly because different frequency distributions overlap at the margins—speak of whites, blacks, etc., but this does not turn them into 'races', nor does it carry any extra-physical implications. On the other hand, since the term 'Caucasian' coined by Blumenbach was used by him in a 'racial' sense, we shall continue to use that term in quotes when reference is made to its use in that sense.

However, from the early eighteenth century onwards, some other European scholars began to re-evaluate Egypt. Thus, Kristophe August Heumann (1681–1764) declared that Egyptians could not have been philosophical due to Egypt's hot climate. He was followed by historian Jacob Brucker (1696–1770) who said real philosophy began only with the Greeks. In 1767, Scottish priest William Duff similarly put the Greeks above the Egyptians (Bernal 1987, pp. 196–216). This devaluation of Egypt gathered pace in the nineteenth century. In 1893, Egyptologist Gaston Maspero said that the Egyptians 'always remained semi-barbarian...Thothemes III and Rameses II resemble Mtesa of Central Africa more than they do Alexander or Caesar'; in 1904, Egyptologist Wallis Budge said: 'The Egyptians, being fundamentally an African people...no African language is suitable for giving expression to theological and philosophical speculations' (Bernal 1987, pp. 250–66, the quotes from Maspero and Budge are on pp. 260–1). In consequence to this progressive devaluation of the Egyptians, sometime during the nineteenth century, the library of the University of Gottingen transferred Egyptian mythology from the 'Western' section to the 'Oriental' section (ibid., p. 479). In other words, *the Egyptians were 'racially' re-classified while they were being intellectually devalued.* In this process of the devaluation of the Egyptians and the elevation of the Greeks, Greece's intellectual debts to Egypt came to be progressively denied. The trend was first set by the British classicist Samuel Musgrave who declared in 1782 that ancient Greek culture was 'autochthonous' (ibid., p. 188). That this subsequently became the predominant view can be seen from the vehement reaction to Bernal's attempt to restore these debts, an attempt which, with reference to the case of mathematics that we use as an illustration, is made on solid scholarly grounds.

Denying the Egyptian roots of ancient Greek intellectual development is only part of the story. By the eighteenth century, France had become Europe's pre-eminent power and French the official and high-society language. German intellectuals hence suffered from an identity crisis leading to the rise of romanticism. The romantics believed that a 'people' and the land that 'belongs' to it interact to produce a particular culture, whose 'purest' expression is in sound (language and folk songs). German romantics sought to retrieve this 'pure' German culture. Contemporaneously, Scotland's union with England in 1707 greatly undermined the former's Celtic culture, which also engendered a romantic movement among the Scots. Both German and Scottish romantics keenly searched for ancient folk songs. In the 1730s, Aberdonian Thomas Blackwell declared ancient Greece to be 'Europe's childhood' and *The Iliad* to be an ancient folksong of the whole of Europe (Bernal 1987, pp. 204–15)—as noted in the previous section, during the eighteenth century, the notion of Europe had shed its previous restriction to the Latin world and became understood in a vague geographical sense. It should be noted that while the previous call for a purported 'return' to Greece, engendered by the hatred of Islam, had already marked an initial drifting away from the position that the Greeks constituted, as stated by Roger Bacon, a 'them', Blackwell's was the first explicit statement that the Greeks were one with Europe, and, indeed, its origins ('childhood'). This marked the birth of *the notion of Europe as a 'trans-historical' entity*, which quickly became accepted generally over the course of the century. With 'racial' notions already predominant, such a transformation of the Greeks, who were white, was also 'racially' grounded.

As mentioned above (see Sec. 4.2), Athenians ridiculed Thebes and Argos for having been founded by 'barbarians'. From Homer and Hesiod onwards down to the Athenian tragedians of the early -fifth-century, Herodotus, Isocrates, Plato, and so on, all reported that Thebes was founded by the Phoenician Kadmos (Europa's brother, sometimes rendered as Cadmus) who had been sent by the King of Tyre to search for Europa, while Argos was founded by Danaos (of either Egyptian or Syrian origin). The Dorian royalty ruling the Peloponnese claimed to have descended from Danaos and were proud of it. In fact, there were also reports of Athens itself being founded by Kekrops coming from the Egyptian city of Sais. From antiquity until the end of the eighteenth century, nobody ever doubted that these colonizations of Greece by Phoenicia and Egypt did take place. But in the 1820s, Karl Otfried Muller suddenly denied their authenticity. His denial with regard to Kadmos did not immediately gain approval, but those concerning Danaos and Kekrops were instantly endorsed (Bernal 1987, Chap I; *idem.* 2001, pp. 296–8). Since then, that ancient Greece had never been colonized by the Egyptians became orthodoxy in European academia, a position steadfastly maintained by Bernal's critics. We are not in a position to adjudicate between Bernal and his critics with regard to this, and there is no need for us to do so as it is irrelevant to our concern. What does interest us in relation to it is the following.

Firstly, colonization is a most effective means of cultural transmission, and it is interesting to note that Muller's about-turn occurred when the process of the intellectual devaluation of the Egyptians which eventually led to their being re-categorized from white to 'Oriental' was underway among substantial numbers of European scholars. Secondly, the differential treatment received by Muller's arguments in European academia concerning Egypt on the one hand, and Phoenicia on the other, is intriguing. Perhaps it was grounded purely on good scholarly grounds. Or perhaps not? Christianity came from the Hebrews, while both Phoenicians and Hebrews were Semites. Was this the reason why Muller's about-turn with regard to west Asia was initially rejected? Whatever the reason, between then and the 1960s, this rejection was reversed as Greece's debts to west Asia came progressively to be denied, and it was, as already seen, not until the 1990s that those debts were restored. What happened in between? As is well-known, anti-Semitism began to rise in the mid-1800s until European governments repudiated it after WWII, and none other than Walter Burkett himself acknowledges that the denial of Greece's debts to west Asia was a product of anti-Semitism.

Concerning the denial of Greece's debts to Phoenicia, there is one interesting issue that is worth mentioning. The biggest problem facing that denial is Greece's borrowing of writing from west Asia, especially the alphabet from Phoenicia. However, if the latter borrowing itself cannot be denied, attempts were made to minimize its significance, by pushing the date of borrowing as late as possible, so as to imply that ancient Greek culture had already been formed by the time of the borrowing, thereby minimizing the latter's significance. The first to do so was the philologist Friedrich August Wolf, who proposed the date of the borrowing to be in the early -700 s, which incidentally would make Homer and Hesiod illiterate. Archaeologist Rhys Carpenter pushed the date further forward to –720 in 1933. If Carpenter were correct, the spread of the alphabet in Greece would have been extraordinarily rapid, when so questioned, Carpenter replied, 'Truly the Greek climate does miracles to a young alphabet; we can almost see it growing' (Bernal 1987: 327, 393–9, the quote is from p 397). The dating of the borrowing to the -eighth century has until today been taken as the 'standard' date by Eurocentric scholars (Goody 2006, p. 27). With regard to this, it is worth taking a look at the view on the matter of contemporary historical linguists, who are unaffected by the debate between Eurocentrism and its opponents.

The Greeks borrowed writing from west Asia twice. Around −2000, the Minoan Greeks of Crete borrowed the moraic system from Byblos,[19] a thriving west-Semitic trading centre, establishing a script known as Linear A. Linear A was subsequently replaced by a script known as Linear B, developed by the Mycenaean Greeks, as the Mycenaeans overtook the Minoans politically. Before that, Linear A had spread to Cyprus, where it developed into a script called Linear C. Linear B continued to be used until ca. -1100, and Linear C even after that; in the meantime, the Canaan itself had switched to the abjad, a purely phonetic consonantal alphabet unmixed with non-alphabetic symbols (see note 19) with completely stylized symbols (Fischer 2004, pp. 75–82; Rogers 2005, pp. 148–51).

During the period ca. -1225 to −1175, late-Bronze Age society in west Asia was in general upheaval. From ca. -1100, with the exception of Linear C used by Cypriote Greeks, writing appeared to have vanished in the rest of Greece altogether for more than a century. Then, Cypriote Greeks again borrowed writing from west Asia, this time from the Phoenician abjad, which they transformed into their own alphabet by also symbolizing vowels.[20] Despite the development of this alphabet, Cypriote Greeks retained Linear C until ca. -220 but only as a consciously archaic script used on monuments and the like. Alphabetic writing diffused to various parts of Greece. Different local variations came into being, but in ca. -403, the Athenians made the Ionian script, in which Homer's *Iliad* and *Odyssey* had been written, compulsory in all documents. By the middle of the -fourth century, all competing versions had given way to the Ionian script (Fischer 2004, pp. 75–82). While all scholars agree that the Greek alphabet was derived from the Phoenician alphabet, there is disagreement concerning the timing of Greece's borrowing. Some scholars date it to as

[19] A moraic system is a largely phonetic writing system that, in the development of writing in the Mesopotamian to east Mediterranean areas, preceded Egypt's mixed consonantal alphabet—a consonantal alphabet is one in which only consonants are symbolized, this suited the characteristics of the languages concerned, as the context can always unmistakably show to native speakers of the language what the unsymbolized vowels are; Egypt's consonantal alphabet was mixed, because it combined alphabets with non-alphabetic symbols retained from the previous system, a result of institutionalized resistance to complete change as can be expected from the path-dependency perspective.

[20] This was done because Phoenician and Greek belonged to two different language families, the former to the Semitic branch of the Afroasiatic family, the latter to the Indo-European family; in consequence, all Phoenician words began with consonants, whereas many Greek words began with vowels. This difference forced the Greeks to also symbolize vowels in borrowing the Phoenician abjad (O'Connor 1996, p. 94; Fischer 2004, pp. 124–5).

early as the -eleventh century (Rogers 2005, pp. 153–5); most scholars, however, date it to the period from the -tenth century to, at the latest, ca. -850 (Fischer 2004, pp. 122–9; Swiggers 1996, pp. 267–8); much earlier than the 'standard' date adopted by Eurocentric scholars until today.

With both Egyptian and Phoenician roots severed or minimized, ancient Greek civilization and its intellectual development became the 'autochthonous' achievements of a white people, the purported originators of the 'trans-historical' Europe.

To summarize, 'Europe' (limited to the Latin world) became Latin Europe's secular collective identity in consequence to the Ottoman's conquest and incorporation of Byzantine, and the split of Latin Christianity into two. As racism built upon the success of the colonial enterprise rapidly rose, 'Europe' and, following it, Europe (understood in a vague geographical sense from the eighteenth century onwards) became imbued with connotations of 'superiority', 'progress' and 'civilization'. Earlier, the hatred of Islam had induced a call for a purported 'return' to Greece, this purported 'return' eventually consummated in retrospectively transforming Greece into the 'childhood' of Europe, thereby constructing the latter into a 'trans-historical' entity. By making ancient Greek civilization 'autochthonous' and reducing Arabic learning to the role of 'preserving' Greek learning in its original state, the rise of modern science, to speak only of it in relation to our subject matter, was turned into a purely 'internal' development of the 'trans-historical' Europe stemming from its 'superior' 'essence' understood in 'racial' terms.

4.7 The Invention of the Notion of the 'West'

The notion of the 'trans-historical' Europe as analysed in the previous section had acquired dominance by the turn of the nineteenth century. In this notion, while the 'trans-historical' entity was conceptualized as humankind's elite in cultural-civilizational terms, the term Europe itself was understood in a vague geographical sense. In this latter sense, Europe included many states (the eastern Europeans and Russians) whose cultural achievements were far below that attained by the elitist French, British, Germans, and so on.[21] This created a conceptual inconsistency, and it was

[21] In relation to this, it is well to note that in the early days after Russia's invasion of Ukraine in late February 2022, an American television journalist covering the exodus of Ukrainians from the country gained instant global fame (or infamy) by calling the Ukrainians a 'semi-civilized people'.

in the context of this inconsistency that the French philosopher of positivism and conceptual founder of sociology as a discipline Auguste Comte (1798–1857) advanced the notion of the 'West' to bypass having to use the imprecise word of Europe.

In Comte's view, humankind goes through three stages of evolution, dependent on the development of the human mind in particular—this was typical of the unilinear view of history engendered by the Enlightenment. Comte called his three stages the 'theological', the 'metaphysical' and the 'positive'. We need not bother about what he meant by these characterizations, suffice it to say that Comte was of the view that only five population groupings of western and central Europe, namely, the French, the Italian, the Iberian, the British and the Germanic, which constituted what he called the 'elite of humanity', had reached the threshold of the 'positive' stage.[22] He argued that this 'elite of humanity' should be organized in such a way in accordance with the laws of positivism as to form a 'Western Republic', which would set an example for the rest of humanity to follow. In this supranational entity, France would occupy the pre-eminent position, and its capital should, hence, be located in Paris. Unsurprisingly, taking up by the then-dominant view of Europe being a 'trans-historical' entity tracing its origins to Greece, Comte added Greece, though located in eastern Europe, as an 'accessory member' of the 'elite of humanity'. The 'Western Republic' was not to be the final stage of Comte's plan, its role was merely to lead the whole of humanity to the final and permanent state of society governed by the scientific laws of sociology prescribed in his positive philosophy, a process which Comte reckoned would take seven centuries. In a tacit acknowledgment of Greece being the origins of the 'elite of humanity', Comte stipulated that in the just-mentioned final and permanent state of the whole of humanity, the capital would be moved from Paris to Constantinople.[23]

In formulating his views, Comte originally referred to the entity of the 'five great nations' constituting the 'elite of humanity' mostly as Europe, however, as time progressed, he increasingly employed the term 'Western Europe'. Then, in the late 1840s, Comte came round to employing the terms 'West' (*l'Occident*) and 'Western' (*occidental[e]*), and their cognates

[22] The following account of Comte's views and his advancing of the notion of the 'West' is based upon Varouxakis (2019).

[23] The city had also been known as Istanbul for centuries by Comte's time, Comte's preference to call it Constantinople was to refer to its Greek origins.

such as 'Westernness' (*l'occidentalite*). From July 1848 onwards, 'all books and circulars published by the rue Monsieur-le-Prince publishing industry were headed "REPUBLIQUE OCCIDENTALE"' (Varouxakis 2019, p. 434). Varouxakis (ibid., p. 429) observes: 'Thus Comte's "West" was a complex "sociological" notion, and certainly not primarily a geographical unit'. Varouxakis calls Comte's notion a sociological one because Comte advanced it as part and parcel of his socio-political project of organizing a future perfect state on the basis of his positive philosophy. In the context of the analysis of the present chapter, however, it is clear that Comte's notion of the 'West' was based upon the hitherto existing notion of Europe as analysed in the previous section. As noted at the beginning of this section, calling this entity Europe creates a conceptual inconsistency in the notion. In his 'law' of the three stages of development, Comte's 'elite of humanity' that traces its origins to Greece is understood, of course, in cultural-civilizational terms. By advancing the non-geographical notion of the 'West' to refer to this 'trans-historical' entity in place of the vague notion of Europe, the just-mentioned inconsistency can be overcome.

In sum, the notion of a 'trans-historical' elite entity 'autochthonously' responsible for its civilizational achievements stemming from its 'essence' was a *construction* occurring over more than a century, which had been accomplished by Comte's time, while it was not exactly Comte's intention to do so—his motive was his above-mentioned socio-political project—it was Comte who gave this elitist entity the new name of the 'West', thereby *inventing* the notion of the 'West'. We call Comte's substituting of one word for another an invention instead of a construction because the substance of the notion of the 'West' had already previously been constructed under the name of Europe. Comte's notion was introduced into Britain in the 1850s and 1860s by British Comtists (Varouxakis 2020). It gained wider general popularity in Europe as a result of a debate that occurred in Russia in the 1860s between Slavophiles and Westernizers; during this debate, the leader of the Slavophile movement Nikolay Kanilevsky referred to the 'West' as a civilization which, he claimed, was in decline (Neumann 1996; Bassin 1992, 1998; Heller 2010).

Our analysis in the previous and present sections has established the present chapter's argument that the notion of the 'West' that underpins the Eurocentric position in relation to the NQ, as best exemplified in Weber's 'forms of rationality' thesis, is a construction-cum-invention, and that the entity that this notion refers to, the existence of which is taken for granted as 'self-evident' in both popular and scholarly discourses, is fictitious. In the late 19th to early 20th centuries, the construction of the notion of the 'West' underwent further developments (GoGwilt 1995; Bonnett 2003, 2004); however, since they are not directly relevant to our argument, we bypass these further developments.

4.8 The Ongoing Importance of Deconstructing the Notion of the 'West' for Analysing the Positive NQ

The construction of the notion of the 'West' as a fictitious 'trans-historical' entity (originally under the vague notion of Europe up to the early 1840s—for simplicity, this qualification will be dispensed with hereinafter) is part and parcel of the construction of the narrative that modern science, to confine ourselves to it, is the product of the '***internal***' and 'organic' development of this 'trans-historical' entity on an '***autochthonous***' basis *throughout the entire process* from Greece until the seventeenth century and thereafter. In a word, the notion of the 'West' is a thoroughly AWD-paradigm-type Eurocentric concept. However, in our discussion with some non-Eurocentric scholars, we have noticed a great reluctance to cast doubt on the notion of the 'West' and to not regard Greece as the origins of such a 'trans-historical' entity.[24] Given the immense implications of our deconstruction of this notion for the non-Eurocentric case, we find this a little puzzling. Why this insistence, even among non-Eurocentric scholars, on the notion that there exists a 'trans-historical' entity originating with Greece, developing through Latin Europe up until the present day?

In section 1 above, we noted that for something to exist trans-historically, it must possess some immutable attribute that does not change over time; we pointed out that neither culture, language, nor geography (including climate) can serve as this immutable attribute. We have now seen that in the construction of the 'West' as a 'trans-historical' entity, 'race' served as the underpinning immutable attribute that enabled such a construction. With racism now being abandoned by most scholars, what can replace 'race' as such an immutable attribute? There can only be one candidate, namely, whiteness. As said (see note 18), it is legitimate to speak of whites, blacks, and so on, but without racist connotations. So it

[24] In contrast, we find sociologists much more ready to respond positively to our genealogical deconstruction of the notion.

is, indeed, possible for non-racists today to speak of Greece, Rome, the Middle Ages and so on as constituting a trans-historical entity consisting of whites. But *without* racist connotations, is it *meaningful* or *logically consistent* to do so? To answer this question, consider the common term 'Western' culture or civilization which is taken to have begun in Greece. A moment's reflection will show that to speak of 'Western' culture or civilization in this sense *necessarily* assumes that white people, *because they are white*, have produced over a period of more than two millennia a particular culture or civilization that stands in contradistinction to other cultures or civilizations, that is to say, *assumes that there are some extra-complexion implications of whiteness.* For racists, such an assumption is exactly what their racist view consists of. But for non-racists who reject extra-complexion implications of whiteness, to think of whites as constituting a 'trans-historical' entity that has, *just because* they are white, produced such a particular culture or civilization can make no sense and is logically inconsistent with their rejection of extra-complexion implications of complexion. Perhaps non-racist non-Eurocentric scholars who insist on speaking of the 'trans-historical' 'West' have never given any thought to this? *From the dialogical perspective*, the correct way of talking about modern Euro-American civilization, as a distinct civilization in comparison to other civilizations, is to see it as having been produced *not* by a 'trans-historical' white people *on the basis of their being a white people*, but as an outcome of Latin Europeans, with their own particular cultural elements, *also* borrowing and adapting numerous cultural elements from various other civilizations, both past and present, through a lengthy historical process.

In our above-mentioned discussions with non-Eurocentric scholars, we have come across the argument that since Latin Europe had taken up Greek learning, there is no need to insist on distinguishing between them. This involves *a conflation between scholarly tradition and society/culture*, a point that we have also made in another context in Chap. 3. What Latin Europe took up from Greece was a *scholarly tradition*, while we distinguish between them as different *societies/cultures.* The Abbasids also took up the Greek scholarly tradition, but would anyone make the mistake of saying that the Muslims and the Greeks constituted one and the same 'trans-historical' entity? That no one would make such a mistake in this case is clearly because the Arabs were not white; this also goes to illustrate our point that it is because both the ancient Greeks and Latin Europeans were white that even non-racist scholars today are so reluctant to problematize the notion of the 'West'. Other than scholarship, Latin Europe and its successor also *adopted* various other aspects of Greek culture such as sculpture and so on, but adoption does not make the adopter and the party who is being adopted one and the same trans-historical entity.

Today, even Eurocentric scholars admit that Arabic scholarship took Greek learning to new heights instead of merely 'preserving' it in its original state as well as admit to the scholarly debts that Latin Europe owed to Arabic scholarship. Hence, the AWD paradigm in its original form is no longer being supported by Eurocentric scholars. Given that, is it still necessary for supporters of dialogism to deconstruct the notion of the 'West'? The answer is an emphatic 'yes', because without that deconstruction, it is still possible for Eurocentric scholars to maintain the AWD paradigm close to its original form, as a result of which, dialogism faces the danger of being reduced to a sideshow even given the above admissions. We illustrate this with reference to two instances, the first concerns the physicist Chen Fangzheng, who takes it upon himself to demolish Needham's suggestion that modern science is a multi-civilizational outcome, the second concerns Floris Cohen who adopts the comparative approach as he understands that approach (see the last part of Chap. 2) with regard to the rise of modern science.

Chen Fangzheng, a physics academic and later administrator of the Chinese University of Hong Kong, met Needham when the latter stayed at that university on a visiting trip sometime in the late 1970s or early 1980s. He takes great exception to Needham's view that modern science is a sea into which the river of Chinese science also flowed, and subsequently wrote a thick treatise in Chinese on the rise of modern science using mathematical astronomy as illustration, in order to refute Needham's view. Arguing that modern science is the 'product' of the 'Western scientific tradition', he traces the centre of this tradition through 'Croton [first base of Pythagoras], Athens, Alexandria [Ptolemy's base], *Baghdad*, *Iran* and *Central Asia*, *Cairo*, *Cordoba* [a centre of learning in Islamic Spain], *Toledo* [another centre of learning in Islamic Spain], Paris, Oxford, North Italy, Cambridge, etc.' (Chen Fangzheng 2009, p. 599, our emphases). As can be seen from this passage Chen fully acknowledges the advances achieved by Arabic scholarship over Greece and the role it played in this process, though the role of Indian scholarship in the development of Arabic scholarship is absent in his discussion—for the moment, we leave aside Chen's failure to cover this and other non-European inputs. At the end of his study, he remarks (ibid., p. 634, our emphases):

> *Western* science is not only a branch of its [the '*West's*'] many scholarly fields, it is the embodiment of the *spirit* of its entire *civilization*. To truly know *Western* science and the *spirit* behind it, requires a comprehensive understanding of *Western* philosophy, religion and the entirety of its *civilization*.

Chen has purportedly arrived at this 'conclusion' on the basis of more than six hundred pages full of detailed technical discussion of mathematical astronomy, but that is actually not the case, because there is one small problem with Chen's 'conclusion', which is that in his own analysis, the centres through which the 'Western scientific tradition' migrated over the centuries include those that we have underlined in the above first quote from him. He does belatedly realize that he has fallen into a self-contradiction. To 'resolve' the problem, he makes a Popperian ad hoc modification of the most blatant kind. In his whole study, he has not for one moment sensed the need to define what he means by the 'West'—undoubtedly because, like most other people, he takes that notion for granted—but now, at the very end, he proposes the notion of '"the West" in the broad sense' (ibid., p. 631, quotes around *the West* are Chen's). Without defining what this 'broad sense' is supposed to mean, by means of it, he turns Arabic scholarship into 'Western' scholarship. His making of an ad hoc modification to try to salvage his so-called 'conclusion' shows clearly that the latter is actually his a priori starting-point.

Let us dissect Chen's logic by explaining on his behalf what he, without himself being aware of it, means by the 'West' 'in the broad sense': (a) There is a 'trans-historical' entity known as the 'West' which began with Greece, hence, the Greek tradition was the beginning of the 'Western' tradition; (b) Arabic development of the Greek tradition was, in his scheme, a further development of the 'Western' tradition ('in the broad sense'); (c) Latin Europe belonged to the 'trans-historical' 'West', given (b), its debt to Arabic scholarship was an 'internal' debt owed by one stage of the 'Western' tradition ('in the broad sense') to an earlier stage; ergo, (d) Newton was the final outcome of one single 'autochthonous' 'Western' tradition. This is the AWD paradigm in pure form, though with the twist of transforming Arabic scholarship into 'Western' scholarship 'in the broad sense'.

While introducing the notion of '"the West" in the broad sense' constitutes an ad hoc modification, one can ask the question as to how Chen could justify to himself the turning of Arabic scholarship into 'Western' scholarship, if only 'in the broad sense'. Chen was able to do this because he conflates scholarly tradition with civilization (or society/culture), a conflation that we have referred to shortly above. Chen begins with the fact that the Greeks established a particular tradition, by seeing Greece to be the origins of the 'West', Chen takes this to be the tradition of 'Western' civilization; Chen gives full acknowledgement to the fact that the Arabs took up the Greek scholarly tradition and developed it to new heights, but due to *Chen's conflation of scholarly tradition with civilization*, for him, despite the Arabs not being 'Westerners', Arabic achievements constituted achievements of 'Western' *civilization* simply by virtue of their being achievements made in the Greek *scholarly tradition*. Hence, the contributions of Arabic scholarship notwithstanding (i.e., *the dialogical argument notwithstanding*), for Chen, on the 'common sense' assumption of Greece being the origins of the 'West', modern science remains the adventure of 'Western' civilization. Even if the contribution of Indian mathematics to Arabic scholarship is taken into account, the logic of Chen's conception remains to be that modern science is still, largely though not entirely, the adventure of 'Western' civilization. To fully overcome Chen's fallacy, pointing out that scholarly tradition must not be conflated with civilization is not enough, for it is clear that *without deconstructing the notion of the 'West', dialogism faces the danger of being reduced in large measure to a sideshow*.

Floris Cohen is a leading contemporary light in the history of the rise of modern science. In his highly sophisticated analysis of the rise of modern science (Cohen 2011),[25] he makes various observations on historical explanation and analysis that are refreshing and in line with our own thinking. To give a few instances: 'the past, as also the present, displays a blend of situational logic and contingency…events are neither *wholly* predetermined [we would say deterministically explicable] nor *just* a matter of chance [let us add: since deterministic factors are always present, despite the importance of contingency, developments are never just, i.e., wholly, chance]' (p. xxvii). 'In the writing of history, pattern must be discerned, not imposed…As far as I can tell, the procedure here followed is an uncommon one, and it seems advisable to mark it. Wherever in the book that follows I have solidified my factual material into concepts and theories by means of this partly inductive, partly deductive process…' (p. xxii). Reading this passage reminds us of our own explanation of how we arrived at our theoretical framework mentioned at the beginning of Chap. 3.

[25] All references to Cohen hereinafter are, with one exception, to this work, hence, where only pagination is provided, it refers to this work.

'Due primarily to the vicissitudes of military-political history, such stimulating and transformative experiences were to keep happening to the Greek corpus of nature-knowledge but never to its Chinese counterpart...' (p. 46); regardless of the substantive accuracy of this comment (which is not to say that we necessarily disagree with it), it coincides with the path-dependency view though Cohen lacks the precise conceptual and theoretical tools of that view.

Often, Cohen seems to give the impression that he is distancing himself from Eurocentrism (see shortly below), yet in his attempt to undermine the dialogical perspective in relation especially to China and, to a lesser extent, India, he seems to take it upon himself to defend the Eurocentric position in its original form (see Chap. 6). And as we show in Chap. 9, in Cohen 2011, in trying to explain the breakthrough of ca. 1600, his resort to certain purported European 'qualities' belongs to the neo-Eurocentric variety mentioned in Chap. 2.

In relation to the argument of the present chapter, Cohen is also fully aware that Greece's intellectual development was not autochthonous (p. 46); further, he even seems to take exception to the notion of the 'trans-historical' 'West':

> Here is the easy answer to that hoary question [concerning the history of science]. The adventure of science started in the West, namely, in ancient Greece, and (but for a holding action by the Arabs) the adventure has always *remained* of the West, namely, of *Europe, where what the Greeks began was destined to come to full fruition* (our emphasis). This broad view of things has been reiterated often...ever since the 19th century, not only has some sort of *identification* with Greece been made in European...historical thought, but this identification has been carried one step further: the Greeks came to look very much like 19th-century Germans or Britons. *Almost*, the Greeks were already us...Historians of ancient Greece have since broken with this *still deeply ingrained* (our emphasis) picture[26] and have sought to draw a different one that stresses, not so much the extent to which ancient Greece was already like Western Europe, but rather its radical *otherness*' (pp. xxii-xxiii).

If we were to try to find a major science historian to endorse this chapter's deconstruction of the notion of the 'trans-historical' 'West', this passage from Cohen would come close to our requirement. This notion, Cohen

[26] To our awareness, this is certainly true of some historians and classicists, whether or not it is true generally, we are not so sure.

points out, is 'still deeply ingrained'—let us add, even among non-Eurocentric scholars. Yet, as is to be shown below, while *verbally* he opposes the view that 'the adventure [in science] has always *remained* of the West, namely, of *Europe, where what the Greeks began was destined to come to full fruition* (our emphasis)', *in practice*, in his analysis of the rise of modern science, he *precisely* presents the picture that it is in 'Europe, where what the Greeks began was destined to come to full fruition'. This divergence between talk and practice is part and parcel of Cohen's general pattern of apparently distancing himself from Eurocentrism, yet in the end adhering steadfastly to it (whether in its original or neo forms).

In his analysis of the rise of modern science, Cohen speaks, firstly, of three 'cultural transplantations', and then of 'three revolutionary transformations' that were responsible for the breakthrough of ca. 1600. In discussing the 'transplantations', he speaks of 'Athens' as the code for natural philosophy, and 'Alexandria' as the code for the mathematical sciences. As mentioned in the previous chapter, 'cultural transplantation' is a misnomer, for what was 'transplanted' was scholarly tradition, not culture. The three 'transplantations' form Chaps. 2 to 4 of his work, which are titled respectively 'Greek nature-knowledge transplanted: The Islamic World', 'Greek nature-knowledge transplanted in part: Medieval Europe', and 'Greek nature-knowledge transplanted, and more: Renaissance Europe'. By 'Greek nature-knowledge transplanted: The Islamic World', he is referring, of course, to the taking up of the Greek tradition by Arabic scholarship. He states that in both natural philosophy and mathematical sciences, the Greek corpus was 'enriched' by Arabic scholars, and that 'there is nothing a priori inconceivable in some Galileo- and/or Kepler-like figure coming forward' in Arabic scholarship (pp. 59, 72). So, like Chen above, Cohen gives Arabic scholarship full recognition.

Cohen has a low opinion of the scholastics, thus, in the third chapter on the 'transplantation' to Medieval Europe, there is a section titled 'Alexandria scholasticized, submerged, and somewhat enriched'. In other words, he argues that the scholastics 'submerged' Greek mathematical sciences in general, though Buridan and Oresme did 'somewhat enriched' them. In discussing the 'transplantation' to Renaissance Europe, he speaks of 'Athens replayed in full' ('in full' because besides Aristotelianism, extra-Aristotelian works of atomism, (neo-)Platonism, and so on were also made available to Latin Europe as Byzantine scholars fled from Constantinople under Ottoman menace), and 'Alexandria: A replay with a difference', referring to the advances that late Renaissance scholars made over Greek mathematical sciences.

The three 'revolutionary transformations' occupy chaps. 5 to 7, titled respectively 'The first transformation: Realist-mathematical science', 'The second transformation: A kinetic-corpuscularian philosophy', and 'The third transformation: To find facts through experiment'. We examine Cohen's arguments concerning these 'transformations' in Chaps. 8 and 9. For now, it needs only be noted that he describes the first transformation as transforming Alexandria into Alexandria-plus, the second Athens into Athens-plus, and concerning the third, he describes it variously as 'specifically European-colored mode of investigation' (p xvi), 'culturally marked pieces of nature-knowledge in Renaissance Europe' (p 114), 'Nature-knowledge colored the European way' (p. 134), and 'the culture-bound European mode' (p. 139).

Whereas Cohen does acknowledge Latin Europe's debts to Arabic scholarship which had taken Greek scholarship to a higher level, in a work of some eight hundred pages, we count only the following times when such an acknowledgement is made: 'Was there revolutionary transformation due to the importation of foreign knowledge [referring to Arabic scholarship]?...The answer to these and numerous questions is undoubtedly 'yes'. But...' (p. 207)[27]; 'earlier transplantations (Greek→Arabic, Arabic→Latin, and Greek→Latin)...' (p. 549). He even mentions India and China on several occasions: 'These developments in the possibilities of travel, preservation, and communication go some way toward explaining how the number zero and the decimal place system could be adopted from India, how the Taoist idea of achieving longevity through the *elixir* could reach the Islamic world' (p. 55); 'Mathematicians in the Islamic world made a new start, which was aided by...developed in previous centuries in India' (p. 57); '[al-Biruni] examine[d] and record[ed] with sympathy the state of learning in Hindu India' (p. 61). As can be seen, these acknowledgements are always made *in passing* and *very briefly*. In his historiographical work concerning the rise of modern science (Cohen 1994, p. 438), an acknowledgement of India's role is made within parenthesis only: 'of the Muslim world (enriched in its turn by Hindu mathematics)...'. The above-mentioned acknowledgement of an 'Arabic→Latin' 'transplantation' is made pretty late in the book on page 549, almost as an afterthought, and in no more than just two words and the '→' symbol.

[27] We discuss his 'but' in Chaps. 8 and 9.

As is common knowledge now, the 'Arabic→Latin' transmission involves whole branches of knowledge as well as methodologies, hence, one wonders why, in addition to a chapter each in the Greek→Arabic, Greek→Medieval Europe, and Greek→Renaissance Europe 'transplantations', *there is no chapter devoted to the Arabic→Latin Europe (inclusive of both Medieval Europe and Renaissance Europe*[28]*) 'transplantation'*? Further, Galileo's debts to Buridan and Oresme, highlighted by leading historians such as David Lindberg and Edward Grant, can nowhere be found in Cohen's schema of his three 'transplantations', a point that we leave until Chap. 8.

In the argument *embodied in the very structure of Cohen's book* from the three 'transplantations' of Chaps. 2 to 4 to the three 'revolutionary transformations' of Chaps. 5 to 7 which were responsible for the breakthrough of ca. 1600, the Greek→Arabic 'transplantation' eventually petered off, the Greek→Medieval Europe 'transplantation' reached a dead-end, and it was the Greek→Renaissance Europe 'transplantation' that *directly* (since there is no chapter on the Arabic→Latin Europe transmission) produced the 'revolutionary transformation' of Alexandria into Alexandria-plus (mathematical science becoming realist, which is supposed to have occurred only in Renaissance Europe), and the 'revolutionary transformation' of Athens into Athens-plus (from Greek atomism to 'kinetic corpuscularian philosophy'), and these two 'revolutionary transformations' combined with a third one, which is a 'specifically European-colored mode of investigation', were responsible for the breakthrough of ca. 1600. In short, *it was in Renaissance Europe, to quote from Cohen himself, 'where what the Greeks began [Athens and Alexandria] was destined to come to full fruition [Athens-plus and Alexandria-plus]'*, which, in combination with Renaissance Europe's 'specifically European-colored mode of investigation', produced the breakthrough into modern science. In other words, *the very structure of Cohen's book paints a picture of and embodies an argument concerning the rise of modern science that he verbally opposes.* In this picture, *if the notion of the 'trans-historical' 'West' originating from Greece is taken for granted*, there is, then, no room for the cross-fertilization between different civilizations to play any role; and from Greece to Renaissance Europe, modern science was all along an adventure of the 'West'. Now, Cohen does verbally recognize that Greece was not 'like

[28] With regard to Arabic→Renaissance Europe, one needs only mention Copernicus' copying from al-Shāṭir.

Western Europe, but rather [marked] by its radical *otherness*'. However, given his just-mentioned divergence between talk and practice, one wonders whether or not there is the same divergence with regard to this too. Further, this verbal recognition is made in a book review, not in Cohen's magnum opus, and only a reader who is keenly aware of the stakes involved will recognize the issue for what it is; for the reader who takes the notion of the 'West' for granted, this passing remark of Cohen's probably means little, if anything, at all.

To conclude, in Chen's case, the rise of modern science is the 'internal' development of the 'spirit' of 'Western' civilization notwithstanding the contributions of the Arabs; in Cohen's case, it is a development 'directly' from Greece to Renaissance Europe. Without deconstructing the notion of the 'trans-historical' 'West', the Eurocentric view in close to its original form is capable of sustaining a robust survival, and dialogism is severely undermined. Deconstructing this notion can not only serve to put the final nail in the coffin of Eurocentrism, by the same token, the dialogical perspective is immensely strengthened *precisely* by the fact that the Greek tradition played a key role in the rise of modern science. To say that, however, is not to imply that such a deconstruction constitutes a deliberate tactical manoeuvre, artificially separating what is originally one—for it is precisely the notion of the 'West' that ideologically constructs what are originally two into a fictitious one; it is simply to underline what the stakes involved are for the non-Eurocentric case.

Bibliography

Bala, Arun. 2006. *The Dialogue of Civilizations in the Birth of Modern Science.* Palgrave Macmillan.

Bassin, M. 1992. 'Russia between Europe and Asia: The ideological construction of geographical space'. *Slavic Review* 50(1), pp 1-17.

Bassin, M. 1998. 'Asia'. In N. Rzhevsky (ed.), *The Cambridge Companion to Modern Russian Culture.* Cambridge University Press.

Bernal, Martin. 1987. *Black Athena: The Afroasiatic Roots of Classical Civilization, Volume 1 The Fabrication of Ancient Greece 1785-1985.* New Brunswick: Rutgers University Press.

Bernal, Martin. 2001. *Black Athena Writes Back: Martin Bernal Responds to His Critics.* Durham: Duke University Press.

Bonnett, Alastair. 2003. 'From White to Western: "racial decline" and the idea of the West in Britain, 1890-1930'. *Journal of Historical Sociology* 16(3), pp 320-348.

Bonnett, Alastair. 2004. *The Idea of the West: Culture, Politics and History*. Palgrave Macmillan.

Campbell, John, and Philip Sherrard. 1965. 'The Greeks and the West'. Pp. 69-86 in Raghavan Iyer (ed.), *The Glass Curtain Between Asia and Europe*. London: Oxford University Press.

Chen Fangzheng. 2009. *Inheritance and Rebellion: Why Modern Science Emerged in the West*. 陳方正，《继承与叛逆:现代科学为何出现于西方》(北京:三联，2009)。

Cohen, H. Floris. 1994. *The Scientific Revolution: A Historiographical Inquiry*. University of Chicago Press.

Cohen, H. Floris. 2011. *How Modern Science Came into the World: Four Civilizations, One 17th-Century Breakthrough*. Amsterdam University Press.

Cohen, H. Floris. 2012. 'From west to east, from east to west? Early science between civilizations', book review of Arun Bala, *The Dialogue of Civilizations in the Birth of Modern Science*, Avner Ben-Zaken, *Cross-Cultural Scientific Exchanges in the Eastern Mediterranean*, Sylvain Gouguenheim, *Aristote au Mont-Saint-Michel. Les racines grrcques de l'Europe Chretienne*, Toby Huff, *Intellectual Curiosity and the Scientific Revolution. A Global Perspective*, and Jonathan Lyons, *The House of Wisdom. How the Arabs Transformed Western Civilization*'. *Early Science and Medicine* 17, pp 339-350.

de Santillan, Georgio. 1963. 'On forgotten sources in the history of science'. Pp 813-28 in A.C. Crombie (ed.), *Scientific Change: Historical Studies in the Intellectual, Social and Technical Conditions for Scientific Discovery and Technical Invention from Antiquity to the Present. Symposium on the History of Science Held at the University of Oxford, 9-15 July 1961*. London: Heinemann.

Delanty, Gerard. 1995. *Inventing Europe: Idea, Identity, Reality*. Basingstoke: Macmillan.

Dionisotti, Carlo. 1971. *Europe in the Sixteenth-Century Italian Literature*. Oxford: Clarendon.

Fischer, Steven Roger. 1999. *A History of Language*. London: Reaktion.

Fischer, Steven Roger. 2004. *A History of Writing*. London: Reaktion.

Geary, Patrick. 2002. *The Myth of Nations: The Medieval Origins of Europe*. Princeton: Princeton University Press.

GoGwilt, Christopher. 1995. *The Invention of the West: Joseph Conrad and the Double-Mapping of Europe and Empire*. Stanford University Press.

Goody, Jack. 2004. *Islam in Europe*. Polity.

Goody, Jack. 2006. *The Theft of History*. Polity.

Grant, Edward. 2004. *Science and Religion 400B –AD 1550*. Baltimore: John Hopkins University Press.

Hall, A. Rupert. 1962. 'General Introduction'. In Marie Boas, *The Scientific Renaissance 1450-1630*. London: Collins.

Hall, Edith. 1989. *Inventing the Barbarians: Greek Self-Definition through Tragedy*. Oxford: Clarendon.

Halsall, Guy. 2007. *Barbarian Migrations and the Roman West, 376-568.* Cambridge: Cambridge University Press.

Hart, Roger. 2013. *Imagined Civilizations: China, the West, and Their First Encounter.* Baltimore: John Hopkins University Press.

Hay, Denys. 1968. *Europe: Emergence of an Idea*, 2nd ed. Edinburgh: Edinburgh University Press.

Hay, Denys. 1980. 'Europe revisited: 1979'. *History of European Ideas* 1(1), pp 1-6.

Hegel, Georg Wilhelm Friedrich. 1975. *Lectures on the Philosophy of World History: Introduction: Reason in History*, trans. by H.B. Nisbet, edited by Duncan Forbes. Cambridge University Press.

Heller, Peggy. 2010. 'The Russian Dawn: How Russia Contributed to the Emergence of "the West" as a Concept.' Pp 33-52 in Christopher Browning and Marko Lehti (eds.), *The Struggle for the West: A Divided and Contested Legacy.* Routledge.

Hobson, John M. 2006. *The Eastern Origins of Western Civilisation.* 5th printing with corrections. Cambridge: Cambridge University Press.

Joseph, George G. 2011. *The Crest of the Peacock Non-European Roots of Mathematics*, 3rd edition. Princeton: Princeton University Press.

Kabbani, Rana. 1986. *Europe's Myth of the Orient.* Indiana University Press.

Katz, Victor and B. Lumpkin. 1995. 'The relationship of Egyptian to Greek mathematics'. *History and Pedagogy of Mathematics Newsletter* 35, pp 10-3.

Keay, John. 2005. *The Spice Route.* London: John Murray Publishers.

Keay, John. 2011. *India: A History*, revised and expanded edition. Grove Press.

Leyser, Karl. 1992. 'Concepts of Europe in the Early and High Middle Ages'. *Past and Present* no. 137 (Nov), pp 25-47.

Lindberg, David. 1978. 'The transmission of Greek and Arabic learning to the West'. Pp 52-90 in Idem. (ed.) *Science in the Middle Ages.* Chicago: Chicago University Press.

Lloyd, G.E.R. 1973. *Greek Science After Aristotle.* London: W.W. Norton.

Malik, Kenan. 2008. *Strange Fruit: Why Both Sides Are Wrong in the Race Debate.* Oxford: Oneworld.

Needham, Joseph. 1981. *Science in Traditional China.* Hong Kong: The Chinese University of Hong Kong Press.

Neumann, I. 1996. *Russia and the Idea of Europe: A Study in Identity and International Relations.* Routledge.

Noble, Thomas (ed.). 2006. *From Roman Provinces to Medieval Kingdoms.* London: Routledge.

O'Connor, Michael. 1996. 'Epigraphic Semitic scripts'. Pp. 88-107 in P. Daniels and W. Bright (eds.), *The World's Writing Systems.* Oxford: Oxford University Press.

Palter, Robert. 1993. 'Black Athena, Afro-centrism, and the history of science'. *History of Science* 31, pp 227-87.

Pomeranz, Kenneth, and Steven Topik. 2006. *The World that Trade Created*, 2nd edition. Armond: M.E. Sharpe.

Rogers, Henry. 2005. *Writing systems A Linguistic Approach*. London: Blackwell.

Ronan, Colin. 1983. *The Cambridge Illustrated History of the World's Science*. Cambridge: Cambridge University Press.

Saliba, George. 2007. *Islamic Science and the Making of the European Renaissance*. Cambridge, Mass: MIT Press.

Southern, R.W. 1962. *Western Views of Islam in the Middle Ages*. Cambridge, MA: Harvard University Press.

Stock, Brian. 1978. 'Science, technology, and economic progress in the early Middle Ages'. Pp. 1-51 in David Lindberg (ed.), *Science in the Middle Ages*. Chicago: Chicago University Press.

Swiggers, Pierre. 1996. 'Transmission of the Phoenician script to the West'. Pp. 261-70 in P. Daniels and W. Bright (eds.), *The World's Writing Systems*. Oxford: Oxford University Press.

Todd, Malcolm. 1992. *The Early Germans*. Oxford: Blackwell.

Ullmann, Walter. 1969. *The Carolingian Renaissance and the Idea of Kingship*. London: Methuen.

Varouxakis, Georgios. 2019. 'The godfather of "occidentality": Auguste Comte and the idea of "the West"'. *Modern Intellectual History* 16(2), pp 411-441.

Varouxakis, Georgios. 2020. 'When did Britain join the Occident? On the origins of the idea of "the West" in English'. *History of European Ideas* 46(5), pp 563-581.

Wang Shanbo. 1996. *In Search of the Scientific Spirit: Comparing Chinese and Western Science from a Philosophical Perspective*. 王善博，《追求科学精神:中西科学比较与融通的哲学透视》(南宁:广西人民出版社，1996).

Weber, Max. 1992. 'Author's introduction'. Pp xxviii-xlii in Max Weber *The Protestant Ethic and the Spirit of Capitalism*, trans. by T. Parsons, Introduction by A. Giddens. Routledge.

West, Martin. 1997. *The East Face of Helicon: West Asiatic Elements in Greek Poetry and Myth*. Oxford: Clarendon.

Williams, Raymond. 1976. 'Civilization'. Pp. 48-50 in Idem., *Keywords: A Vocabulary of Culture and Society*. London: Fontana.

Wolf, Eric R. 1997[1982]. *Europe and the People Without History*. Berkeley: University of California Press.

Woolf, Stuart. 1989. 'French civilization and ethnicity in the Napoleonic Empire'. *Past and Present* 124(1), pp 96-120.

CHAPTER 5

Social Essentialism and Sociological Reductionism Concerning Greece in Comparison to China

Given the role of the Greek tradition, a non-essentialist understanding of how the Greeks came to establish their particular tradition is most crucial. This is especially because this issue has always been given essentialist interpretations. In Chap. 3, we have already addressed this issue in a preliminary way, here we would like to pursue it further from another angle. In the literature, there is a thesis concerning Greece in comparison to China which is of special significance because, firstly, it is the most sustained and systematic social-essentialist analysis concerning intellectual development available; secondly, it reproduces Weber's AWD paradigm and DML view in pure form; and, thirdly, it exactly matches the conceptually flawed dual form in which Needham raised the NQ. This thesis argues that a whole range of scholarly developments in Greece was due to a particular social factor that was unique to Greece, and that these developments were lacking in China because of the absence of that social factor. It was advanced by the classical scholar G.E.R. Lloyd and the sinologist Nathan Sivin, in a collaboration that lasted almost three decades until Sivin's passing away in 2022. The main works of Lloyd's and Sivin's—for convenience, reference to them collectively will hereinafter be made in the short form L&S—collaboration were published in the 1990s and early 2000s (Lloyd 1990; Sivin 1995; Lloyd 1996; Lloyd and Sivin 2002), but they have continued to advance the same argument in works on the same or overlapping subject matters published in recent years (Lloyd 2015, 2018; Sivin 2018).

R. W. K. Lau, *Rethinking the Needham Question*,
https://doi.org/10.1007/978-981-97-9472-0_5

The social factor concerned is the institution of the agora debate that existed in Greece but not in China; instead, China is said to be suffering from an 'inhibitory' social institution, namely, scholars' alleged complete submission to political patronage.[1] Hence, we call their thesis the 'agora debate' thesis.

In Chap. 3, in discussing social explanation, we pointed out that even for non-essentialist social explanation, it is important to understand what social explanation can and cannot do. To repeat what we said, what social explanation can provide is to identify the conditions arising from social factors that impact upon intellectual development, either positively or negatively or in other ways, but in science and mathematics social factors are unable to explain the ideational aspects of that development (i.e. the emergence and development of ideas). A common trap of social explanation is to fall into sociological reductionism, which reduces ideas to social factors, that is to say, to see ideas as the direct effects or direct reflection of social factors. As will be seen as we proceed, not only is L&S's thesis social essentialist, it also constitutes a classic case of sociological reductionism.

To critique social-essentialist arguments, there are, broadly, two approaches that can be taken, depending on the specific circumstances of the individual cases concerned. The first is either to show that the social factor concerned has not arisen from purely internal dynamics of the society concerned, but, for instance, as a result of its interactions with other societies; or that while that factor has, wholly or largely, arisen from internal dynamics, no one would imaginably ever claim that it constitutes an 'essence' of the society concerned that marks it off from other societies—the Chinese social institution of mathematics becoming the hereditary specialism of particular lineages, discussed in Chap. 3, is an instance of this. The second approach applies when the social factor concerned, such as the institution of the agora debate in the present case, has indeed arisen through internal dynamics, and which does mark the society concerned off from other societies in a quite distinct way, i.e. there is indeed considerable justification to speak of it as an essence[2]; in such a case, the approach is to show that in another society, even in the absence of such a or an

[1] Scholars' autonomy or the lack of it in China varied, in degree instead of in an either-or way, greatly over the centuries. L&S simply assume that there was none throughout China's history.

[2] Clearly, there can be no hard and fast rule to determine whether or not a social institution that has arisen from internal dynamics does or does not constitute an essence, it's all a matter of judgment and good commonsense.

equivalent social institution, similar intellectual developments can also be found. This is the approach that applies in our present case.

Before proceeding to our analysis, it is worth pointing out that L&S's 'agora debate' thesis is a form of what we refer to as the 'civil society' thesis. This thesis argues that intellectual developments 'unique' to the 'West' and purportedly absent in other civilizations were due to the former's possession of a particular social institution, namely, civil society, that is not found elsewhere. In a work titled *The Rise of Early Modern Science. Islam, China, and the West*, Huff (2003, pp. 2–3, our emphases) states the following in its opening pages, applying the 'civil society' thesis to the rise of modern science:

> What are the sociological foundations—…—that enable us to carry on this enterprise of freely pursuing inquiry wherever it leads?…As we enter the global world as never before, these are surely questions of fundamental importance. To create a truly global order there must be a set of fundamental shared principles—…—which enables us to communicate freely…Perhaps *the conditions that allowed the development of modern science* can tell us something about how societies (and civilizations) can and should be ordered…*the study of the sociological foundations of modern science may have much to tell us about the ingredients that go into the making of the 'open society', freedom of expression…*

If one is reminded by this passage of the political rhetoric of a bygone era,[3] one would not be far wrong. But political rhetoric aside, Huff states that there are certain 'conditions that allowed the development of modern science', and that these 'sociological foundations of modern science' concern '"open society", freedom of expression'. Huff apparently has a very short memory, his book from which the above quote is taken is concerned with 'early modern science', in relation to which need one be reminded of the various European inquisitions from the 1100s right up to the mid-1700s? Or the suspension of Copernicus' *On the Revolutions* by the Congregation of the Index of Prohibited Books in March 1616? Or the referral of Galileo's *The Assayer* of 1623 to the Roman Inquisition? Or the conviction of Galileo of 'vehement suspicion of heresy' in 1633 for his *Dialogue*? Or the need for the manuscript of Galileo's *Discourses* to be smuggled out of Italy for publication in Holland?

[3] Unfortunately, since 2021, we seem to be witnessing a second coming of such an era.

With regard to the 'civil society' thesis, Harbsmeier (1998, p. 414) has observed the following in relation to logical thinking, his special area of expertise:

> No simple distinction between the different roles of logic in despotic versus civil societies will help; occidental despotism (...) was rife in many places where logic flourished. Republican Rome with its freedoms was not noted for its advances in theoretical logic. The story here, when it is fully told, will be extremely complex. It will not be a simple matter of occidental civil society in Greece or Rome versus oriental despotism in China.

Despite being contradicted time and again by historical facts, the 'civil society' thesis continues to be made—Huff's political rhetoric quoted above may allow us a glimpse into why this is so. In any case, with regard to Huff, despite his above-quoted programmatic statement of the 'civil society' thesis in the opening pages of his book concerned, unlike L&S's highly systematic thesis, that work is an eclectic mixture of different kinds of argument, making it as a whole irrelevant to our focus (social essentialism and sociological reductionism) in the present chapter.

5.1 Greek Intellectual Developments Which China Purportedly Lacked

According to L&S's thesis, China lacked the following which were found only in Greece:

1. Real schools of thought.
2. Real public debate between scholars.
3. Independent-minded scholars whose objective was to seek truth instead of political patronage.
4. The search for ultimate foundations.
5. The equivalent of Greek element theories in physics.
6. A philosophy that is 'meant to be natural philosophy'.
7. The concept of nature—it was, it is claimed, only in 1881 that China acquired such a concept from Europe through a Japanese translation of the word 'nature'.
8. A clear analysis of the concept of causation.

For our present purposes, only the last five claims are relevant.[4] Let us examine how these claims lead to one another in a chain springing from the presence or absence of agora debate. In agora debate, Lloyd contends, one argues with other people as 'enemies', 'Ideally, to win the competition [in agora debate], you needed not just a theory you could claim as your own, but one you could present as true, better still as certain' (Lloyd 1996, p. 3). It is this purported quest for certainty in agora debate[5] that is supposed to be *reflected directly* in scholarship into a quest for ultimate foundations by Greek thinkers in the form of Greek element theories in physics, i.e. the doctrines of Empedocles, Anaxagoras, and Democritus—elements being the ultimate foundations. The absence of agora debate in China, on the other hand, translated into the purported absence of the search for ultimate foundations, and hence the purported lack of the equivalent of Greek element theories: 'That leads us to draw on aspects of Greek thought that have no counterpart in China...An obvious example is Greek element theories' (Sivin 1995, p. 2). As is to be seen, L&S believe that a correlative cosmology was all that the Chinese had instead. With element theories, the Greeks established a mature natural philosophy; by contrast, with only a correlative cosmology, the Chinese are said to lack a philosophy that is, in their wording, 'meant to be natural philosophy', i.e. a philosophy that properly constitutes natural philosophy. According to L&S, the conception of nature was a Greek 'invention'; presumably, the Greeks were able to 'invent' such a notion as an outcome of natural-philosophical investigation; by contrast, purportedly without a philosophy that is 'meant to be natural philosophy', the Chinese are said to also lack a similar concept until 1881 as mentioned above. Natural philosophy investigates into the physical causal relations between natural phenomena, presumably, this engendered a clear analysis of the concept of causation by the Greeks; by contrast, possessing only a correlative cosmology that did not investigate into physical causal relations, the Chinese purportedly lacked a clear analysis of the concept of causation.

[4] There are other comparisons besides these eight made by L&S which we do not cover, for reasons to be mentioned at the end of this chapter.

[5] Actually, agora debate is about persuasion, not certainty or seeking ultimate truths, as both Plato and Aristotle decried. As a leading classicist, Lloyd is, of course, fully aware of this, but there is no need for our present purposes to discuss how he proposes to get round this problem.

5.2 How Not to Compare by Selecting a Particular Period for Comparison

Let us begin with a crucial methodological issue, namely, the period of comparison chosen by L&S.[6] Selecting the period of comparison is methodologically equivalent to defining the core terms of an analysis—as seen in the previous chapter in relation to Chen Fangzheng's term of the '"West" in the broad sense', this pretty well determines the kind of conclusions that can be drawn. Sivin (1995, pp. 3–4) states:

> We [Sivin and Lloyd] have looked for a single period in which both cultures seem to be passing through more or less analogous transitions...we believe that there is one. It happens that between roughly 300 B.C. and A.D. 200...By the end of that time both were obsessed instead with preserving what had already been done...

With regard to Greece, setting the start-date at −300 excludes even Aristotle from the comparison, but, in practice, L&S routinely refer to Greek developments prior to −300. To take the most obvious instance with reference to Greek element theories in physics, Empedocles, Anaxagoras, and Democritus all lived in the -5th to, in the case of the last, the early -4th centuries. It is amazing how something so obvious could have escaped their, and especially Lloyd's, notice. The end-date of 200 for Greece covers everybody up to and including Ptolemy, so it is of no particular significance with regard to Greece. It is with regard to China that the implications of setting the period from −300 to 200 are most telling. To see how that is the case, a brief account of China's intellectual developments up to 200 is necessary.

Documented natural-philosophical theorizing dated back to −780. Around −500, a qualitative leap in development began which lasted until −221—the period before −221 is generally referred to as the pre-Qin period, since the Qin Dynasty was established in that year. '[T]his period [-500 to -221] provides clear evidence of an indigenously Chinese and entirely autochthonous intellectual movement in the direction of a critical rationalism and of systematic logical reflection' (Harbsmeier 1998,

[6] In mid-2021, we engaged in private exchanges with Lloyd for over a month on our critique of him and Sivin, for which we are most grateful to him; however, since at the end of our exchanges, he preferred to keep these exchanges private, we are unable to quote from them in the following.

p. 413). As explained in Chap. 3, since the founding of the Zhou Dynasty in −1046, mystical thinking had generally remained insignificant among intellectuals as a naturalistic scholarly tradition was established. However, in the latter Warring States period (from the second half of the -fourth century to −221), it rapidly made a comeback by infiltrating into the natural philosophy of the following two circles: Firstly, the Huanglao (黃老) scholars who founded the Yin-Yang School (*Yinyang jia*陰陽家)[7] and greatly influenced the thinker Zou Yan (鄒衍), generally regarded as the representative figure of the Yin-Yang School—hereinafter, we refer to this strand of thoughts as Huanglao-cum-Yin-Yang thoughts; secondly, the unidentified thinkers responsible for parts of what were subsequently compiled into the *Commentaries on the Book of Changes* (*Yizhuan* 《易傳》) that combined natural philosophy with divination—the *Book of Changes* is a divination book dating from the early Zhou Dynasty.

Under the brief Qin Dynasty (−221 to −206), a wholesale destruction of learning occurred, in which all schools of thought other than the state-sponsored Legalist school were suppressed and private collection of books forbidden. The only non-Legalist thinking that managed to survive was the just-mentioned branch of mysticism-laden natural philosophy that married with divination because divination books were allowed. In the early Han Dynasty (−206 to 220), this married couple continued to flourish in conjunction with Huanglao-cum-Yin-Yang thoughts. In −134, this combination became Han's state-sponsored philosophy while all other schools of thought were suppressed, as a result of which, the most rationalistic works of the pre-Qin period, namely, the works of the Later Mohists (後期墨家) and Gongsun Long (公孫龍) disappeared,[8] while the equally rationalistic propositions of Hui Shi (惠施) and his disciples[9] preserved in the *Book of Zhuangzi*[10] had become incomprehensible to Han

[7] *Yin* and *yang* originated as natural-philosophical concepts of great value, but in the hands of the just-mentioned scholars, they became heavily tainted with mysticism.

[8] The Later Mohists and Gongsun Long specialized in systematic logical investigation and, in the case of the former, optical experimentation as well.

[9] Hui Shi and his disciples engaged in proto-scientific investigation, some of their propositions in the form of paradoxes have been compared (favourably by some scholars) to those of Zeno.

[10] The *Book of Zhuangzi* (hereinafter the *Zhuangzi*) is a collection of essays by the philosopher Zhuangzi (to be discussed presently in the main text) and his disciples. Due to the Qin Dynasty's wholesale destruction of knowledge, these essays had to be recovered and compiled into a collection in the early Han Dynasty; most fortunately, this was able to occur before Han's suppression of non-state-orthodoxy.

scholars. Further, the completely naturalistic natural philosophy of Laozi (ca. mid -6th to early -5th centuries) and that of his successor Zhuangzi (fl. mid -4th to early -3rd centuries) were reduced to the margins of scholarly concerns (there will be a brief discussion of Laozi's and Zhuangzi's natural philosophy in Sects. 5.5.3 and 5.5.4 below). In sum, the line of pre-Qin intellectual development was abruptly and completely broken by the brief Qin Dynasty, and under the Han, a new intellectual tradition, which represented a reversal of the pre-Qin tradition, was imposed and entrenched by the state. By contrast, despite the re-location of the centre of learning on several occasions, the Greek tradition was never once broken, not to mention never suffering from any reversal.[11]

Thus, with regard to China, setting the start-date at −300 eliminates all pre-Qin developments before that date. Further, by 200 (the end of Sivin's period in the above quote), twenty years from the fall of the Han Dynasty, China was not 'obsessed with preserving what had already been done', but had for more than three hundred years been suppressing all non-state-sponsored schools of thought, to call it 'consolidation of what had already been done' is exactly the reverse of the truth.

After Han's fall, there occurred a period of intellectual renaissance in the third century, during which the brilliant mathematician Liu Hui (劉徽, ca. 225–295), to be mentioned in the following two chapters, appeared. Among other things, Liu independently invented what was later known as the exhaustion method. It seems that after agreeing on the end-date of 200 with Sivin, Lloyd then learned about Liu's work from other sinologists (see Lloyd 1996, p. 152 n 20), and hence, a year later, he (Lloyd 1996) effectively extended the end-point in practice by including a fair and objective discussion on Liu. This extension was later formally endorsed in Lloyd and Sivin (2002) and Lloyd (2018) which revise the end-date to 300. However, despite this, L&S effectively retain the end-date at 200 with regard to overall philosophical development. This is because during the post-Han third-century intellectual renaissance, the above-mentioned works of the Later Mohists and so on re-surfaced briefly while the natural philosophies of Laozi and Zhuangzi were taken up in earnest again, but none of this is covered by L&S.

[11] One reason for this contrast is that whereas the centre of Greek learning was able to relocate under changed political circumstances, the same opportunity did not exist in China. We can think of multiple non-essentialist reasons, including contingency, for this difference itself, which, however, are outside our present scope.

Further, setting the start-date at −300 artificially cuts the Warring States period (−475 to −221, the pre-Qin period with the most intense intellectual developments) into two (unequal) halves which defies all logic. The effects of this are clear. During the last century of the Warring States period, the above-mentioned Zou Yan came into prominence. Zou's works have not survived, but parts of them were collected in a compilation of essays and materials put together by a collective of intellectuals under the patronage of the premier of the Qin state Lü Beiwei (呂不偉), the state that later founded the Qin Dynasty, known as *Lü's Spring and Autumn Annals* (*Lüshi chunqiu*《呂氏春秋》hereinafter LSAA), which appeared in ca. -239. LSAA not only includes materials from Zou, it is full of, though not exclusively comprised of, Huanglao-cum-Yin-Yang thoughts. Key parts of LSAA served as the model of and basis for the premier philosopher of the Han Dynasty Dong Zhongshu (董仲舒) in his work titled *Luxuriant Dew of Spring and Autumn Annals* (*Chunqiu fanlu*《春秋繁露》), which represented the highest embodiment of Huanglao-cum-Yin-Yang thoughts and constituted Han's state-sponsored philosophy. Significantly, LSAA and *Luxuriant Dew* are the only two philosophical works specifically mentioned by Sivin (1995, pp. 5–7) for the purpose of comparing with Greece, and to which L&S often refer. In sum, the start-date of −300 enables the coverage of the strand of thoughts documented in LSAA, which constitutes the focus of L&S's coverage for the period from −300 to −221, and it is this strand of thoughts that became Han's state-sponsored philosophy.[12] The pre-eminent historian of ancient Chinese philosophy Lao Siguang (Lao 1981, Introduction), voicing a commonly shared view, describes Han's state-sponsored philosophy as the nadir of indigenous Chinese philosophy. While Han's reversal of the pre-Qin tradition was to have lasting effects beyond Han, this philosophy itself constituted such a low point that it was immediately abandoned after Han's fall. But it is precisely this philosophy that L&S have chosen for comparing with Greece.[13] The effects of this on their various claims concerning what China purported lacked will become obvious as we proceed.

[12] For simplicity, we skip the role of the *Commentaries on the Book of Changes* here.

[13] Inevitably, L&S do occasionally refer to developments before −300 and to non-Huanglao-cum-Yin-Yang thoughts between −300 and −221, but such occasional references are either skewed by their focus on Huanglao-cum-Yin-Yang thoughts (as shown in Sect. 5.4 below), or are insufficient to alter the overall picture presented by their 'agora debate' thesis.

5.3 China's Alleged Lack of the Greek Search for Ultimate Foundations and the Equivalent of Greek Element Theories

L&S's central claim—the claim that, as seen in Sect. 5.1 above, leads to the other claims made by them—is that Chinese thinkers lacked the search for certain, ultimate foundations as evidenced in their lack of the equivalent of Greek element theories. Besides being social-essentialist in nature, their belief that Greek element theories in physics were directly engendered by, i.e. being a direct reflection of, the *social* institution of the agora debate (both sides in the debate allegedly seeking certainty to defeat the other side) constitutes a case of sociological reductionism—the reduction of the *substance* of scholarship, at one level of social practice, as being directly the effects of what occurs at a different (supposedly more basic) level of social practice.

In order to address L&S's central claim with regard to China, it will be useful first to take a brief look at the Greeks concerning speculations on what Aristotle calls *archê*, i.e. 'that from which a thing first comes-into-being and into which it is finally destroyed'. *Archê* is the ultimate foundations that L&S talk about. The various *archê* substances proposed by the Milesians (Thales, Anaximander, and Anaximenes), Empedocles, Anaxagoras, and Aristotle (water for Thales, and so on) are all based upon empirical concepts.[14] In contrast, as explained in Chap. 3, induced by a combination of factors among which non-naturalistic ideas figure prominently, Parmenides asserted that the empirical world is illusory; reality is, instead, an absolute, unborn, indestructible, uniform, changeless, and motionless unitary Being. Parmenides did not address the issue of where its abode is. Inspired by Parmenides, Plato posited that empirical things are imperfect copies of divine Forms which exist outside time and space. While, unlike Parmenides, he did not reject the empirical as completely

[14] An empirical concept is a concept derived directly or indirectly from observation—to illustrate, by contrast, 'black hole' is a non-empirical concept, for it is deduced theoretically, not derived from observation, since black holes are unobservable. Incidentally, Anaximander's *archê* is *apeiron*, which is sometimes misleadingly translated as 'boundless' or 'infinite', however, it actually means 'without boundary/differentiation' between what subsequently became separated into distinct empirical things (Guthrie 1962, p. 86; Kirk et al. 1983, pp. 109–10), hence its reference is to empirical substances. Anaxagoras' 'seeds' ('seeds' of bone, etc.) are so fine as to be imperceptible, but the notion of imperceptible empirical things is still based upon an empirical concept.

illusory, he nonetheless regarded it as less real than the divine Forms. The posited reality that exists outside time and space is subsequently known as the *transcendent* in European philosophy. The atomists were likewise inspired by Parmenides, arguing, like Plato, that the empirical is less real than atoms. They did not explicitly address the issue of the abode of the atoms, but since for them empirical things are formed by atoms, by implication, the latter must, by default, exist within time and space. One focus of L&S's reference to element theories is the distinction between the empirical and non-empirical:

> Greek element theories, which claim that things are made up of minute ultimate parts that usually do not look like the parts that are big enough for us to see. Element theories build on the idea that reality is hidden, and direct experience is in some ultimate sense not real. (Sivin 1995, pp. 2-3; cf. Lloyd and Sivin 2002, pp. 142-57)

Atomism was a minority view throughout the ancient period, instead, it is Plato's view of the transcendent divine Forms which constitutes the most distinctive view of the Greeks on the matter.

Contra Plato, the non-empirical need not be transcendent. In modern scientific realism,[15] there is a category called the *unobservable*, which refers to the non-empirical that exists within time and space. The concept of the unobservable was advanced in the late nineteenth century as scientists were positing the existence of sub-atomic particles such as the electron. Pre-Qin Chinese natural philosophers did not regard the empirical as either illusory or less real, neither did they posit the transcendent. But they drew a distinction that, in spirit, corresponds to the modern distinction between the empirical and the unobservable, with the latter being the ultimate foundation of the former. This began with Laozi, who is eliminated, except on one occasion for reasons to be mentioned in the next section, from L&S's comparison because he lived long before −300[16]:

[15] On modern scientific realism, see Hacking (1983) and Chalmers (2013). Scientific realism is challenged by instrumentalism, which need not concern us.

[16] Laozi lived from ca. mid -6th to early -5th centuries. The original version of the *Laozi*, by Laozi himself, is known as the Guodian (郭店) *Laozi*—so called because it was discovered in an archaeological find at the Guodian archaeological site. We follow the text of this version, but we also cross-reference where our quotes can be found in the transmitted version of the *Laozi* (sometimes with differences in wording), since most people are familiar only with the latter. It should be noted that Chinese experts on Laozi all regard the Guodian ver-

There are four versions of the *Book of Laozi* (hereinafter the *Laozi*), the earliest one is known as the Guodian *Laozi*, written by Laozi himself, while the version that most people are familiar with is the transmitted version (see note 16). In passages in the Guodian *Laozi* that correspond to Chaps. 25 and 40 of the transmitted version, Laozi stated that there is a thing which

> exists prior to Heaven and earth [*tiandi*天地, the natural world]...it is independent and immutable, it can be the mother [giving birth to] all under heaven. Its name is unknown, [I] style it as Dao, I arbitrarily name it as the primal. It is in non-stop motion [over time], it moves far away but then returns.
>
> (先天地生......獨立而不改，可為天下母。未知其名，字之曰道，吾強為之名曰大。大曰逝，逝曰遠，遠曰返。)
>
> [To and] fro is the way in which Dao moves.
>
> (返者道之動。)[17]

Laozi's Dao, hence, refers to what Aristotle called *archê*. In pre-Laozi times, Chinese thinkers had arrived at the notion that nature operates according to a cyclical pattern, which they called 'the way of operation of nature' (*tian dao* 天道—see Chap. 2, on which further below). The statements 'moves far away but then returns' and '[To and] fro is the way in which Dao moves' show that Dao obeys nature's 'way of operation', thereby implying that it is a natural thing existing within space; and it follows from 'It is in non-stop motion [over time]' that Dao exists within time. But Laozi's Dao is not part of the empirical world. In a passage in the Guodian *Laozi* corresponding to Chap. 40 of the transmitted version, Laozi stated:

> The things of all under heaven are born from *you* (有, meaning that which possesses real existence), born from *wu* (無, meaning that which is without form).
>
> (天下之物生於有，生於無。)[18]

sion as the authoritative one when it comes to interpretation; the transmitted version is much longer than the Guodian version and clearly contains materials added to Laozi's original text by transmitters of the text.

[17] Our translations reference Chen (2012b, pp. 197–203, 259).

[18] Readers closely familiar with Laozi will be aware that there is a small but crucial difference in the text of the quoted passage between the Guodian and transmitted versions of the *Laozi*, which has a major impact on its interpretation. This, however, cannot be discussed

We have explained the meanings of the terms *you* and *wu* as they were used by Laozi to facilitate the reading of the quoted passage. Our authority is Wang Bi (王弼), the third-century annotator of the *Laozi* usually taken as authoritative by both Chinese and overseas scholars—Wang was a central figure in the post-Han third-century intellectual renaissance, which, as previously mentioned, L&S do not cover at all except for the mathematician Liu Hui:

> If it is to be said to be *wu*, but things are born from it; if it is to be said to be *you*, but it is without form. (Wang and Lou 2008, p. 31)
> (欲言無邪，而物由以成。欲言有邪，而不見其形。)

In other words, Laozi's Dao is *wu* because it is without form, it is at the same time *you* because as that which gives birth to empirical things, it possesses real existence. From Laozi onwards, the term *wuxing* (無形), that which is without form, was commonly used to refer to what we call non-empirical today. Thus, Laozi's Dao is something that possesses real existence within time and space but is non-empirical, in a word, it is, in modern terminology, unobservable. In the understanding of mid-Warring States thinkers, the unobservable ultimate foundation of the empirical constitutes both the origins and the constituent of the empirical, though whether or not this was also Laozi's understanding is uncertain. Sometimes they continued to use Laozi's term of Dao to refer to it, but mostly they referred to it as *qi* (氣). The term *qi* originally meant cloud gas, but it subsequently acquired various other meanings as used in different contexts, in the context of the speculation on the ultimate foundations of empirical things, it refers to this non-empirical ultimate foundation. The reason why post-Laozi pre-Qin thinkers extended the term *qi* to be used in this sense is a long story that cannot be explained here. There are different versions of *qi* doctrines during the Warring States period, which, however, need not concern us.

Thus, from Laozi onwards, pre-Qin thinkers, like the Greeks, also searched for the ultimate foundations of the empirical. While it is not our concern to go into any detailed comparison of Greek and Chinese *archê* doctrines, there is one general difference between them which is crucial. Right from the beginning, China's non-empirical ultimate foundations

here. Since the Guodian version is the one written by Laozi himself, it represents his original view. On this point, see Ding (1998, p. 218) and Chen (2012b, p. 168).

correspond to the naturalistic concept of the unobservable. As analysed in Chap. 3, as an unforeseen consequence, China developed a naturalistic intellectual tradition from the early -1st millennium onwards which lasted until the Qin Dynasty (−221 to −206) engaged in a wholesale destruction of learning. This was why in searching for the ultimate foundations of the empirical, pre-Qin scholars arrived at a naturalistic concept. In contrast, in Greece, on the one hand, the *archê* substances of most scholars were derived from empirical concepts, while, on the other, Plato's notion of the transcendent represented the most distinctive Greek view on the issue of the relationship between the non-empirical and the empirical. It is only the notion of atoms that provided something equivalent to China's Dao/*qi* in this regard. But, unlike Laozi, who explicitly theorized about Dao's attributes as something unobservable, the unobservable status of atoms was implied by default only.

L&S seem completely unaware of the above,[19] making them believe that China did not have the equivalent of element theories. This has everything to do with their focus on Huanglao-cum-Yin-Yang thoughts as China's frame of reference. The signature concept of this strand of thoughts is *wu xing* (五行), which should be translated as 'five phases' in the present context. The history of the notion of *wu xing* is too complicated to be discussed here, suffice it to mention that in the hands of the Huanglao scholars who founded the Yin-Yang School, underpinned by notions of cosmic sympathy, *wu xing* became used as the basis for drawing up correlations with the correlated groups locked into 'mutual procreation' (*xiang sheng*相生) and 'beginning and end' (*shizhong*始終) relations with one another. Taking up these ideas, the above-mentioned Zou Yan added the notion of 'mutual conquest' (*xiang ke*相剋), and applied the whole correlative schema to the succession of dynasties (Sun 1995, pp. 219–20; Hu Jiacong 1995, pp. 107–8, 121). The *correlative cosmology* of *wu xing* has nothing to do with pre-Qin's speculations on the ultimate foundations of the empirical world, as is obvious from our above discussion on the latter. But, if with regard to ancient Chinese thoughts pertaining to natural phenomena, one's focus is solely on Huanglao-cum-Yin-Yang

[19] Lloyd and Sivin (2002, pp. 196–7) mention the use of the term *qi* in various contexts, but there is no mention of its use to denote the non-empirical ultimate foundations of the empirical.

thoughts, as it is for L&S,[20] then one is easily misled into believing that China had no equivalent to Greek element theories.

5.4 Was Natural Philosophy Unique to the Greeks?

One can imagine that from L&S's point of view, with the element theories of Empedocles, Anaxagoras, and Democritus, the Greeks' speculations on *archê* developed into a mature natural philosophy. Since the Chinese are believed to lack the equivalent of Greek element theories, but only possessed a correlative cosmology instead, it follows that they could not possibly possess a philosophy that properly constitutes natural philosophy. On the other hand, anyone with even a casual acquaintance with ancient Chinese philosophy is aware that the philosopher Laozi is commonly regarded as the key progenitor of natural philosophy in China. Hence, while Laozi lived some two centuries before −300, and hence is outside of L&S's chosen period of comparison, not to discuss him would be inexcusable. At the same time, according to the revered late Angus Graham (1989, pp. 215–7), the *Laozi* is a forgery dating from ca. -250, which makes it fall within L&S's period of comparison. For either or both of these reasons, L&S have provided a discussion of Laozi—but in no more than a single paragraph.

To begin with, it should first be pointed out that Graham's belief that the *Laozi* is a forgery of ca. -250 is mistaken. There are, in fact, as mentioned in the previous section, four known versions of the *Laozi*, only three of which need to concern us. The earliest is the Guodian *Laozi*, authored by Laozi himself and dating from the early -fifth century or even earlier (Ding 1998). Graham was unaware of the existence of this version since its discovery came after his untimely death. The second version is referred to as the silk version, which circulated during the late Warring States period in the -third century (Han Wei 2013). The final version is the transmitted version dating from the third century, annotated by the post-Han third-century scholar Wang Bi mentioned in the previous section. The Guodian version is about one-third of the length of the

[20] Cf. Raphals (1998, p. 1132, our emphases) in a review of Lloyd 1996: 'He [Lloyd] explores Greek and Chinese...underlying values...(Chinese *yin-yang* and *Five Phase* language reflected the consensus in Han *correlative cosmology*)'.

transmitted version,[21] while the silk version is similar in length to the later, but the order of the chapters is different and, occasionally, there are some critical differences in wording between the two.

Lloyd and Sivin (2002, p. 204, our emphasis), as noted, discuss the *Laozi* in one single paragraph, part of which reads:

> The Book of Lao-tzu [Laozi] begins with two Ways (*tao* [Dao])....
> These two are the same
> But diverge in name as they issue forth.
> Being the same, they are called the mystery,
> Mystery beyond mystery
> Gateway of the manifold arcana.
> This is not meant to be natural philosophy...It is about the difference between common experience and mystical breakthroughs.

For the record, L&S's quote is from the second half of the first chapter of the transmitted version. This chapter is not found in the Guodian version[22]; in the silk version, it does not constitute the first chapter and its wording differs from that of the transmitted version. But since the transmitted version is the version most people are familiar with, and in the Chinese original, its first chapter is written with brilliant metrical elegance, except for those with in-depth knowledge of the philosophy of Laozi, it is generally taken (erroneously) to provide a programmatic summary of the philosophy of Laozi. To return to L&S, given that Laozi is commonly regarded as the key progenitor of natural philosophy in ancient China, by dismissing his philosophy as 'not meant to be natural philosophy...', what they are saying is that China did not possess a philosophy that properly constitutes natural philosophy. In other words, natural philosophy is supposed to be unique to the Greeks.

[21] Incidentally, there are disputes as to whether or not the Guodian version is the full text of Laozi's original work. Our own view is that it most probably is; in any case, by means of comparing texts of other thinkers dating from the -5th and -4th centuries, it can be shown that certain parts of the transmitted version not found in the Guodian version incorporate ideas that could not have dated from earlier than the mid -fourth century, and hence must have been added by post-Laozi transmitters of the text. This is probably why Graham, being unaware of the existence of the Guodian version, mistakenly believed that the *Laozi* was a late forgery.

[22] Whereas not all of the views expressed in this chapter of the transmitted version are consistent with the views of the Guodian version, this is not the case with the above-quoted passage.

L&S's interpretation of the quoted passage follows directly from how they translate a couple of words in it. The word they translate as 'mystery' is *xuan* (玄), and the word they translate as 'arcana' is *miao* (妙). According to the *Shuowen Jiezi* (《說文解字》), the first comprehensive dictionary in ancient China completed ca. 100,[23] the word *xuan* means 'deep and remote' (*youyuan*幽遠); according to the *Etymology of Words* (1987, p. 1093), the word means 'deep and profound' (*shenao*深奥). In an essay titled 'The gist of *Laozi*' (*Laozi zhilue*老子指略),[24] Wang Bi, who was responsible for the transmitted version of the *Laozi* from which L&S quote, stated: '"*xuan*", [it's] what's called deep' (*Xuan, weizhi shen ye* '玄'，謂之深也). As discussed in the previous section, Wang Bi explained that Laozi's Dao is non-empirical, hence, what Wang was saying with regard to the word *xuan* is, of course, that Dao is deep because it is non-empirical. Finally, in modern times, the word 'metaphysics' as a disciplinary name was initially translated as *xuan xue* (玄学the study of the deep and profound), a clear indication of what this word means in the scholarly context.

The word *miao* 妙did not exist in pre-Qin and early Han times, we know this because there is no entry for it in the above-mentioned *Shuowen Jiezi*. In the silk version of the *Laozi*, which circulated before the *Shuowen Jiezi*, the word used instead of 妙 is 眇 (the two words differ in tone in pronunciation) which means 'the finest' (*Etymology of Words* 1987, p. 1197). The word 妙 that came into being after the *Shuowen Jiezi* carried (and still does) several different meanings, and one of which is interchangeable with 眇 to mean 'the finest' (ibid., p. 400). In 'The gist of *Laozi*', Wang Bi used the word 妙precisely in the sense of 'the finest' by using it in combination with the word, which carries the same meaning, *wei* (微), and used the combined term (*weimiao*微妙) in conjunction with the term 'that which is without form' (*wuxing*無形) to speak of 'the finest which is without form' (*weimiao wuxing*微妙無形). As explained in the previous section, the term 'that which is without form' refers to the non-empirical, the non-empirical is, by virtue of being non-empirical, imperceptible, which is what the combined term 'the finest' (*weimiao*微妙) refers to.

Are L&S aware of the above? In daily parlance, the words *xuan* and *miao* (妙) are sometimes used to refer to something that is difficult to fathom without implying why it is so, it is clear that L&S are keen, perhaps

[23] The text is accessible online in full.

[24] This short essay is missing in most library collections, but available online.

a little too keen, to read 'mystery' and 'arcana' into what is difficult to fathom. Why is that? The answer should be clear. In addition to regarding the Huanglao-cum-Yin-Yang correlative cosmology as the be all and end all of ancient Chinese thoughts that had something to do with natural phenomena, it is only by doing so that they can claim that the philosophy of Laozi amounts to nothing more than 'about the difference between common experience and mystical breakthroughs', whereby the conclusion can be drawn that China lacked a philosophy that 'is meant to be natural philosophy'.

5.5 Is the Concept of Nature a Greek 'Invention' That China Lacked?

Lloyd and Sivin (2002, pp. 199–200) state:

> We have looked at a very few of the technical terms that early Chinese used when they defined the characteristics of animals…[25] What the various terms refer to sometimes overlap certain denotations in English of 'nature' and words allied to it. But before modern times, Chinese did not need a word that meant 'nature' (…). The classical word most commonly so translated, *tzu-jan* [*ziran* 自然]… did not mean 'nature' until 1881…

There is a reason why L&S say 'did not need' instead of 'did not have' straightforwardly, which will be explained in due course. For now, since 'did not have' follows from 'did not need', and in the book's index (p 342), under 'nature', there is an entry for 'no concept in China before 1881', L&S's claim is, hence, that China lacked the concept of nature before 1881. According to L&S, it was only in 1881 that China *acquired* the concept of nature from Europe, when in that year Chinese scholars followed the practice of Japanese modernizers who had used the Chinese term *ziran* to translate the term 'nature' in European textbooks (ibid., pp. 200, 292 n 16).[26] Before proceeding, let us define the concept of

[25] What these terms are is not mentioned by L&S in this context. This passage was probably written by Sivin.

[26] As is well-known, Japanese writing was borrowed and adapted from Chinese writing, with borrowed Chinese characters (called *kanji*) figuring heavily in the Japanese script. The Chinese borrowing from Japanese modernizers in 1881 is according to the German sinologist Michael Lackner, L&S's source with regard to this matter; we take Lackner's view as reliable.

nature as referring to the domain of natural phenomena which operates or functions in accordance with naturalistic physical relationships; such a definition is unremarkable and should be non-controversial.

The word *tian* (天) is one of ancient Chinese philosophy's core terms/concepts. In the early Zhou Dynasty, it referred to the supreme anthropomorphic deity (Xu 1975, Chap. 2), as in the term *tianming* (天命), usually translated as Mandate of Heaven, with Heaven ('H' in upper case) meant in the non-naturalistic sense of the supreme deity. But it acquired various other meanings afterwards. As the twentieth-century Chinese philosopher Feng Youlan points out (Feng 1993, p. 55), by early Han, the term was used in four other different senses as well, only two of which need concern us. The first means sky (translated in English as 'heaven'—'h' in lower case—in the naturalistic sense) as *tian* in *tianxia* (天下 literally all under heaven) and *tiandi* (天地 literally heaven and earth). As is generally known, the term *tiandi* refers to nature or the natural world generally. The reason why L&S do not regard *tiandi* as constituting the concept of nature will be explained shortly.

In another naturalistic sense of the term, *tian* means nature all by itself, and it is this sense of the term that most precisely and generally denotes the concept of nature. By the Spring and Autumn period (−711 to −476) at the latest and probably earlier, Chinese thinkers had arrived at the notion that nature (*tian*) operates according to a cyclical pattern or mechanism, which they called 'the way of operation of nature' (*tian dao* 天道), in which *tian* means nature and *dao* means way of operation. If one goes through the *Zuo's Commentary* (*Zuozhuan* 《左傳》), a chronicle of historical events between −722 to −468 compiled in the -fourth century, one will find various occasions on which the term *tian dao* is used in this sense, for instance, in the years of −690 and −484 (Zuo et al. 1996, pp. 101, 1259).[27] It is likely that 'way of operation' was originally understood in the sense of pattern alone, but textual evidence shows that by the Warring States period (−475 to −221) at the latest, it was also understood in the sense of mechanism. In the chapter 'The operating of *tian*' ('*Tian yun*'天運) of the *Zhuangzi*, it is stated: 'Who is in charge of it [the operating]? Who is sustaining it? … Or is there a mechanism (*jijian* 機緘) that is responsible for it such that it operates necessarily? Is it that it goes on

[27] These two years are more conventionally referenced as the 4th year of Duke Zhuang of the *Zuo's Commentary* (左传庄公四年) and the 11th year of Duke Ai of the *Zuo's Commentary* (左传哀公十一年).

operating and cannot stop?' (Chen 2012a, p. 295). Here is a clear statement that nature operates necessarily in accordance with its own naturalistic physical mechanisms. How, then, is it possible to argue that prior to 1881, China did not have a concept of nature, with nature understood, as we have defined it unremarkably above, as the domain of natural phenomena which operates or functions in accordance with naturalistic physical relationships?

Lloyd (1996, pp. 6–7, emphases added) discusses several Chinese terms in connection with L&S's claim that China lacked the concept of nature before 1881:

> 天 (*tian*, heaven), 物 (*wu*, things), 性 (*xing*, character), 理 (*li*, pattern),[28] 道 (*dao*, the way), 自然 (*zi ran*, spontaneity, literally 'self-so'),[29] do service perfectly adequately, *in different contexts*, to express ideas where the English translation would be in terms of nature… Yet, that is still not to say that there is a concept of nature in Classical Chinese, *just the one*.…

This quote clearly shows, firstly, that Lloyd's understanding of the word *tian* in a naturalistic sense is *limited* to its meaning of sky (heaven), that he is *unaware* that the word was also used to refer to nature all by itself, and that he is also *unaware* of the above-discussed key term of *tian dao*, in which *tian* precisely means nature. Secondly, the two phrases that we have stressed in the above quote show that Lloyd's view is more precisely that China lacked a *single* ('just the one') concept of nature applicable to *all contexts* (cf. his 'in different contexts') concerning that concept. This

[28] Translating *li* as 'pattern' is problematic, though quite common among sinologists. While this cannot be discussed here in detail, a simple example will illustrate this. The *li*-philosopher Cheng Yi (程颐) of the Song Dynasty spoke of *li* in two senses, one pertaining to the transcendent realm, the other in a naturalistic sense. The former sense is irrelevant to our present discussion; with regard to the naturalistic sense, Cheng states: 'Everything has its own *li*. Such as why fire is hot, why water is cool' (*wu wu jie you li. Ru huo zhi suoyi re, shui zhi suoyi hanleng* 物物皆有理。如火之所以热，水之所以寒冷) (quoted in Lao 1983, p. 236). To speak of the *why* of fire being hot and the *why* of water being cool is to speak of what necessarily is the case with natural things for physical reasons, which is far beyond talking about pattern. As just seen in the main text, if it is likely to be the case that 'way of operation' in 'way of operation of nature' was originally understood in the sense of pattern only, by the Warring States period at the latest, the concept of mechanism concerning how nature operates had already been formed.

[29] Lloyd is correct to note that *ziran* is literally 'self-so', but his rendering of it as spontaneity is inaccurate. We explain why later on.

becomes clear in his argument in a more recent work that such a single concept was purportedly a Greek 'invention', where, concerning China, he states:

> The Chinese recognized the spontaneous as the spontaneous, heaven and earth as heaven and earth, the different characteristics that different creatures are born with as those characteristics, and so on. But they were not tempted to suppose that they were dealing with the same problem in all those different instances. They had indeed different concepts in each of those contexts, speaking of *ziran* when discussing the spontaneous [see note 29], of *tiandi* when talking of heaven and earth, and *xing* when identifying the key characteristics of things. (Lloyd 2015, pp. 21–2)

To see Lloyd's point in the above two quotes, it should be noted that in English one uses the same *term* 'nature' to speak of nature in general, as well as the nature of, say, lions or peonies, hence, in both contexts (cf. 'in different contexts' in the first of the above two quotes from Lloyd), English-speakers understand that they are dealing with the same problem, i.e. they are referring to the same single *concept*. If in Chinese, one spoke of the *tiandi* (heaven and earth, meaning nature generally) of lions or peonies, then, Lloyd would be satisfied that China also had a single concept of nature that applies to the context of speaking of nature in general, as well as to the context of speaking of the nature of lions or peonies. However, instead, two different *terms*, *xing* (which Lloyd renders as a character or key characteristics) and *tiandi*, are used in the two different contexts (speaking of the *xing* of lions and peonies instead of the *tiandi* of lions or peonies), hence, Lloyd believes that the Chinese did not 'suppose' that they were dealing with the same problem, i.e. they were not referring to the same single *concept* in the two different contexts. This is Lloyd's logic behind L&S's claim that China lacked the concept of nature before 1881. However, firstly, Lloyd's logic actually *contradicts* L&S's own view that China finally possessed the concept from 1881 onwards with the word *ziran*, for, on this logic, *ziran also cannot* constitute such a single concept, since even after 1881, one does not speak in Chinese of the *ziran* of lions or peonies, but continue to speak of their *xing*. Where has Lloyd's logic gone wrong?

Before analysing this, it will be useful to point out that different languages have different characteristics, hence, they often do not match one another perfectly in usage, expression, etc.[30] The most well-known example of this is, perhaps, the use in most Indo-European languages of the existential verb (e.g. I think therefore I *am*) also as a copula (e.g. Lloyd *is* a classical scholar), whereas most other languages use other verbs or particles or other elements to perform the same function of copula (Reding 2004, p. 71). That non-Indo-European languages do not use the existential verb also as a copula does *not* mean that the function of copula is not performed in these languages by other means.

Returning to Lloyd's logic, it is wrong because it *conflates concept* (that which pertains to the same problem) with *term*—using English or Greek as a reference point, wrongly believing that the use of two different terms in two different contexts in *another* language ipso facto precludes their being underpinned by the same single concept in *that* language. This is clear if one realizes that besides *xing*, Chinese also speak of *tian xing* (天性), the inborn *xing endowed by tian*, meaning precisely inborn nature. Hence, to speak in Chinese of the *xing* of lions and peonies, is exactly identical to speaking of the *tian xing* of lions and peonies. This term (*tian xing*) provides a direct *rebuttal* of Lloyd's logic as expressed in his not-dealing-with-the-same-problem argument, and has a long history. Thus, with reference to humans, the late-Warring States philosopher Xunzi contrasted *tian xing* with *jimo* (積靡 meaning social nurturing) (in the '*Ru xiao*' chapter of the *Xunzi*《荀子·儒效》). That, in the present context, the terms *xing* and *tian xing* carry the same meaning and can be used interchangeably is evident in Xunzi's following statement: 'The why it is so of life, that's called *xing*' (*sheng zhi suoyiran zhe, weizhi xing*生之所以然者，謂之性) (in 'The righting of names' chapter of the *Xunzi*《荀子·正名》). The Tang Dynasty annotator of the *Xunzi* Yang Liang (楊倞) explains: 'Humans [are]…hence endowed by *tian* with *xing* [which constitutes] their necessary principles' (*ren…guyou biran zhi li, shi suo shouyu tian zhi xing ye*人…故有必然之理，是所受於天之性也) (quoted in Lau 2013, p. 78). *Tian* in Yang's statement refers to nature in the abstract, while *xing*, as he states in so many words, refers to the nature of humans as *endowed by tian*.

In Chinese, there are countless other terms (see *Etymology of Words* 1987, pp. 369–76) which are formed in the same way as the term *tian xing*, in which the word *tian* is used either in the straightforward sense of

[30] It is perhaps worth stressing in passing that in noting this, we do now wish to imply any endorsement of the linguistic determinism of the Sapir-Whorf thesis.

nature, or in the sense of naturally endowed or given. For instance, to speak of the superior natural endowments of an individual such as an athlete, one speaks of her *tian fu* (天賦), with *fu* meaning endowment; to speak of the special natural gift of an individual, one speaks of her *tian zi* (天資), with *zi* meaning gift; to speak of what someone is naturally endowed with in general, one speaks of *tian sheng* (天生), with *sheng* meaning literally giving birth to, as in the famous line of the poem 'Please keep drinking' (*Jiang jin jiu* 将进酒) of China's greatest poet Li Bai (李白), who was deeply frustrated by prolonged lack of career success: 'The abilities that nature has endowed me with, they must definitely be of some use' (tian sheng *wo cai bi youyong* 天生我才必有用). Among the numerous terms formed with the word *tian* in this way, one is particularly significant in the present context, namely, *tian gong* (天工), which has been in use since pre-Confucius times (ibid.: 370). Originally, it meant the function of nature, subsequently, it acquired the additional meaning of 'work [*gong*] of nature [*tian*]', i.e. what is formed by natural processes, with this latter meaning being contrasted to *ren gong* (人工) which means 'work [*gong*] of humans [*ren*]', i.e. artificial. The reason why we have given these examples is to show that not only is the term *tian* meaning nature a core concept of ancient Chinese philosophy, it also *permeates the daily Chinese vocabulary*. That L&S are unaware of it is both surprising and not what one would expect of major scholars.

Finally, let us return to the term *ziran*. In modern Chinese, this term is used, among other usages, *as a noun* to denote nature, a practice that can be traced back to 1881. Before then, the term had only been used as an adjective or adverb. As seen, Lloyd renders it as 'spontaneity' or 'spontaneous' in pre-1881 times. This is inaccurate. Firstly, the term had never been used as a noun (Lloyd's 'spontaneity') in pre-1881 times, but this is only a minor point. Secondly and more importantly, whereas the term is literally 'self-so' or '[it is] so of its own accord', what it means is 'naturally self-so' or '[it is] so of its own natural accord', i.e. the specification 'naturally' or 'natural' are implied in and integral to the term. Thus, the Eastern Han philosopher Wang Chong (王充 27 to ca. 97) states in the 'Warning' chapter of *Lunheng* (《论衡·谴告》): '*Tian dao*, that's *ziran*' (夫天道，自然也). *Tian dao*, as seen, refers to the mechanism of nature, thus, in describing it as self-so, the meaning can only be naturally self-so. Hence, unlike terms such as *tian xing*, etc., in which the presence of the word *tian* makes it unmistakable that these terms are underpinned by the same single concept of nature, the word *tian* is not present in the term *ziran*; but, like in the case of the term *xing* (on its own, in contrast to the term *tian xing*), the concept of *tian* as nature is integral to it, i.e. to express it in Lloyd's terminology, the Chinese well understood that the terms *tian* (nature) and *ziran* ([it is] so of its own accord) deal with the same problem. It should be noted that the adjectival and adverbial uses of *ziran* have been retained after 1881, despite its also being used since then as a noun to denote nature.

Further, it is interesting to note that during the Western Han Dynasty (−206 to 9), the terms *tian* (nature) and '[it is] so' (*ran* 然) were combined to form the term *tianran* (天然), literally '[it is] so of nature's accord'; in the *Records of the Grand Historian* (*Shiji* 《史记》), it was used in the sense of what an individual is naturally endowed with; by the third century, it had also been used in the same sense as *ziran* (*Etymology of Words* 1987, p. 374). The use of tian*ran* and zi*ran* synonymously is direct confirmation, if such confirmation is needed at all, that *ziran* means it is so of its own naturally given accord. It is not for no reason that Japanese modernizers used *ziran* to translate the European term 'nature', what they did was to extend the term's adverbial and adjectival uses to using it as a noun as well.

Finally, as seen, L&S do not only claim that China lacked a concept of nature prior to 1881, they further claim that it 'did not need' such a concept. This second claim is an intriguing one, and one wonders why they should make it. This has everything to do with the nature of their entire comparison between China and Greece itself as examined in Sect. 5.2 above. In the discussion leading up to L&S's statement that China 'did not need' a concept of nature (Lloyd and Sivin 2002: 198–9), three individuals (Lü Beiwei, Liu An and Dong Zhongshu) and three works (the *Book of Changes*, *Supreme Profundities* [*Tai xuan* 《太玄》][31] and the *Inner Canon*) are mentioned. Lü was the sponsor of the previously mentioned LSAA. Liu was an early-Han aristocratic sponsor of the compilation titled *Huainanzi* (《淮南子》), which, like LSAA, is classified by modern scholars as a work of Huanglao thoughts. Dong, as mentioned in Sect. 5.2, was the author of *Luxuriant Dew*. *Supreme Profundities* is an early-Han work that combines Huanglao-cum-Yin-Yang thoughts with an attempt to mimic the *Book of Changes* (Lao 1981b, pp. 124–34). The *Book*

[31] L&S mis-translate the word *xuan* in the title as mystery. In the previous section, it has been explained, in the context of the first chapter of the transmitted version of the *Laozi*, why such a translation is not serious and irresponsible.

of Changes, as noted previously, is a divinatory manual. The *Inner Canon* (*Huangdi neijing*《皇帝内經》) represents the branch of medicine which attempted to theorize medicine on the basis of Huanglao-cum-Yin-Yang thoughts. In short, all the individuals and works mentioned belonged to the body of thoughts embodying the infiltration of mysticism into natural philosophy and proto-science.

The concept of *tian* was formulated and elaborated over several centuries and was commonly shared among intellectuals. Did Huanglao-cum-Yin-Yang scholars also share it? To simplify, let it just be observed that their mysticism-laden correlative cosmology was underpinned by notions of cosmic sympathy, i.e. beliefs in *mystical* relationships between natural phenomena *instead of* naturalistic, physical ones. Taking their thoughts as their frame of reference for comparing with Greece, and hence, treating them as what ancient Chinese natural philosophy is all about, this is why L&S believe that 'China did not need' a concept of nature.

5.6 Did Only the Greeks Possess a Clear Analysis of the Concept of Causation?

The concept of causation exists in all cultures, even peoples at low levels of development possess the concept, if only laden with mysticism, as in the belief in magic. On top of possessing the concept, peoples in more developed cultures also subject it to analysis, as in Greece. Thus, Aristotle distinguished between four kinds of cause, of which only efficient cause approaches the modern concept of causation. Aristotle did not have the concepts of necessary and sufficient conditions in relation to efficient cause, it was the stoic philosopher Chrysippus (−third century) who first explored some of the issues involved: 'He does not have the terminology of necessary and sufficient conditions but explores some of the distinctions that those terms can be used to express' (Lloyd and Sivin 2002, p. 160). In China, the word *gu* (故) (hence, thus, therefore) means precisely one thing or event being caused by another. One of the earliest documented usages of *gu* is in −780 when a top court official named Bo Yangfu (伯阳父) advanced a hypothesis on the cause of a major earthquake, concluding with the statement 'that's why (*gu*) the earthquake occurred' (*guyou dizhen*故有地震). Did Chinese thinkers subject the concept of causation to analysis? The answer is well-known to anyone with some familiarity with pre-Qin philosophy.

Like Aristotle and Chrysippus, the previously mentioned Later Mohists, who were active from the mid- until the end of the Warring States period, did precisely that. But unlike Chrysippus, they did not only explore 'some of the distinctions' involved, they also had the terminology as well as advanced definitions of the terms concerned. Firstly, cause is defined as 'Cause, that which when present produces something subsequent to it' (*gu, suode er hou cheng ye* 故，所得而後成也).[32] The Later Mohists then introduced the concept of *Xiaogu* (小故 literally small cause) and defined it as 'its being present, the it-is-so is not necessarily entailed; its being absent, the it-is-so is necessarily not entailed' (*Xiaogu, you zhi bubiran, wu zhi biburan* 小故，有之不必然，無之必不然). In today's terminology, *Xiaogu* is none other than necessary but insufficient condition. Then follows the definition of the concept of *Dagu* (大故 literally big cause): 'its being present, the it-is-so is necessarily entailed; its being absent, the it-is-so is necessarily not entailed' (*Dagu, you zhi biran, wu zhi biburan* 大故，有之必然，無之必不然). In today's terminology, *Dagu* is none other than sufficient and necessary condition.

How about sufficient but not necessary condition? In the part of the *Mohist Canons* where the above terms and definitions are found, this is not mentioned. The possibility that this was lost in the process of transmission cannot be discounted. Pre-Qin texts were written on narrow bamboo strips tied together with strings, as previously mentioned, soon after the First Emperor established the Qin Dynasty in −221, he forbade the private collection of books, thus much of pre-Qin texts had been lost by the early Han Dynasty,[33] and these texts had to be re-constructed from surviving materials, sometimes by means of memory of old scholars. As also previously mentioned, after the Qin, the Later Mohists, among others, continued to be subject to state repression under the Han. Hence, there can be no doubt that when their works re-surfaced briefly in the mid-third century, they must have already been much mutilated. In any case, in the

[32] All quotes are taken from the *Mohist Canons*, the compilation of the surviving materials of the Later Mohists. The translations are ours, referencing Tan and Sun (2009). In our translations, we try to be as literal as possible. It should be noted that Classical Chinese is written in highly concise form, which is probably due to its being written on narrow bamboo strips, this and Classical Chinese's different grammatical structure may sometimes make our translations sound unfamiliar to the English speaker.

[33] This loss was also due to a most consequential contingent event, namely, that the Qin Dynasty's imperial library was completely burnt down in a fit of rage by the warrior Xiang Yu (项羽) during the dynasty's overthrow.

methodological chapter known as 'Taking the lesser' (*Xiaoqu* 小取) of the *Mohist Canons*, one can indeed find a discussion of an issue that can be expressed in terms of sufficient but not necessary condition: 'the it-is-so is identical, the why-it-is-so need not be identical' (*qiran yetong, qisuoyiran bubitong*, 其然也同，其所以然不必同). In modern terminology, the 'it-is-so' refers to how something is, the 'why-it-is-so' refers to the cause. Thus, how something is might be entailed by, say, either X or Y individually, with both constituting a sufficient but not necessary condition.

The analysis of the concept of causation is directly related to L&S's claim that Chinese thinkers 'did not need' to investigate into the naturalistic, physical relationships between natural phenomena (see the end of the previous section), so what do they say about such an analysis in China after mentioning Aristotle and Chrysippus?

> Except for some *elliptical* remarks in the Mohist Canons, however, the extant Chinese texts from the Warring States and Han periods do not contain explicit analyses of causation. (Lloyd and Sivin 2002, p. 159, our emphasis)

The first thing to note is that the Later Mohists' analysis is dismissed in a grand total of three words ('some elliptical remarks'), not only without any explanation of why their analysis is supposed to be 'elliptical', but, incredibly, also without so much as informing their reader what these allegedly 'elliptical' remarks are. This passage was probably written by Sivin. In his own work, Lloyd (1996, p. 108, our emphasis) addresses the issue of the analysis of causation in China as follows. Firstly, he remarks that 'There are, for instance, some *elliptical* passages in the Mohist canon'. After that, he quotes from Angus Graham's translation of the definitions of *gu* (故), *dagu* (大故) and *xiaogu* (小故) and Graham's explanations of them. However, he makes no comment on whether or not he agrees or disagrees with Graham's explanations and how these explanations bear upon his preceding 'some elliptical passages' statement. For him, it seems, regardless of well-known contrary analysis by, among others, the highly respected Angus Graham, that China allegedly did not have a clear (non-elliptical) analysis of causation constitutes an article of faith.

5.7 Concluding Remarks

That L&S's thesis is social-essentialist in nature has also been noticed by other scholars, though not in the terminology of essentialism. In a review of Lloyd and Sivin 2002, Cullen (2004, p. 361) observes: 'in the conclusions offered by the authors, it does seem as if the society *did* determine the science in both Greece and China'. The poverty of social essentialism is clear, not only from our demonstration that each and every claim made by L&S with regard to lacks that China was purportedly suffering from is empirically false, it can also be seen with reference to Greece itself. Empedocles, the first of the Greek element-doctrine thinkers, hailed from and lived in south Italy, far from the scene of Athenian agora debate. His doctrine was a *reaction* to the philosophy of Parmenides, who also hailed from and was based in south Italy. Furthermore, Parmenides' philosophy was formulated before agora debate began to thrive in Athens. Further still, his philosophy was itself a *reaction* to the *archê* doctrines of the Milesians of Ionian Greece. In other words, the appearance of element doctrines cannot be understood except by putting it in the context of the logic of intellectual development itself. As we observed in Chap. 3 (note 33), intellectual development has a logic of its own which is *independent of external factors*, such as a particular idea invoking a specific response, a scenario that is well-captured by the concept of *reactive-sequence path dependency*.

In critiquing L&S, it is not our intention to say that there are no differences between China's and Greece's intellectual developments, quite the contrary, despite considerable commonalities, the two developments, even for the pre-Qin period for China, followed markedly different trajectories, some aspects of which have correctly been analysed by L&S.[34] However, explanations of these divergencies cannot be found by taking the social-essentialist approach, especially not in L&S's sociological-reductionist form.

Finally, there is a common impression among scholars unfamiliar with China that Chinese natural philosophy was imbued with mysticism, hence, they may find our above discussion of Chinese natural philosophy out of their expectation. Perhaps a few brief words of elucidation on this issue are in order. As pointed out in Chap. 3 and Sect. 5.2 above, China established a highly naturalistic intellectual tradition in the pre-Qin era from the early

[34] We have not discussed these aspects because they are irrelevant to our critique of social essentialism and sociological reductionism.

Zhou Dynasty until −221. It was during this period that the natural philosophy discussed by us was formed. This tradition was reversed under the impact of the wholesale destruction of learning during the period of the Qin and Han dynasties (−221 to 220).

Before −221, as the socio-political-ethical system established by the Zhou disintegrated, a development that quickly accelerated from the -fifth century onwards, mystical thinking began to experience a revival among certain intellectual circles. In Sect. 5.2, we mentioned that these circles refer to the Huanglao-cum-Yin-Yang (hereinafter HcYY) circle and the *Commentaries on the Book of Changes* circle. The former hailed from what modern scholars refer to as the Yan-Qi cultural area around today's Bohai (the sea half-encircled by today's Shandong and Korean peninsulas) region—Yan refers to the state of Yan of the Zhou Dynasty occupying parts of today's Liaoning and Hebei provinces; Qi refers to the state of Qi occupying today's Shandong peninsula. The Yan-Qi area is precisely the area from where the highly superstitious Shang people originally hailed, before they extended politically to the middle course of the Yellow River to found the Shang Dynasty (ca. -1600 to −1046). It will be recalled from Chap. 3 that after overthrowing the Shang and suppressing their rebellion several years later, the Duke of Zhou invented a legitimation myth, which, as a contingent unforeseen consequence, was what led to the formation of a naturalistic intellectual tradition; and that to implement this legitimation myth, the Zhou established an elaborate system of rites that constituted Zhou's socio-political-ethical system. From this angle, it is comprehensible why when this latter system disintegrated, mystical thinking began to rear its head again in a cultural area from which the Shang originally hailed, where Shang culture must have remained strong. While far from being insignificant, the HcYY circle was outside of the pre-Qin philosophical mainstream, originally influential only in the Yan-Qi cultural area where it was patronized by the state of Qi. Its correlative cosmology became influential generally only towards the end of the pre-Qin era when the doctrine of Zou Yan which applied that cosmology to the succession of dynasties (see Sect. 5.3 above) attracted the attention of the state of Qin, and especially its king who was soon to unite the country and become the First Emperor. As explained, this correlative cosmology, with the five-phase doctrine as its signature notion, became the state orthodoxy under the Han Dynasty, which constituted the nadir of indigenous Chinese philosophy; but precisely because it represented such a low point, it was immediately abandoned after the fall of Han. Thereafter, the five-phase doctrine all but disappeared in mainstream natural philosophy.[35]

The *Commentaries on the Book of Changes* was compiled in early Han from various materials written in the late pre-Qin era by unknown scholars plus further similar materials written in early Han. The *Book of Changes* was a divinatory manual with which all pre-Qin intellectuals were familiar, but it *never* figured in the ideas and doctrines of any mainstream philosopher. The unknown scholars just mentioned were on the absolute margins of pre-Qin philosophy, which is why their identities remain a mystery until today. Nonetheless, through philological analysis, modern scholars are able to identify them to be hailing from today's Sichuan area in the middle course of the Yangzi River known as the Shu (蜀) area, possibly with later inputs from collaborators from the Yan-Qi cultural area. The Shu area was an area where there was a sizable population of migrants from the Shang people after their defeat by the Zhou. Thus, just as in the case of the HcYY circle hailing from the Yan-Qi area, the unknown authors of the *Commentaries* hailed from an area where Shang culture remained strong, which reared its head again infiltrating into natural philosophy as Zhou's socio-political-ethical system rapidly disintegrated.[36] In our discussion in Sect. 5.3 above, for simplicity's sake, we put the *Commentaries* on one side (see note 12) because, though it was influential during the Han Dynasty, the latter's state orthodoxy was dominated by the HcYY correlative cosmology, while the thinking of the *Commentaries* was marginal to it.

HcYY's correlative cosmology, however, cannot compare with the *Commentaries*, in which HcYY's five-phase doctrine plays no role at all, in terms of natural philosophical depth and sophistication. Some texts constituting the *Commentaries* are, indeed, as simplistic and absurd as HcYY's correlative cosmology, but its key texts are on a completely different level.[37] These latter texts contain an intriguing amalgamation of mystical

[35] It continued to survive only, on the one hand, in popular consciousness and discourse, and, on the other, in Chinese medicine. The story of how and why it was incorporated into Chinese medicine, and how and to what extent that incorporation affects the efficacy of Chinese medicine—the only ancient proto-science that has been able to prove its medical value in various ways in an age of modern medicine—are issues that cannot be addressed here.

[36] Hence, in Lau 2020 (Chap. 9), we describe the re-emergence of mystical thinking in natural philosophy in the last century or so of the pre-Qin era as 'The revenge of history'.

[37] It should be remembered that the various texts comprising the *Commentaries* were written by different individuals, and compiled (in most cases posthumously) together by yet other individuals, hence, this inconsistency is as can be expected.

thinking (around the notion of macrocosm-microcosm cosmic sympathy) with key rationalistic and naturalistic ideas taken from pre-Qin natural philosophy plus some profound original insights of their own. For instance, the notion of *taiji* (太极), popularized by Daoist religion's graphical representation of it that has intrigued some leading twentieth-century scientists, originates from one of these key texts. After HcYY's correlative cosmology had been abandoned with the fall of Han, it was the *Commentaries* which constituted the mainstream in post-Han natural philosophy. The *Commentaries* appears nowhere in L&S's discussion, but in modern scholarship, the study of the *Commentaries* constitutes a discipline all by itself, on a par with the study of Confucianism and Daoist philosophy.[38] Because its key texts contain the above-mentioned amalgamation, rationalistic and naturalistic Chinese thinkers could steer it towards the rationalistic-naturalistic direction, while mysticism-oriented thinkers could steer it towards the opposite direction. This is why, on the one hand, the impression is formed by scholars unfamiliar with China that Chinese natural philosophy was imbued with mysticism, while, on the other, modern scholars are also able to discern profound natural philosophical insights in it.

Bibliography

Chalmers, Alan. 2013. *What is This Thing Called Science?*, 4th edition. University of Queensland Press and Open University Press.

Chen Guying. 2012a. *The* Zhuangzi*: Contemporary Annotations and Interpretation.* 陈鼓应，《庄子今注今译》，修订重校本(香港:中华书局，2012)。

[38] The reader may have noticed that here and in the previous but one sentence in the main text, we have separately mentioned Daoist religion and Daoist philosophy. There is widespread confusion among sinologists who are unfamiliar with Chinese philosophy between the two, for instance, both Needham and Sivin (1995) mistook the two to be one and the same. This is most unfortunate. Daoist philosophy refers to the philosophy founded by Laozi as developed by Zhuangzi. Daoist religion was founded by Zhang Daoling (张道陵 ca. 34 to ca. 156) several centuries later. Although Zhang claimed to be inspired by Laozi and Zhuangzi, his main doctrinal source was actually the Huanglao school, the ideas of which he combined with folk superstitious beliefs and practices to found his religion. With regard to Zhang's claim to have been inspired by Laozi and Zhuangzi, Lao Siguang (1981a) has provided a detailed analysis of how Daoist religion actually represents a *distortion* and *subversion* of their philosophies.

Chen Guying. 2012b. *The* Laozi*: Annotations and Interpretation with Introduction and Commentary*, revised and expanded edition. 陳鼓應，《老子注譯及評介》，修訂增補本(香港:中華書局, 2012).

Cullen, Christopher. 2004. 'Book review of G.E.R. Lloyd and Nathan Sivin, *The Way and The Word*'. (See Lloyd and Sivin 2002). *Studies in History and Philosophy of Science* 35, pp 357-62.

Ding Yuanzhi. 1998. *The Guodian* Laozi*: Explication, Analysis and Research*. 丁原植，《郭店竹簡老子釋析與研究》(臺北:萬卷樓圖書有限公司，1998)。

Etymology of Words. 1987. Combined edition compiled by the Editorial Department of the Commercial Press, China. 商务印书馆编辑部，《辞源(缩印合订本)》(香港:商务印书馆香港分馆，1987)。

Feng Youlan. 1993. *History of Chinese Philosophy*, Volume 1, expanded edition. 馮友蘭，《中國哲學史》, 增訂本, 上冊(臺北:商務，1993).

Graham, Angus C. 1989. *Disputers of the Tao*. Chicago: Open Court.

Guthrie, W.K.C. 1962. *A History of Greek Philosophy, Vol. 1, The Early Presocratics and the Phythagoreans*. Cambridge: Cambridge University Press.

Hacking, Ian. 1983. *Representing and Intervening*. New York: Cambridge University Press.

Han Wei. 2013. 'The textual features and scholarly value of the Western Han bamboo-strip *Laozi*'. 韓巍，〈西漢竹書《老子》的文本特徵和學術價值〉，收入陳鼓應(編)，《道家文化研究》，第二十七輯(北京:三聯，2013).

Harbsmeier, Christoph. 1998. *Science and Civilisation in China, Volume 7 Part I: Language and* Logic. Cambridge: Cambridge University Press.

Hu Jiacong. 1995. *New Investigations into the* Guanzi. 胡家聰，《管子新探》(北京:中国社会科学出版社，1995)。

Huff, Toby. 2003. *The Rise of Early Modern Science: Islam, China and the West*, 2nd ed. New York: Cambridge University Press.

Kirk, G.S., J.E. Raven, and M. Schofield. 1983. *The Presocratic Philosophers*, 2nd edition. Cambridge: Cambridge University Press.

Lao Siguang. 1981a. *New Edition History of Chinese Philosophy*, Volume 1. 勞思光，《新編中國哲學史(一)》(台北: 三民書局, 1981a).

Lao Siguang. 1981b. *New Edition History of Chinese Philosophy*, Volume 2. 勞思光,《新編中國哲學史(二)》(臺北: 三民書局, 1981b).

Lao Siguang. 1983. *New Edition History of Chinese Philosophy*, Volume 3(1). 勞思光,《新編中國哲學史(三【上】)》(臺北: 三民書局, 1983).

Lau, Raymond W.K. 2013. 'On the achievements of Xunzi's theory of *Xing* from the social science perspective'. 劉榮錦，〈從社會科學角度看荀子性惡論的成就〉，《新亞學報》第三十一卷[上] 2013年6月 頁65-109.

Lau, Raymond W.K. 2020. *Intellectual Developments in Greece and China: Contingency, Institutionalization and Path-Dependency*. Newcastle upon Tyne: Cambridge Scholars Publishing.

Lloyd, G.E.R. 1990. *Demystifying Mentalities*. Cambridge: Cambridge University Press.

Lloyd, G.E.R. 1996. *Adversaries and Authorities: Investigations into Ancient Greek and Chinese Science*. Cambridge: Cambridge University Press.

Lloyd, G.E.R. 2015. *Analogical Investigations: Historical and Cross-Cultural Perspectives on Human Reasoning*. Cambridge: Cambridge University Press.

Lloyd, G.E.R. 2018. 'Introduction: methods, problems and prospects'. Pp 1-29 in G.E.R. Lloyd and Jenny Jingyi Zhao (eds.) *Ancient Greece and China Compared*. Cambridge University Press.

Lloyd, G.E.R. and Nathan Sivin. 2002. *The Way and the Word: Science and Medicine in Early China and Greece*. New Haven: Yale University Press.

Raphals, L. 1998. 'Book review of G.E.R. Lloyd, *Adversaries and Authorities*' (see Lloyd 1996). *Journal of Asian Studies* 57(4), pp 1132-3.

Reding, Jean-Paul. 2004. *Comparative Essays in Early Greek and Chinese Rational Thinking*. Aldershot: Ashgate.

Sivin, Nathan. 1995. *Medicine, Philosophy and Religion in Ancient China Researches and Reflections*. Variorum.

Sivin, Nathan. 2018. 'Why some comparisons make more difference than others'. Pp. 33-9 in G.E.R. Lloyd and Jenny Jingyi Zhao (eds.), *Ancient Greece and China Compared*. Cambridge University Press.

Sun Kaitai. 1995. 'The relationship between Zou Yan and the Daoists'. 孫開泰，〈鄒衍與道家的關係〉，收入陳鼓應(編)，《道家文化研究》，第八輯(上海:上海古籍出版社，1995)，頁213-225.

Tan Jiajian, and Sun Zhongyuan. 2009. *The* Mozi*: Contemporary Annotations and Interpretation*. 谭家健、孙中原(注译)，《墨子今注今译》(北京:商务，2009).

Wang Bi (Wei Dynasty, annotator) and Lou Yulie (proof and explication). 2008. *The Canon of Dao and De of Laozi: Annotations, Proof and Explications*. 〔魏〕王弼注·楼宇烈校释，《老子道德经注校释》(北京:中华书局，2008).

Xu Fuguan. 1975. *History of Chinese Doctrines of Human Nature: The Pre-Qin Period*. 徐复观，《中國人性論史先秦篇》(台北:商務, 1975).

Zuo Qiuming, Hu Zhihui and Chen Kejiong. 1996. *Zuo's Commentary*. 左丘明、胡志辉及陈克炯，《左传》(长沙:湖南人民出版社, 1996).

CHAPTER 6

The Issue of Evidence of Knowledge Transmission Across Different Civilizations

The issue of the transmission of knowledge across different civilizations is crucial to the dialogical perspective. Because scholars who, to varying degrees, retain the Eurocentric view invariably misinterpret scholar tradition, values, and so on in essentialist terms, it is believed that if as many claims of such knowledge transmission—hereinafter, by knowledge transmission, the meaning is transmission across different civilizations—made from the dialogical perspective can be discredited, this would strengthen the Eurocentric case. Efforts to discredit knowledge-transmission claims come in two broad types, namely, that which concerns the issue of evidence of transmission, and that which concerns the issue of purported 'cultural obstacles' to transmission. The latter issue will be examined in the next chapter, while this chapter concerns the former issue. It will be shown why the Eurocentric effort to discredit transmission claims by faulting them for lacking evidence embodies an impoverished view of evidence in historical investigation, which has been generally rejected by scholars of various historical disciplines.

R. W. K. Lau, *Rethinking the Needham Question*,
https://doi.org/10.1007/978-981-97-9472-0_6

6.1 The Legal Criterion of Onus of Proof and the Principle of the Presumption of Innocence

Let us begin with Needham, who made many claims of knowledge transmission from China to Europe, mostly of a technological nature, but also some of a scientific nature. His main basis for making these claims is to demonstrate chronological priority, i.e. first found in China and subsequently also in Europe. Cohen (1994, pp. 431–9) examines five transmission claims made by Needham that Needham argued enriched European science before the breakthrough of ca 1600.[1] He dismisses all of them for being unsupported by evidence by arguing that chronological priority does not constitute such evidence at all.[2] He wraps up his discussion regarding evidence with a section titled 'The onus of proof', stating: 'Roman law has it (upon him who makes an assertion it is incumbent to prove it)', i.e. the burden of proof lies with the claim of transmission. By mentioning Roman law as well as speaking in terms of the legal criterion of the onus of proof, it is clear that for Cohen, the proper guideline concerning evidence in historical investigation is the *legal* criterion. It is interesting to note that Needham himself also referred to the same legal criterion, though arguing in the opposite direction: 'the onus of proof lies upon those who wish to maintain fully independent [European] invention' (quoted in Cohen 1994, p. 436). In thinking in terms of the onus of proof, both Cohen and Needham have misconceived the issue concerning evidence in historical investigation. To explain why that is so, let us take a look at Cohen's dismissal of knowledge-transmission claims made by scholars from the dialogical perspective.

In his work that gave dialogism its name, Bala (2006) addresses the issue of evidence, arguing for a stricter requirement than chronological priority. For instance, he claims that mathematical developments in India in the 14th to 16th centuries by the Kerala School contributed to the invention of calculus in Europe in the seventeenth century. Bala bases his claim on the fact that Matteo Ricci visited Goa for a few years before he went on to China in 1582, that Goa was close to Kerala, and that one of the objectives of the Jesuits in visiting Asia was to find out about Asian

[1] Some of these claims will be mentioned as we proceed, but there is no need to discuss all of them specifically, or their merits or lack thereof.

[2] In relation to at least one instance, this is disingenuous since Needham did provide one piece of direct evidence of transmission, which Cohen neglects to mention at all. This instance will be mentioned shortly below.

science. Hence, there are reasons to believe that Ricci learned about the Kerala School's developments, which he transmitted to Europe. As can be seen, Bala's case is based upon various pieces of *circumstantial evidence*. Cohen (2012, pp. 342–3) dismisses Bala's reasoning outright: 'Without a shred of evidence...Not only does Bala curtail but little Needham's really *unlimited freedom to speculate*...Bala's criterion really implies the promotion of scholarly sloth'. The final comment is somewhat unfair because Bala does talk about trying to find evidence in Jesuit correspondence, and, if nothing comes up there, in mariners' correspondence.

Bala's mentioning of Jesuit correspondence reminds us of Needham's argument that the notion of the universe as an infinite empty space, one of the earlier-mentioned five transmission claims made by Needham that Cohen examines, was transmitted from China to Europe. It is well-known that in 1595, Ricci wrote a letter back home explaining the above Chinese notion with the remark that it is 'absurd' (Chen 1996, p. 162). Needham used this as a piece of *direct evidence* of transmission, but in his comment on this issue, Cohen *fails* to mention this at all. Returning to Bala, he further argues that 'dominant themes in Indian mathematics...afterwards became major themes in European mathematics—themes so new that they were considered to *break with the classical tradition* of mathematics' (Bala 2006, p. 72, our emphasis). We are not mathematicians, hence, not in a position to judge the validity of the italicized point. Assuming that it is true, then, in view of how deeply entrenched scholarly traditions mould how scholars nurtured in them think (their approach, etc.), this is a piece of *circumstantial evidence* of some force in favour of transmission—this piece of evidence belongs to a particular kind of circumstantial evidence of knowledge-transmission that Roger Hart (see immediately below) refers to as 'evidence from mathematics' or mathematical evidence. Cohen completely ignores this argument of Bala's.

The mathematically trained historian Roger Hart (2011, 2013) has published two monographs concerning ancient Chinese mathematics, respectively titled *The Chinese Roots of Linear Algebra*, and *Imagined Civilizations: China, the West, and Their First Encounter*. In these works, he critiques the traditional account in the history of mathematics that what is called Gaussian elimination was first discovered by Carl Friedrich Gauss (1777–1855), while determinants were first developed by Gottfried Wilhelm Leibniz (1646–1716). Hart argues that methods for solving systems of linear equations with n conditions in n unknowns, with one variant being identical to Gaussian elimination, can be found in the Chinese mathematical work titled *Nine Chapters on Methods of Calculation* (*Jiuzhang suanshu* 《九章算术》) that was compiled sometime during the period from -1st to the 1st centuries. He also argues that:

> In addition to the more general solution using what is now called Gaussian elimination, determinantal-style[3] calculations and solutions were also known for a distinctive class of [Nine Chapters] problems [which contains the 'well problem' to be mentioned below]...it is in commentaries to the Nine Chapters [compiled by later mathematicians] that records of determinantal calculations and solutions are preserved. (Hart 2013, pp. 167-8)

Corroboration for Hart's above arguments can be found in Joseph (2011):

> This method of solving simultaneous linear equations is essentially the same as the one we use today, the development of which is attributed in the West to the famous German mathematician Karl Friedrich Gauss (1777-1855). But, over fifteen hundred years before Gauss, Chinese mathematicians were using a variant of one of Gauss's methods. (p. 240)
>
> So, as early as the beginning of the Christian era, a variant of Cramer's rule for solving two equations in two unknowns was known in China...we can detect in this method an early hint of the concept of a determinant. Apparently this escaped the notice of Chinese mathematicians, but it was taken up by the Japanese mathematician Seki Takakazu (...) in 1683—ten years before Leibniz, to whom historians of mathematics usually attribute the discovery of determinants. (pp. 235-6)

With reference to Seki Takakazu who discovered determinants prior to Leibniz, Joseph (ibid., pp. 240, 299) notes that he was steeped in Chinese mathematics and worked within that tradition, a point also noted by Hart.

Hart further notes that in Europe, Gaussian elimination, identical to the ancient Chinese methods, appeared ex nihilo in the fifteenth century in the work of, among others, the French monk Johannes Buteo (born Jean Borrel, ca. 1492 to ca. 1572). Concerning the above-mentioned determinantal-style calculations and solutions in *Nine Chapters*, he also notes:

[3] By 'determinantal-style' or 'determinantal', Hart means conceptually akin to determinants.

> Problems of the form of augmented matrix (5.8) [this refers to the 'well problem' mentioned in the previous quote from Hart]...are *so distinctive that they can serve as a 'fingerprint', a kind of unique identifier*...Many examples of problems equivalent to the form given in augmented matrix (5.8) can be found in the work of Leonardo Pisano (c. 1170-c. 1250), more commonly known today by the name of Fibonacci. Not only are these problems themselves distinctive, the solutions to these problems are even more so. The solutions to these problems are without analog in modern mathematics, and in fact these solutions are esoteric enough that they have not been adequately analysed in previous studies of Fibonacci's mathematics. To put it another way, it was because of my familiarity with Chinese [Nine Chapters] problems and their solutions that the problems and solutions I found in Fibonacci made sense. (Hart 2013, pp. 173–4, our emphasis)

Hart places great stress on what he calls 'evidence from mathematics', that is, 'mathematical evidence preserved in extant treatises' (Hart 2013, p. 29). The most striking illustration of this is his just-quoted argument concerning problems of the form of a particular augmented matrix serving as 'a fingerprint, a kind of unique identifier'.

In sum, in this case, we have a classic scenario of chronological priority: Developments (Gaussian elimination and determinantal-style calculations and solutions) first appearing in China, later, they appeared elsewhere (Europe in the works of Buteo and Fibonacci; Japan where Seki Takakazu discovered determinants). In the case of Seki Takakazu, since he was steeped in Chinese mathematics and worked within that tradition, there can be little doubt that his discovery of determinants was inspired by Chinese sources—*Nine Chapters* and the determinantal-style calculations and solutions preserved in commentaries to that work. As is well-known, Leibniz was a Sinophile, could it be that he also had knowledge of Chinese mathematics and was similarly inspired by it in his discovery of determinants? Since this is not material to our present discussion, we leave that question at that.

In *The Chinese Roots of Linear Algebra*, Hart (2011, p. 192) claims only that 'the essentials of the methods used today...in modern algebra were not first discovered by Leibniz or by Gauss'. In *Imagined Civilizations*, he goes further and speaks of 'the circulation of mathematical practice across the Eurasian continent prior to the West's "scientific revolution"...The diffusion of these practices across Eurasia was likely effected by merchants, missionaries, and travelers, following the routes by which commerce, art, and religion circulated' (Hart 2013, p. 192).

Cohen (2017, p. 228, our emphases) has published a review of *Imagined Civilizations*, in which in relation to Hart's argument with regard to Fibonacci, he remarks:

> His [Hart's] straightforward claim is that *similarity* allows no other explanation than direct or indirect adoption…Hart does not take the trouble to produce even one shred of *empirical evidence* for any tangible route of transmission at all. Practice is just tacitly taken to spread more or less everywhere and to do so without leaving written traces—two presuppositions that conveniently absolve the historian from seeking any *empirical evidence* at all for his wide-ranging claims.

Thus, instead of addressing Hart's argument concerning mathematical evidence ('fingerprint' and 'unique identifier'), it is simply brushed aside with a single word 'similarity'—the careful reader will have noticed that whereas Hart speaks in *professional terms* of the precise mathematical form in which problems are stated and their solutions, Cohen uses the *non-professional, daily term* 'similarity', which hides the force of Hart's argument. Expectedly, 'empirical evidence' of 'tangible route of transmission' is demanded. As seen, Hart does suggest such possible routes of transmission, but what is demanded by Cohen is surely *proof as in onus of proof*, as he does with regard to Bala.[4] The demand for proof is not limited to Cohen. The same demand was made of Bernal by his critics: 'At a methodological level, my debate with [Robert] Palter is the same as with most of the other contributors [to the compilation *Black Athena Revisited* directed against Bernal mentioned in Chap. 2]: Is *proof* required to justify seeing cultural loans from Egypt and the Levant to Greece?' (Bernal 2001, p. 248).

In law court, circumstantial evidence is admissible but generally it is insufficient to establish full proof, only under very strict conditions can conviction be established on the basis of circumstantial evidence alone. Thus, in demanding proof in accordance with the legal criterion of onus

[4] According to Joseph (2011, p. 464–6): 'The young Fibonacci grew up in North Africa, where his father was in charge of a customhouse. He was first introduced to Indian mathematics by his Islamic teachers there. As a young man, he traveled extensively around the Mediterranean…observing the various computational systems used by merchants, particularly merchants in the Islamic world'. Fibonacci's upbringing and his travels constitute a piece of circumstantial evidence that lends credence to the possible routes of transmission suggested by Hart. On the issue of circumstantial evidence, see the next section.

of proof, Cohen and his kindred spirits demand ***direct evidence*** of knowledge transmission or cultural loans in general, lacking which claims of such transmission and loan are dismissed, and ridiculed as 'promoting scholarly sloth' or 'absolving the historian from seeking empirical evidence'. That this demand for direct evidence is what Cohen requires is evident in his dismissal of Bala's argument based upon circumstantial evidence as engaging in 'unlimited freedom to speculate' while ignoring his argument based upon mathematical evidence, as well as in his brushing aside, by resorting to the unprofessional term of 'similarity', of Hart's argument concerning mathematical evidence.

It is interesting to note that whereas Cohen and his kindred spirits demand proof in the form of direct evidence, ***when such evidence does exist but which is unfavourable to their argument, they refuse to accept it.*** As seen in Chap. 4, that the Greeks had initially learnt geometry from the Egyptians who invented it was reported by one Greek thinker after another, yet, Eurocentric scholars refuse to admit this direct evidence by means of restricting geometry to the axiomatic-deductive system. Similarly, as mentioned above, Ricci wrote a letter back home in 1595 in which he mentioned the Chinese notion of the universe as an infinite empty space, this piece of direct evidence was mentioned by Needham, but in dismissing Needham's claim of China-to-Europe knowledge transmission in this case, Cohen does not mention this at all. Incidentally, in relation to this astronomical notion with reference to the issue of evidence, another Eurocentric scholar has even tried to deny that such a notion existed in ancient China by trying to discredit a piece of direct evidence that it did. This case is so instructive that it is worth some explanation.

In the 'Explanation of sources' of the 'Astronomical materials (first part)' of the *The Book of Later Han* (《續漢書·天文志上》注引),[5] it is reported that according to Cai Yong (蔡邕) of the Eastern Han Dynasty (25–220), one of the three ancient schools of astronomy called the Infinite-Universe School (*Xuanye shuo* 宣夜説 hereinafter IU School) had disappeared due to the lack of succession—it should be noted that though

[5] Ancient Chinese texts with the title of *Book of [dynasty name]* (sometimes, as in this case, with further specifications) are official histories and records compiled by high-level court scholar-officials on the basis of primary documents, both official and non-official. Chinese courts were extremely meticulous in compiling such documents, hence, such official histories and records are authoritative sources of historical data. The *Book of Later Han* referenced here is the first of two similar works compiled at different times that cover the Eastern Han Dynasty (25–220), this version was compiled during the second half of the third century.

it had disappeared as a school, its core ideas were taken up by the astronomical school that was to become dominant in ancient China from the Eastern Han Dynasty onwards. In the 'Astronomical materials (first part)' of the *Book of Jin* (《晉書·天文志上》),[6] it is reported that materials of the IU School had been preserved by the early Western Han Dynasty (−202 to 9) court librarian Xi Meng (郗萌). Besides the materials preserved by Xi Meng, some other surviving materials from the IU school were reported by the Eastern Han Dynasty scholar Huang Xian (黄宪) who lived shortly after Xi.[7]

Briefly, the IU School argued that the universe is an infinite empty space; celestial bodies are not attached to any non-existent celestial spheres, instead they float freely in empty space at different distances from earth; each of them moves independently in accordance with its own natural physical mechanism—in this chapter and elsewhere in the present work, when reference is made to the notion of the universe as an infinite empty space, its other details just mentioned are included as well. As can be seen, the IU School's highly naturalistic[8] physical model of the universe is the most advanced in the entire ancient world.[9]

The existence of the IU School and its physical model, far more advanced than anything Greece had to offer, prove to be an inconvenient embarrassment to the 'agora debate' thesis of G.E.R. Lloyd and Nathan Sivin, examined in the previous chapter. As mentioned, although the IU School subsequently disappeared as a school, its core ideas were taken up by what later became ancient China's dominant astronomical school. Faced with the above, what Lloyd and Sivin have done is, on the one

[6] The Jin Dynasty was established in 266 and lasted until 420. The *Book of Jin* was compiled in 648 in the early Tang Dynasty (618 to 907), when political stability was restored after a lengthy period of political turmoil after the fall of Jin.

[7] For an analysis of the IU School and quotations from its surviving materials, see Chen Cheng-I (1996) and Chen Meidong (2008). Both of these sources take for granted that the IU School was formed during the latter Warring States period (−475 to −221); elsewhere (Lau 2020), we have provided an analysis to establish the correctness of this dating of the school.

[8] Cf. our analysis of the establishment of a naturalistic intellectual tradition in pre-Qin China in Chap. 3.

[9] Despite this advanced physical model, the IU School was abandoned during the Western Han Dynasty mainly because it lacked a mathematical model to complement it, as a result of which, it lacked operational capacity which was much emphasized in ancient Chinese astronomy.

hand, to ignore the surviving notion of the universe being an infinite empty space, and, on the other, to claim that the IU School never existed.

Needham had discussed the IU School and mentioned that its existence was reported by the polymath Ge Hong (葛洪283–343),[10] who correctly mentioned that materials of the school had been collected by the above-mentioned Xi Meng. Without even bothering to investigate into the mentioning of Xi Meng by Ge, Lloyd peremptorily remarks that Ge's report 'does not inspire confidence', and in a footnote, thanks Sivin for 'stressing to me the difficulties of accepting this [Ge's] report', whereby he summarily denies the existence of the school altogether by describing it as 'so-called' (Lloyd 1996, p. 151). Thus, the existence of an entire astronomical school is denied in a grand total of two blank assertions ('does not inspire confidence' and 'difficulties of accepting [Ge's] report'), period.

In sum, the cases of the dismissal of the Greeks' own repeated reports on having learnt geometry from the Egyptians who invented it, Cohen's failure to mention Needham's mentioning of Ricci's letter of 1595, and Lloyd's and Sivin's summary denial of the existence of the IU School, all go to show *a consistent pattern*: When direct evidence does exist which are unfavourable to their argument, Eurocentric scholars do not hesitate to dismiss, deny, or ignore it, even for those who at the same time demand, as a matter of legal principle, direct evidence.

In the criminal law court, if the prosecution cannot prove guilt beyond reasonable doubt, the accused is automatically taken to be innocent.[11] This is, of course, the well-known principle of the presumption of innocence. Cohen believes that this principle also applies to historical investigation. In relation to Needham's China-to-Europe transmission claim concerning magnetism, after noting Needham's problems with evidence, he states:

> It is striking that Needham...does not appear to give a thought here to the one inference that, for the uncommitted reader, cries out to be made..., namely, that...knowledge of the directive properties of the magnet...were

[10] While Ge was a polymath, as a priest of the Daoist religion, he is most famous for his researches into herbal medicine and alchemical pursuits.

[11] The onus of proof and proof beyond reasonable doubt are the criteria of criminal law, hence, in demanding proof, the reference point for Cohen and Bernal's critics is criminal law. In contrast, in civil law, instead of proof and proof beyond a reasonable doubt, circumstantial evidence is given a much greater role and the dominance of probabilities replaces proof beyond reasonable doubt.

independently discovered in Europe...this [is the] commonsense solution. (Cohen 1994, pp. 432-3)

In other words, where there is a lack or insufficiency of evidence on the criterion of criminal law, one should, according to Cohen, ***automatically assume*** the absence of transmission, as in law court. He claims that making such an automatic assumption constitutes 'commonsense' for the 'uncommitted reader'. We shall shortly see whether or not this is so.

6.2 The Criterion of the Balance of Probabilities or Plausibility in Historical Investigation

It would not be a wild claim to note that in history, direct evidence often simply does not exist or has not survived. Classical scholars precisely face this problem of the non-existence of direct evidence with regard to the early Pythagoreans, because the Pythagoreans strictly observed a code of secrecy typical of mystical sects and did not write anything until Philolaus broke that code ca. mid -fifth century. How should historians deal with this then? W.K.C. Guthrie and Charles Kahn are two leading classicists whose approach to the Greeks cannot be more divergent, with one stressing the continued influence of mystical thinking on the Greeks and the other preferring a rationalist reading of them, but both have addressed the above issue concerning the early Pythagoreans in a similar way. Guthrie (1962, p. 217, our emphases) states: 'one can only demand *proof* in so far as the subject admits of it—and if we are to speak of Pythagoreanism at all, we must at many points remain content with *probabilities*'—assessed, let it be mentioned, on the basis of indirect (such as comments made by contemporaries on them) and circumstantial evidence. Kahn (2001, pp. ix, 38, our emphases) remarks: 'the history of Pythagoreanism before Philolaus...must remain an area for ***informed speculation***', an assessment of what is 'more ***plausible*** than' the alternative. It should be noted that in Kahn's view, the assessment of what is more plausible than the alternative does not amount to, to quote from Cohen's critique of Bala mentioned above, engaging in 'unlimited freedom to speculate', but to speculate—or, as we would rather put it, hypothesize—on an informed basis. Thus, both Guthrie and Kahn are of the view that in historical investigation, proof is often not the appropriate criterion, and in speaking of balancing probabilities or plausibility, it is safe to assume that their requirements are less stringent than those in civil law (see note 11).

This recognition is not limited to Guthrie and Kahn, but is widely shared, if we may use this word, commonsense among scholars of different historical disciplines. Thus, in response to his critics, Bernal (2001, p. 248) makes the following methodological point: 'Thus, I believe that all that is required to hypothesize Greek borrowings in science as well as other subjects from Egypt and Syro-Palestine is, once again, not proof but competitive plausibility'. Bernal's illustration of this in relation to a particular issue is worth quoting at length:

> For example, Palter [the science historian mentioned above] refers to the disagreement between the Egyptologist J.H. Breasted and the medical historian Henry Sigerist on whether the clear similarity between the Egyptian and the Greek methods of treating dislocated jaws was the result of diffusion from Egypt or independent invention (...). Sigerist argued for the latter, as he saw 'really only one anatomically possible way to do this.' Therefore, as Palter puts it, 'Egyptians and Greeks *may well* [Bernal's emphasis] have independently discovered the procedure' (...)...If the treatment had been used in ancient Peru, for example, we should undoubtedly see it as an independent invention. In this instance, however, substantial cultural contact between Egypt and Greece not only took place during the relevant millennia but, as Sigerist himself maintained, these contacts were particularly close in medicine (...). (Bernal 2001, p. 250)

We have quoted this long passage because it is pregnant with implications with regard to our present discussion. Firstly, it should be noted that neither Bernal nor Palter, his critic, deals with proof here; the same applies to Breasted (who argues for diffusion) and Sigerist (who argues for independent Greek invention). This is simply because, in Guthrie's words, 'the subject [does not] admit of it'. Both parties in each case are engaging in the balance of probabilities or plausibility. Secondly, while we are not familiar with the details of the debate between Breasted and Sigerist, it would seem to us that there is a problem with Sigerist's argument. According to him, there is 'really only one anatomically possible way to do this', this argument is a little similar, though in reverse direction, to the argument made by Bala and Hart on the basis of mathematical evidence, and should be taken seriously. However, Sigerist appears to undermine his own argument by admitting that there was a lot of contact between Egypt and Greece, which was especially close in medicine. In other words, one piece of circumstantial evidence that he admits to (close contact especially in medicine) *cancels out* another piece of circumstantial evidence ('really only one anatomically possible way') that his entire argument for independent invention relies on. That being the case, since both parties are engaged in balancing what is more plausible, Sigerist's argument is much undermined by none other than himself.

Thirdly, in critiquing Bernal in relation to this case, Palter himself, based upon Sigerist, states only that 'Egyptians and Greeks *may well have* independently discovered the procedure' (our emphasis). In other words, in the absence of direct evidence, he simply *prefers to believe in* independent invention, *despite* the above-mentioned admission by Sigerist himself. This brings us to a most crucial *methodological* point made by Bernal:

> I believe that given the closeness in space and the overlap in time, it would be astonishing if substantial cultural exchange had not occurred around the East Mediterranean [cf. his contrast with the hypothetical case of Peru in the previous quote from him]. Furthermore, given the greater age, wealth, and elaboration of Egyptian and Levantine civilizations, it would be surprising if the predominant cultural flow during the Bronze and Early Iron Ages had not been from southeast to northwest rather than the other way round...In the case of ancient studies generally, I find *the presumption of cultural isolation* around the Eastern Mediterranean held by most Hellenists, ancient historians, and archaeologists of the Aegean extremely implausible. (Bernal 2001, pp. 248–50)

In the above, we quoted Cohen to the effect that according to him, the presumption of independent invention based upon the legal principle of the presumption of innocence constitutes 'commonsense' for the 'uncommitted reader'. We can now see why this is absolutely not the case. In the case concerning Bernal and his critics, since proof is out of the question, *both parties begin with presumptions*, one the presumption of substantial cultural exchange, the other the presumption of cultural isolation. Which presumption is more plausible? The reader can judge for herself. In the case of Palter with regard to the treatment of dislocated jaws, as said, he prefers to believe in independent invention despite Sigerist's admission. Thus, contra Cohen, the argument for independent invention is absolutely not any 'commonsense' of the 'uncommitted reader', but *the preferred presumption of someone committed to a particular standpoint.* It should be noted that in coming to this conclusion, we are only making *a general methodological point* against Cohen's legal-principle-derived general claim of 'commonsense' and 'uncommitted reader', and not to defend Needham with regard to magnetism, which is of no interest to us in the present context—the balance of probabilities or plausibility in each specific case needs to be assessed individually.

Needham is certainly wrong to say that the onus of proof lies in the claim of independent invention. This is because the onus of proof is the wrong criterion, and if one adopts such a criterion, then its concomitant principle of the presumption of innocence works in favour of the opposing claim. What the above analysis tells us is that *if the balance of what is more probable or plausible in any particular case tilts towards one direction, then it is up to the opposite claim to find more indirect and/or circumstantial evidence to support its own case.* In other words, *contrary to* the legal principle of the presumption of innocence, the criterion of the balance of what is more probable or plausible *does not favour any one side in an* a priori *way.* For instance, in the above case of treatment of dislocated jaw, it would seem to us that, with regard to the case of Greece (in contrast to the hypothetical case of ancient Peru), the balance tilts towards diffusion, thus, to dispute that, it is up to the claim for independent invention to strengthen its own case. We return to this point shortly below.

Let us illustrate further the poverty of Cohen's criteria of the onus of proof and the presumption of innocence with reference to some other historical disciplines. The bronze age appeared in China much later than in Mesopotamia, some Chinese archaeologists such as Xu Hong (2021, pp. 29–30) argue that bronze had been diffused to China from Mesopotamia despite the absence of any such evidence, direct or indirect. As another example, a basically complete morphographic writing system in China appeared during the Shang Dynasty (ca. -1600 to –1046) apparently out of nowhere, hence, linguists such as Fischer (2001) argue that writing had not been invented independently in China but was diffused from Lower Mesopotamia.[12] Again, there is no direct or indirect evidence of diffusion to support Fischer's argument. On Cohen's legal conceptualization of historical evidence, both Xu Hong's argument and Fischer's argument should be dismissed out of hand without further ado for lacking any direct evidence at all, let alone in the amount sufficient to prove beyond reasonable doubt. However, despite the absence of direct or

[12] Fischer is not the first one to advance this argument, C.J. Ball and A. Ungnad had already done so in the early twentieth century (see Fischer 2001, p. 328 n 4 of Chap. 5).

indirect evidence, neither Xu nor Fischer is engaging in 'unlimited freedom to speculate', for both base their arguments on credible circumstantial evidence. To illustrate with Fischer: 'The earliest Chinese inscriptions display the characteristic writing in columns, reading from top to bottom and from right to left, that distinguishes inscriptions on the reverse sides of Mesopotamian monuments (until about 1500 BC). Chinese writing also incorporates the Mesopotamian principle of…' (Fischer 2001, p. 168). These are serious scholarly arguments that present a case with considerable plausibility, and it is up to scholars who argue for independent invention to counter such arguments with circumstantial evidence supporting their own case.[13]

To sum up, contrary to Cohen, scholars of historical disciplines ranging from archaeology (Xu Hong), historical linguistics (Fischer), classical studies (Guthrie and Kahn), Egyptology (Breasted), to history of medicine (Sigerist), not to mention Bernal, all well understand that historical investigation is often not about proof or the presumption of innocence, but about the balance of what is more probable or plausible on the basis of circumstantial evidence, if neither direct nor indirect evidence is available. To give one final, simple example closer to the issue of the rise of modern science, Descartes denied ever having been influenced by Galileo, but, as Cohen (2011, p. 241) notes himself, in view of his travels through Italy in the early 1620s (a piece of circumstantial evidence), historians generally do not find his denial *plausible*.

With the above in mind, let us return to Bala and Hart, and Cohen's dismissal of their arguments. As mentioned, Bala hypothesizes that

[13] Some recent archaeological discoveries do provide strong circumstantial evidence arguing for independent invention. At the Tao Si (陶寺) archaeological site in southern Shanxi province, a broken ceramic pot was unearthed in 2000 which bears two individual characters pre-dating Shang by several centuries or more which were unquestionably direct precursors of the Shang system (Ge Yinghui 2007). This provides evidence, though insufficient, of a 'missing developmental stage'. More significantly, in 1992, a discovery was made at the Ding Gong (丁公) archaeological site in Shandong province of a writing system that pre-dated Shang's by some eight centuries or so, but it was based upon a different system from Shang's (Feng Shi 2007). Since Shandong province is where the Shang people originally hailed from before they moved towards north-central China to found the Shang Dynasty, this indicates that the Shang people had had centuries of development of writing before arriving at the morphographic system found on Shang's oracle bones. Fischer was probably unaware of these latest developments when he published his book in 2001. That, however, does not affect the point that we are making, which is that Fischer made his argument on the basis of circumstantial evidence that he was aware of at the time of his writing.

mathematical developments in India in the 14th to 16th centuries by the Kerala School contributed to the invention of calculus in Europe in the seventeenth century. He bases his claim on the fact that Matteo Ricci visited Goa for a few years before he went on to China in 1582, that Goa was close to Kerala, and that one of the objectives of the Jesuits in visiting Asia was to find out about Asian science. Hence, there are reasons to believe that Ricci learned about the Kerala School's developments, which he transmitted back to Europe. Bala further argues that 'dominant themes in Indian mathematics...afterwards became major themes in European mathematics—themes so new that they were considered to *break with the classical tradition* of mathematics' (Bala 2006, p. 72, our emphasis). Thus, what Bala does is to try to argue for a plausible hypothesis on the basis of various pieces of circumstantial evidence. Our concern here is not whether the degree of plausibility is sufficiently convincing, instead, our focus is on the last-mentioned piece of circumstantial evidence on the assumption that the italicized point in the quote is true—'on the assumption' because we do think that Bala's argument is made all too briefly in the form of a pure claim, and needs to be substantiated much more thoroughly such as is being done by Hart. As previously mentioned, Cohen ignores this piece of evidence in his critique of Bala; similarly, he brushes aside Hart's argument, which is made with great elaboration, concerning mathematical evidence with the non-professional and misleading word 'similarity' without addressing Hart's professionally based point about 'fingerprint' and 'unique identifier'.

Mathematical evidence (and other evidence analogous to it, such as the evidence used by Sigerist mentioned above) is an especially strong form of circumstantial evidence. Confronted with such evidence, what those who oppose claims based upon such evidence should do is to strengthen their own case with countervailing circumstantial evidence. For instance, as said, Sigerist cancels out his own 'really only one anatomically possible way' argument by admitting to the close contacts in the field of medicine between Egypt and Greece. However, we would like to illustrate this with a more striking and illuminating instance. The mid-third century Chinese mathematician Liu Hui (刘徽) developed what later became known as the exhaustion method in estimating a circle's area, a method that Archimedes had previously invented. The Graecophile and anti-Needham physicist Chen Fangzheng (2002, pp. 489–90), discussed in Sect. 4.8 of Chap. 4, claims that Liu must have plagiarized it from Archimedes. Unsurprisingly, Chen has no evidence, either direct or indirect, to support his claim, he simply claims that Archimedes' method had come to Liu's attention probably through its transmission in Buddhist documents containing it which were diffused to China (Buddhism first diffused to China during the 1st and 2nd centuries). As noted in Chap. 4, Chen places great stress on the uniqueness of the Greek tradition, which he believes to be an embodiment of the 'soul' or 'spirit' of the Greek people, and his argument is precisely that, given China's different mathematical tradition, Liu Hui lacked the required conditions for inventing the exhaustion method independently. As can be seen, Chen's claim is based upon an argument of 'evidence from mathematics' similar to Bala's concerning Europe lacking the tradition to develop calculus independently.

For those who disagree with Chen's claim, they might, in a similar way to Cohen, simply dismiss it on the basis of Chen's failure to produce any direct evidence of transmission; and in Chen's case, such a dismissal actually has much more justification in comparison to Cohen's. This is because Liu Hui lived in a period in which east-west communication lines had barely begun to be established—as noted, diffusion of Buddhism to China began only during the 1st and 2nd centuries, while Liu lived in the third. In view of this, Chen's claim is really rather frivolous.[14] For anyone to make such a claim on a more serious basis, one would expect the conducting of at least some research into the literature analysing the contents of Buddhist tracts transmitted to China over different periods, to see if it is possible that tracts transmitted in the early days of Buddhism's diffusion to China, the amount of which is unlikely to voluminous to begin with, contained more than purely or largely religious contents.

Despite the above, however, from the perspective of our above analysis, it would still be necessary to take the bull of Chen's 'evidence from mathematics' argument by its horns. To do so, the question that needs to be asked is whether or not Liu possessed the scholarly-tradition-derived conditions for independently inventing the exhaustion method. As the facts

[14] This is all the more so in view of the following. In making his assertion, Chen references two works (Li Yan and Du Shiran 1987; Martzloff 1997). Actually, Martzloff does not make any such assertion at all, which Chen misrepresents as that Martzloff 'does not rule out' his (i.e. Chen's) assertion. Concerning Li and Du, Chen states 'Li Yan [why only Li is mentioned is not known] seems [why "seems" only is also not known] to hold this view [i.e., the view that Liu Hui plagiarized from Archimedes]' (see note 34 on p 490 of Chen's work); however, what Li and Du (ibid.: 68) actually say is exactly the opposite: 'It should also be noted that Liu Hui's method was not influenced by Archimedes. He obtained it independently'.

are, what fails Chen is his unfamiliarity with ancient Chinese philosophy and proto-science. Conceptually, the exhaustion method is based upon the notion of infinite divisibility. As is well-known, in Greece, this issue was debated between Zeno and his opponents such as Gorgias. In China, the same debate, which some scholars regard to be at an even more sophisticated level than the Greek one, occurred during the -4th and -3rd centuries, in which the concept of the infinitesimal approaching zero as limit was stated in terms analogous to the modern notion (Sun Hong'an 2008, p. 42; Chen Cheng-I 1987). This debate was terminated by the First Emperor after his founding of the Qin Dynasty in −221 when he engaged in a wholesale destruction of learning, and remained suppressed under the Han Dynasty (−206 to 220). However, after the fall of the Han Dynasty, there occurred a vigorous intellectual renaissance in the third century, during which the above debate was revived. Guo Shuchun (1984) has provided an analysis of the conceptual sources of Liu Hui, showing how the revival of that debate was central to Liu's work. Liu, in fact, provided a methodological explanation of his method, and Chen Cheng-I (1987, p. 27) has shown how it echoes that debate with 'a remarkable statement of the limit concept'.

Chen's case shows that if Bala's and Hart's arguments are to be critiqued, the proper way to do so is to show, in the case of Bala, that the antecedent mathematical tradition in Europe was conducive to, or at least not unfavourable to, the invention of calculus, and, in the case of Hart, that Buteo's and Fibonacci's works concerned had roots in the antecedent European mathematical tradition. This, in our view, is especially required with regard to Hart, whose case, argued in great detail, is particularly strong. To *not* do so, but either ignore (in relation to Bala) or dismiss in a single unprofessional word (in relation to Hart) Bala's and Hart's arguments, and instead fault their claims on the criterion of onus of proof, precisely amounts, if we may use Cohen's own words, to 'scholarly sloth'.

Finally, notwithstanding our critique of Cohen's insistence on using the legal criterion, his warning against 'scholarly sloth' certainly needs to be taken seriously. This can be illustrated with reference to China's notion of the universe being an infinite empty space. As noted, there is hard direct evidence of its transmission to Europe in the form of Ricci's letter of 1595. However, we are unaware of any research being done to trace the notion's trail of spreading in Europe. In Galileo's *Sidereal Messenger* of 1610, he listed seven observations which contradicted the solid-sphere model: 'However, he failed to recognize the observation as an indication that stars are not attached on a crystalline sphere with a fixed ratio. Consequently, he did not initiate the process to abandon the solid-sphere world-view' (Chen Cheng-I 1996, p. 236). Was Galileo then aware of China's notion, which had reached Europe some fifteen years earlier? Needless to say, to trace the notion's spreading trail in Europe requires very painstaking archival research, and most supporters of dialogism probably lack the training and expertise, not to mention the language requirements to do so. But it must be acknowledged that without efforts in that direction, their case remains, as it were, incomplete. The same applies to similar cases.

Bibliography

Bala, Arun. 2006. *The Dialogue of Civilizations in the Birth of Modern Science.* Palgrave Macmillan.

Bernal, Martin. 2001. *Black Athena Writes Back: Martin Bernal Responds to His Critics.* Durham: Duke University Press.

Chen Cheng-I. 1987. 'A comparative study of early Chinese and Greek work on the concept of limit'. Pp. 3-52 in Idem. (ed.), *Science and Technology in Chinese Civilization.* Singapore: World Scientific.

Chen Cheng-I. 1996. *Early Chinese Work in Natural Science.* Hong Kong: University of Hong Kong Press.

Chen Fangzheng. 2002. 'On the development of Chinese mathematics and its relationships with the rise and fall of dynasties and external influences'. 陈方正，'论中国数学发展与皇朝盛衰和外来影响的关系'，收入氏著，《在美妙的新世纪的门上》（香港：中文大学中国文化研究中心，2002），页479-92。

Chen Meidong. 2008. *Ancient Chinese Astronomical Thoughts.* 陈美东，《中国古代天文学思想》(北京:中国科学技术出版社, 2008).

Cohen, H. Floris. 1994. *The Scientific Revolution: A Historiographical Inquiry.* University of Chicago Press.

Cohen, H. Floris. 2011. *How Modern Science Came into the World: Four Civilizations, One 17th-Century Breakthrough.* Amsterdam University Press.

Cohen, H. Floris. 2012. 'From west to east, from east to west? Early science between civilizations', book review of Arun Bala, *The Dialogue of Civilizations in the Birth of Modern Science*, Avner Ben-Zaken, *Cross-Cultural Scientific Exchanges in the Eastern Mediterranean*, Sylvain Gouguenheim, *Aristote au Mont-Saint-Michel. Les racines grrcques de l'Europe Chretienne*, Toby Huff, *Intellectual Curiosity and the Scientific Revolution. A Global Perspective*, and Jonathan Lyons, *The House of Wisdom. How the Arabs Transformed Western Civilization*'. *Early Science and Medicine* 17, pp 339-350.

Cohen, H. Floris. 2017. 'Review of *Imagined Civilizations: China, the West, and Their First Encounter* by Roger Hart'. *East Asian Science, Technology, and Medicine* 2017, pp 225-31.

Feng Shi. 2007. *New Theses on Ancient Writing and Ancient History.* 冯时，《古文字与古史新论》(台北:台湾书房，2009).

Fischer, Steven Roger. 2001. *A History of Writing*. London: Reaktion.

Ge Yinghui. 2007. 'Deciphering the name of King Yao, advancing investigation of the sources of civilization'. 葛英会，'破译帝尧名号，推进文明探源'，北京大学震旦古文明研究中心编，《古文明研究通讯》，总32期(2007)，页1-6.

Guo Shuchun. 1984. 'Investigation into the origins of the thoughts of Liu Hui'. 郭书春，〈刘徽思想探源〉，《中国哲学史研究》，1984年2期，页57-62.

Guthrie, W.K.C. 1962. *A History of Greek Philosophy, Vol. 1, The Early Presocratics and the Phythagoreans.* Cambridge: Cambridge University Press.

Hart, Roger. 2011. *The Chinese Roots of Linear Algebra*. Baltimore: John Hopkins University Press.

Hart, Roger. 2013. *Imagined Civilizations: China, the West, and Their First Encounter*. Baltimore: John Hopkins University Press.

Joseph, George G. 2011. *The Crest of the Peacock Non-European Roots of Mathematics*, 3rd edition. Princeton: Princeton University Press.

Kahn, Charles H. 2001. *Pythagoras and the Pythagoreans A Brief History.* Indianapolis: Hackett.

Lau, Raymond W.K. 2020. *Intellectual Developments in Greece and China: Contingency, Institutionalization and Path-Dependency.* Newcastle upon Tyne: Cambridge Scholars Publishing.

Li Yan and Du Shiran. 1987. *Chinese Mathematics A Concise History*, translated by John N. Crossley and Anthony W.-C. Lun. Oxford: Clarendon.

Lloyd, G.E.R. 1996. *Adversaries and Authorities: Investigations into Ancient Greek and Chinese Science.* Cambridge: Cambridge University Press.

Martzloff, Jean-Cluade. 1997. *A History of Chinese Mathematics.* Translated from the French original of 1987 by Stephen S. Wilson. Berlin: Springer.

Sun Hong'an. 2008. *Ancient Chinese Mathematical Thoughts.* 孙宏安,《中国古代数学思想》(大连:大连理工大学出版社, 2008).

Xu Hong. 2021. *The Earliest China: The Rise of the* Erlitou *Civilization.* 许宏，《最早的中国:二里头文明的崛起》，(北京:三联书店，2021).

CHAPTER 7

Alleged 'Cultural Obstacles' to Knowledge Transmission and Huff's 'Intellectual Curiosity' Argument

Besides demanding proof (sometimes inconsistently, as when unfavourable direct evidence is either ignored, dismissed, or denied) and insisting upon the legal principle of the presumption of innocence, another way sometimes employed to undermine dialogism is to resort to a form of *cultural-essentialist* argument, namely, to claim that there are 'cultural obstacles' to knowledge transmission across different civilizations. Such a claim may come as a surprise, since is it not generally acknowledged, even by Eurocentric scholars, that Arabic scholars successfully took up Greek learning as well as Hindu mathematics? Huge cultural gulfs existed between Islamic society and polytheistic Greece as well as Hinduist India, yet, they did not pose any 'cultural obstacles' to knowledge transmission.

Nonetheless, the claim of 'cultural obstacles' to knowledge transmission remains alive and well—unfortunately, even among scholars who argue for dialogism. As mentioned in Chap. 3, none other than George G. Joseph has remarked that 'in a number of other areas of human knowledge, notably in philosophy and the arts, the [cultural] barriers [to knowledge transmission] are often insurmountable unless filters can be devised to make foreign "products" more palatable'. Could Joseph have been unaware of the taking up of Greek philosophy by Arabic scholars who developed it further?

In this chapter, we show, firstly, why claims of 'cultural obstacles' betray a deficient conceptual-theoretical understanding of how and why

R. W. K. Lau, *Rethinking the Needham Question*,
https://doi.org/10.1007/978-981-97-9472-0_7

obstacles to transmission of knowledge across cultures do sometimes occur, and, secondly, that a cultural-essentialist interpretation of such cases is contradicted by historical facts. This is done by means of a detailed examination of the reception of Euclid's *Elements* in China over the centuries, and of the introduction of the Tycho Brahe model and the telescope to China in the late 16th to early 17th centuries. By coincidence, the introduction of the telescope to China is used by the historical sociologist of science Toby Huff to advance a cultural-essentialist neo-Eurocentric 'intellectual curiosity' argument. We show how Huff's own analysis actually cancels out his own argument, and how his argument betrays an extraordinarily simple-minded conception of how scholarship develops, according to which scholarship develops like blank calculators, each and every one of which should automatically produce the same output given identical inputs.

7.1 Needham as Cohen's Authority for Claiming 'Cultural Obstacles' to Knowledge Transmission

It comes as no surprise that a Eurocentric scholar such as Cohen should advance claims of 'cultural obstacles' to knowledge transmission. What is surprising is that in making such claims, he is able to find support with a quote from none other than Needham:

> The mutual incomprehensibility of the *ethnically*-bound concept systems did severely restrict possible contacts and transmissions in the realm of scientific ideas. This is why technological elements spread widely through the length and breadth of the Old World, while scientific elements for the most part failed to do so...The sciences of the medieval world were in fact tied closely to their *ethnic* environment, and it was difficult if not impossible for people of those different environments to find any common basis of discourse (Needham 1969, p. 16, 1981, p. 9, our emphases, quoted twice by Cohen. [1994, p. 438; 2012, p. 350])

To see the claim, repeatedly contradicted by historical facts, that 'scientific elements for the most part failed [to spread across cultures]' coming from Needham is completely unexpected. What is even more so is that Needham's point is made in terms of ethnic, i.e. bio-psychological-essentialist, obstacles. Given the enormous service that Needham has done to the non-Eurocentric cause, we would not want to dwell on this point further. Anyway, it is clear that, in quoting Needham in support, Cohen is talking about 'cultural obstacles': 'Many culture-bound attitudes of either acceptance or rejection...may stand in the way of a smooth transmission' (Cohen 1994, p. 470); 'conceptions of nature [are]...linked up with their own cultural environment' (Cohen 2012, p. 349).

Cohen's 'cultural obstacles' claims sometimes degenerate into irrelevant caricatures. For instance, concerning the reception of the European mechanical clock during the Qing Dynasty,[1] he states: 'Courtiers and concubines took their clocks as nothing but funny toys, and none of all these timepieces (...) ever left the imperial palace' (Cohen 1994, p. 470). One wonders whether or not courtiers and concubines are appropriate illustrations concerning scientific transmission.[2] Incidentally, China had developed a mechanical clock long before the Qing Dynasty, which, however, was later forgotten as a result of the intervention of a contingent event, namely, the invasion of the Mongols. The substantive merit or lack of it of Cohen's discussion in this particular case is of no concern to us,[3] it is the above telling caricatural remark that we want to draw the reader's attention to.

Caricature aside, Cohen refers more seriously to the instance, mentioned by Needham, of the translation of fifteen chapters of Euclid in the twelfth century that was left untouched in China's Imperial Library to illustrate how culture is supposed to pose obstacles to transmission. Why did Euclid not strike a chord with Chinese scholars? In Chap. 3, we analysed the great resistance on the part of scholars nurtured in a particular deeply entrenched scholarly tradition to accept a significantly different tradition. We also mentioned that a scholarly tradition includes not only substantive knowledges, but also formal aspects; and that some substantive knowledges are tradition-specific because the formal aspects are embodied in them, and we gave the example, precisely, of Euclidean geometry. It is specifically in relation to these substantive knowledges that the above resistance is fiercest. For instance, in Chap. 3, we also mentioned, with reference to the mathematician Li Ye, China's development of a computational mathematical tradition, that even brilliant, theoretically minded

[1] The mechanical clock is one of the five China-to-Europe transmission claims made by Needham which Cohen examined in detail mentioned in the previous chapter.

[2] It is worth mentioning that many cultural-essentialist claims concerning China are made by making reference to the Qing Dynasty (1644–1911), which is most unfortunate and misleading. We discuss this later on in the present chapter.

[3] Hobson (2006, pp. 130–2) provides a discussion that contradicts Cohen's.

mathematicians such as Li who had been nurtured in it were unable to break through its limits. That is why Euclidean geometry failed to strike a chord in China since Euclid is the very embodiment of a very different tradition, namely, Greece's theoretical tradition. The obstacles that Cohen talks about with reference to this case are, to repeat in the terminology of Chap. 3, intellectual, not cultural, in nature.[4]

Moreover, as also discussed in Chap. 3, from the angle of path-dependency theory, *timing* is a crucial factor, i.e. the effects of a certain event or factor can be very different if it occurred or intervened at a particular point in time in the developmental process instead of another. Thus, the introduction of a particular tradition-specific knowledge from one tradition to another may receive different receptions depending on the timing of that introduction. The simplest illustration of this is the degree of entrenchment of the receiving tradition, the higher this degree, the greater the resistance, and vice versa. This is an internal factor. External factors can also affect reception, such as the political environment obtaining at a particular time. In the case of the translation of Euclid in the twelfth century, its external circumstances are not well-understood, which might also have played a role, and can add further to our understanding.

Other than scholarly tradition, we may, perhaps, hazard an informed guess on another possible internal factor which is related to the influence of timing. We make this informed guess, not so much as an argument in its own right, but in order to illustrate the complexity of the matter that a simplistic 'cultural obstacles' argument can only fail to recognize. As is well-known, the Song Dynasty (960–1279) was a high point in China's development in many respects, among which is learning. In philosophy (including natural philosophy), beginning with Zhou Dunyi (周敦颐 1017–1103), a series of thinkers elaborated a philosophical system that

[4] In Sect. 3.4 of Chap. 3, it was mentioned that Cohen's notion of 'cultural transplantation' is a misnomer (because what was transplanted was scholarly tradition, not culture), and that Cohen's misnomer results from his conceptual confusion between scholarly tradition and culture. In speaking of 'cultural obstacles' to knowledge transmission between different civilizations, Cohen suffers from the same confusion, for although he speaks repeatedly of 'cultural obstacles', he occasionally *seems* to show an awareness that the obstacles are really intellectual in nature, as when he speaks (in Cohen 2024) of 'people reared in…different conceptual frames' being unable to engage in dialogue. We stress 'seems' because, due to his above-mentioned confusion, Cohen apparently believes that conceptual frames are *culture*-specific. That is why he believes that resistance to different conceptual frames constitutes a 'cultural obstacle'.

came to be known as neo-Confucianism, reaching a high point with Song's premier philosopher Zhu Xi (朱熹1103–1200), who lived precisely in the twelfth century. Prior to the Song period, for centuries, for reasons that need not concern us here, indigenous Chinese philosophy had atrophied while Buddhist philosophy became dominant. The neo-Confucian movement was a conscious attempt to challenge the influence of Buddhist philosophy by reviving indigenous Chinese philosophy on a new basis (Lao Siguang 1983). Could this stress on retaining indigenous Chinese learning, which constituted a value among the intelligentsia, have affected scientists and mathematicians as well? If so, could it be that the cold-shouldering of Euclid in the twelfth century also had something to do with this? We do not know and do not want to push this further. In any case, much of that cold reception is explicable in terms of scholarly-tradition-derived intellectual resistance.

Does it follow from our argument concerning such resistance that all transmissions between significantly different scholarly traditions are impossible? Not so. Just as a scholarly tradition is non-culture-specific in the sense explained in Chap. 3, some individual substantive knowledges are non-tradition-specific, which can fit into, and hence be shared by, different scholarly traditions without much difficulty. A good example is the notion of the universe being an infinite empty space previously mentioned, which can easily be shared by both the ancient Chinese and the Greek-derived astronomical traditions. In such cases, transmission is possible even between deeply entrenched scholarly traditions which are significantly different from one another, and it cannot be doubted that the just-mentioned notion was indeed transmitted from China to Europe, since, as explained in the previous chapter, direct evidence of this exists in Matteo Ricci's letter back home of 1595. The issues of transmission and, if this does occur, obstacles to transmission are more complex than we have discussed so far, as will be seen as we proceed.

7.2 The Second and Third Episodes of Euclid's Reception in China: Xu Guangqi and Matteo Ricci, and Li Shanlan

The story of Euclid's reception in China did not end with the twelfth century; there were a second episode and then a third episode of this story. To begin with the second episode, barely three centuries after the 12th,

Euclid was again translated into Chinese in 1606–1607 by a scholar-official named Xu Guangqi (徐光启1562–1633) in collaboration with Matteo Ricci, for which Xu wrote a preface which eulogized European learning. This eulogization is not only contrary to what one would expect from the cultural obstacles standpoint, it also seems to cast serious doubts on our perspective of scholarly-tradition-derived intellectual resistance to transmission. Hence, how this happened is worth consideration in detail.

Originally an ordinary Chinese scholar brought up in the Chinese scholarly tradition, Xu Guangqi was subsequently converted to Christianity. The conventional account of this is epitomized by the French sinologist Jacques Gernet who states: 'Virtually all the literate converts of the early seventeenth century were attracted to the missionaries by their scientific teaching' (Gernet 1985, p. 22). This narrative of Xu and his colleagues becoming Catholics because they had been attracted by the 'superiority' of European learning, that comes perfectly naturally from the Eurocentric point of view, has been repeated time and again until the present day. Let us examine whether or not the historical facts bear this narrative out with reference to the most famous of these Chinese literati, namely, Xu.

Born in 1562 in Shanghai, Xu was educated in the traditional manner and went through the various levels of civil service examinations.[5] At the age of eighteen, having successfully passed through the two lowest levels of examination, he achieved *shengyuan* (生员) or *xiucai* (秀才) status, and became a teacher; for the next fifteen years, he failed at the next level of examination, and it was only in 1597 that he finally passed it to achieve *juren* (举人) status. *Juren* status entitled one to become an official, but only at the local level; those who harboured the ambition of becoming a capital official in Beijing had to go through further examinations including the final metropolitan examination. Before he achieved *juren* status, Xu went to Guangdong to teach in 1595, it was there that he met the Jesuit missionary Lazaro Cattaneo. Xu was able to achieve *juren* status because the chief examiner at the examination concerned was appreciative of his learning. In 1600, Xu went to Nanjing to pay a visit to this person on his way to Beijing to attend the metropolitan examination in order to become a *jinshi* (进士)—achieving that status is a prerequisite of becoming a capital official—which he failed on that occasion. While in Nanjing,

[5] For a comprehensive study of Xu's life and career, see Jami, Engelfriet and Blue (eds. 2001). Part of the following is according to the second chapter of that collection by Timothy Brook titled 'Xu Guangqi in his context'.

he also met Ricci. The latter went to Beijing the following year to be invited to become a court adviser by the then-reigning Emperor Wanli (万历). In 1603, eight years after he had first come into contact with Christianity, Xu was baptized in Nanjing. The next year, in 1604, at the age of forty-one, Xu finally achieved *jinshi* status and, thanks to the abdication of the person ahead of him in the examination, became a capital official. Why did Xu become a Christian? 'It is tempting', Brook (2001; 96) observes, 'to explain Xu's interest and conversion in psychological terms, as arising from the frustration of long years of uncertainty and a blocked career. Such frustration may well have opened Xu to Christian influence, but it cannot fully account for the choices he made'. Nor, apparently, would conversion be helpful to his career at that time. Was it, then, as the conventional account has it, because he had, *out of intellectual conviction*, become awed by European science?

In an analysis of extant Yuan, Ming and Qing civil service examinations, Elman (2000) notes that during the Ming, the triennial provincial-level examination for *juren* status sometimes featured questions requiring a basic competence in astronomy, calendrics, and music. Thus, we can assume that Xu possessed some basic technical training in those areas. A work from 1603 titled 'Measurement of river works' is Xu's earliest extant work related to practical matters. In this work, 'there is in fact little beyond the most elementary mathematics. Xu does show familiarity with the terminology and calculations of the sides of right triangles (....), taken from the chapter on right triangles (...) in popular mathematical manuals'. In a letter titled 'Reply to a fellow townsman', Xu spoke about the merits of Christianity. The date of this letter is uncertain, but it was certainly sometime during the period after he had become attracted to Christianity up to his baptism or shortly thereafter; on the other hand, it was before he became conversant with Ricci's works, for in the letter he stated that '[t]oday [I] have not yet obtained [Ricci's] writings in translation' (Hart 2013, pp. 236–7). From the above two extant texts by Xu, it is clear that by the time of his baptism or shortly thereafter, he had not yet read Ricci's works on European science, and that his knowledge of mathematics was only elementary, despite possessing some basic technical training. This contradicts the above-mentioned conventional account as exemplified by Gernet of his conversion.

Brook observes the 'growing place of Buddhism in gentry life during the latter half of the sixteenth century, when the gentry elite of Shanghai [Xu's native place] patronized Buddhist monasteries and absorbed themselves in religious ideas and commitments…Xu's own life thus coincides almost perfectly with the period of deepening religious devotion among the gentry'. In Brook's view, Xu's turn to religion needs to be understood in this context. Before proceeding, it should be noted that, contrary to popular misunderstanding, Confucianism is not a religion, but a thoroughly this-worldly doctrine in which the supernatural plays no role at all. The devotion to Buddhism among some members of the gentry, however, created a reaction among other members who regarded Buddhism as undermining the authority of Confucianism and the socio-political-ethical order based upon it. These people 'revived the long tradition of anti-Buddhist polemics within the statecraft strain of Confucianism, the strain with which Xu felt most sympathetic. Christianity had the advantage of being untainted by this tradition. It was a religion, but it was not Buddhism and it was not condemned by Confucianism' (the above quotes are from Brook *op.cit.*, pp. 96–7). In sum, nurtured intellectually in the traditional way, Xu was committed to Confucianism, his turn to religion needs to be understood in the above-mentioned context of the general trend among the gentry. In turning to religion, there were three options open to Xu. He rejected Buddhism and the Daoist religion. The reason for this is obvious. This is because both Buddhism and Daoist religion promoted other-worldly concerns, whereas as a Confucian, Xu's concerns were this-worldly. It is from this angle that we are able to understand why he was attracted to Christianity. This is because although Christianity was also other-worldly-oriented,[6] it was new to China and Xu, with as yet unknown implications for the Confucian socio-political-ethical order, which is in contrast to the cases of Buddhism and Daoist religion which had long been queried for being inconsistent with that order.

Evidence supporting our above interpretation can be found in Xu's 'Reply to a fellow townsman', in which he stated:

> Buddhism has entered China for one-thousand eight-hundred years. The minds of the people and the way of the world, they cannot compare with antiquity with each passing day…If we reverently believe in the Lord of Heaven (天主referring to the Christian God in this context),[7] it will definitely occur within several years that people will become sages and noble

[6] Without implying any endorsement of Weber's protestant-ethic thesis, let it be noted that the Jesuits were Catholics.

[7] This term had three meanings in China (see Hart 2013, p. 238 n 159; cf. *Etymology of Words* 1987, p. 370). In Xu's context, it undoubtedly refers to the Christian God.

gentlemen. The way of the world could then be compared to that of Yao, Shun, and the Three Ages,[8] even greatly exceeding it...The Lord of Heaven [the Christian God] is just what the Confucian texts call the Emperor on High (Shangdi 上帝).[9] Once its existence is believed in, then all the established teachings and admonishments cannot but be observed, all the principles concerning what is right and just discussed cannot but be followed.... (Hart 2013, pp. 237-8, translation slightly altered by us)

More than a decade later, Xu remarked as follows in a document of 1616:

All the schools of Buddhism and Daoist religion (*shidao*释道[10]], their arts are impure, and their teachings are incomplete...If we were to believe in the

[8] The Three Ages (*sandai*三代) refer to the Xia (夏), Shang (商) and Zhou (周) (more accurately the Western Zhou period from −1046 to −771) dynasties. The latter two dynasties have been attested to in written documents dating from them; the existence of the Xia has been subject to dispute, but archaeological discoveries in recent decades have convinced most scholars of its existence. Yao (尧) and Shun (舜) are two pre-Xia kings said to have existed before the Xia, with Yao passing the throne to Shun. Previously dismissed as mere legend, recent archaeological discoveries have also lent credence to reports of their existence in later documents, though both likely overlapped with early Xia instead of pre-dating it. During the period beginning in −771, the Zhou Dynasty's socio-political-ethical system slowly disintegrated, as a result of which, ancient Chinese intellectuals of all schools (which began to come into being from the latter -sixth century onwards) revered the period of Yao, Shun, and the Three Ages as China's golden age, and yearned for a return to it. This article of faith was held by the Confucians throughout the period of dynastic China.

[9] *Shangdi* (or simply *Di*帝) is the term the Shang people gave to the supreme anthropomorphic deity, the Zhou people inherited it but mostly referred to that deity as *tian* (天) (this term was used in several senses, some of which we have examined in Chap. 5, which could be misleading to those unfamiliar with ancient Chinese thoughts). Xu was misleading his addressee in stating that the Lord of Heaven is none other than *Shangdi*, since he could not have been unaware that, unlike the Christian God, the latter was a non-Creator.

[10] As previously mentioned (note 38 of Chap. 5), while Daoist religion claims inspiration from Daoist philosophy (the philosophies of Laozi and Zhuangzi), it actually represents a severe perversion of the latter, hence the two should be strictly distinguished. The phrase *shidao* is part and parcel of the phrase *rushidao* (儒释道). This longer phrase refers to Confucianism, Buddhism, and Daoism as religions, which is incorrect and misleading insofar as Confucianism is not a religion. Despite this, this longer phrase as well as the term 'Confucian religion' (*rujiao*儒教) have remained in general use, even among scholars. Thus, in the phrase *shidao*, *dao* refers to Daoist religion. In relation to this, it is worth noting that in the same document from which the present quote is taken, Xu accused *Chan* Buddhism (the branch of Chinese Buddhism that had risen to dominance) of perverting the intent of Laozi and Zhuangzi, showing that he held the *philosophies* of these two thinkers in great respect.

> Sovereign on High (*Shangzhu* 上主)[11] in the manner in which Buddha and Lao (*folao* 佛老[12]) are believed in, if we accommodate the tributary officials [referring to the Jesuit missionaries[13]] the way we accommodate [Buddhist] monks and Daoists [of Daoist religion], then we would revitalize civilization and deliver governance, and in doing so would definitely exceed the times of Yao, Shun, and the Three Ages.... (Hart 2013, p. 249, translation slighted altered by us)

To sum up so far, Xu's turn to religion was part and parcel of a general trend among the gentry in the latter sixteenth century; as a Confucian at heart, Xu rejected Buddhism and Daoist religion because he saw them as having corrupted China over many centuries; Christianity was something new altogether, with as yet unknown implications for the Confucian socio-political-ethical order that Xu supported. In order to make sense of his conversion to himself as a Confucian, Xu took Christianity to be recovering the 'teachings and admonishments' and the 'principles of what is right and just' of China's past golden age that had been lost.[14]

If Xu's conversion to Christianity had nothing to do with first having been, out of intellectual conviction, awed by the 'superiority' of European science, he did, as *part and parcel of* his conversion, begin to investigate into European learning. Today, in contrast to Xu and his fellow-converts of the late 16th to early 17th centuries, converts to Christianity in the non-Euro-American world do not do so. To understand why there is such a difference, it is worth recalling that in Latin Europe, due to the handmaiden tradition, there had been established the institution of the theologian-natural philosopher, an institution that no longer exists today. In this tradition, investigating into science was said to be in order to better

[11] The term *Shangzhu* was usually used to refer to the sage ruler, but here, it clearly refers to the Lord of Heaven (the Christian God).

[12] The phrase *folao* refers to Buddhism and Daoist religion.

[13] As Hart (2013, p. 252) points out, Xu referred to the Jesuit missionaries as 'tributary officials' in a deliberate attempt in the document concerned to mislead, i.e. to misrepresent them as coming to pay tributes to the Chinese throne, for the purpose of promoting these missionaries and their teachings. By the time of this document (1616), Xu had become most deeply aligned with the missionaries.

[14] The idea that European learning was recovering for China what it had lost since the past golden age was not limited to Xu, but shared by many Confucian scholar-officials of the period. Most Euro-American scholars are unaware of the existence of the article of faith held by Confucians since the pre-Qin era (see note 8), hence, they fail to correctly contextualize this idea, and, more often than not, treat it with ridicule (see, for e.g. Huff 2010, p. 98).

understand God's design. There can be no doubt that, educated in the theologian-natural philosopher tradition, Ricci and his fellow missionaries brought this belief with them to China.

On the other hand, Ricci held Chinese science in contempt. As seen, he regarded China's notion of the universe being an infinite empty space as 'absurd'; likewise, he believed that Chinese mathematics lacked proofs and instead proceeded through the 'wildest imagination' (Hart 2013, p. 198). Let us take a little look at the latter issue. With regard to proofs, it is now well understood that mathematical proofs can take different forms and proof in Chinese mathematics had existed right from the beginning. It is clear that other than the Greek form of proof, Ricci was unable to comprehend other forms existing in different mathematical traditions. With regard to 'wildest imagination', let us refer to a couple of examples. Pascal's triangle is a triangular array of binomial coefficients named after the seventeenth-century French mathematician Blaise Pascal (1623–1662), but 'it received detailed treatment in the hands of [Song Dynasty mathematicians] Yang Hui [杨辉ca. 1238-1298] and Zhu Shijie [朱世杰 1249-1314], some 350 years earlier' (Joseph 2011, p. 247). 'Interest in [indeterminate analysis] arose in both China and India in connection with calculations in calendar making and astronomy. The basic problem was one of finding a procedure for solving a system of *n* equations with more than *n* unknowns', concerning which the Song Dynasty mathematician Qin Jiushao (秦九韶1202–1261) developed methods that predated those developed in Europe by Euler (1742) and Gauss (1801) some 500 years later (Joseph 2011, p. 248; Chen Cheng-I 1996, p. 224). The above-mentioned Chinese mathematicians attained their achievements several centuries before Ricci arrived in China, and the corresponding developments in Europe took place after Ricci had passed away. Little wonder then that Ricci, educated in one particular mathematical tradition, was unable to comprehend a different mathematical tradition. What Ricci failed to understand, he derogatorily dismissed, in the present case as the 'wildest imagination'. Ricci's dismissal of Chinese culture even extended to his claim that Catholic exorcism was 'superior' to Chinese exorcism (Hart 2013, p. 199).

Hence, it is safe to assume that Xu, when he was preached to by Ricci and his fellow Jesuit missionaries, must have been told by them that Chinese science (including mathematics) was not up to scratch, and, in accordance with the hand-maiden tradition, that to belief in the Christian God, he must also learn 'superior' European science. This is why in converting to Christianity, Xu also began to study European learning. Is this assumption of ours merely a reasonable deduction, or is there textual evidence to support it? And if our assumption is valid, how would Xu make sense of what he had been told by Ricci to himself as someone intellectually brought up as a Confucian? The answers to these questions can be found in the preface that he wrote for the Chinese translation of the first six chapters of Euclid, completed by him and Ricci in collaboration, that was published in 1607, in which he eulogized European learning:

> In the times of Yao and Shun…without measurement and numbers, no tasks could be accomplished[15]…Of the Six Arts in the Offices of Zhou,[16] numbers constitute the first; if, when engaging in the [other] five arts, measurement and numbers are not used, then those too cannot attain accomplishments. Therefore, in the aforementioned Three Ages and before, those who excelled in this endeavour [mathematics] prospered…But this was terminated in the flames of the Ancestral Dragon [referring to the First Emperor's wholesale destruction of knowledge shortly after founding the Qin Dynasty in -221]. Since the Han Dynasty many have arbitrarily fumbled about, like blind men shooting for a bull's-eye…until the present time this Way [referring to the mathematics that was supposed to have existed in the times of Yao, Shun, and the Three Ages] has been obliterated…[Euclid's *Elements*]

[15] In the *Records of the Grand Historian* (*shiji* 《史记》), compiled by Sima Qian (司马迁) in the -first century, it is said that Yao and his top officials were adept at measuring the seasons and time cycles related to cultivation. This is why Xu mentioned that in the time of Yao and Shun, measurement and numbers were important. Sima Qian's report concerning this has always been dismissed as pure legend, especially by Euro-American sinologists. Recent archaeological investigation at the Tao Si (陶寺) archaeological site in southern Shanxi province mentioned in note 13 of Chap. 6, however, has discovered the bases of thirteen columnar pillars placed in close formation. By reconstructing these pillars, and with the help of historical astronomers, a particular observation point has been identified, from which by making observations of the rising sun through the twelve gaps between the thirteen pillars, the two equinoxes, the two solstices, and sixteen other time cycles related to cultivation can be precisely measured. On this evidence and based upon exact carbon-14 dating, most Chinese archaeologists are of the view that this site, which is a site of a massive walled settlement complete with a palace area and so on, was where Yao mentioned in the *Records of the Grand Historian* located his ruling base.

[16] *Offices of Zhou* (*Zhouguan*周官) is an official document dating from the latter Zhou Dynasty; the Six Arts (*liuyi*六艺) are rites, music, archery, equestrianism, calligraphy, and mathematics. During the Zhou Dynasty, children of royalty, the aristocracy and other upper-class elites received an education in the Six Arts, in which mathematics, in the view of modern scholars, referred to basic astronomical and calendrical knowledge, basic calculation skills, divination, and so on. On this issue, see Lau (2020, p. 376).

> is the progenitor of measurement and numbers...He [Ricci] stated only that without having translated this book [Elements], none of the other [mathematical] books [introduced by Ricci] could be discussed...In my mind I said to myself: I did not expect—two thousand years after the learning of antiquity was annihilated—that I could suddenly obtain that which restores the corrupt classics and the lost meanings from the times of Yao, Shun, and the Three Ages...Of the learning of Master [Ricci] there are probably three kinds: the greater is self-cultivation to serve Heaven; the lesser is to investigate objects to fathom principles; one aspect of the principles of objects is also called numerical arts.... (Hart 2013, pp. 242-4, translated slightly altered by us)

Xu's preface is a revealing treasure trove that answers a lot of questions concerning our present subject matter. Firstly, as seen, there is no indication from extant textual evidence that before his engagement with Ricci, he had been trained in the Chinese mathematical tradition while his understanding of right angles was elementary. We have definitive proof of this from the above quote. As has become well-known and well-established for some time by now (Joseph 2011), Chinese mathematics reached a first high point in the *post*-Han Dynasty mid-third century in the works of Liu Hui, mentioned in Chap. 5, and that of another mathematician named Zhao Shuang (赵爽); after that, various individual mathematicians achieved important breakthroughs over the centuries; then, in the thirteenth century, a quartet of mathematicians (Li Ye 李冶 [mentioned in Sect. 3.4 of Chap. 3], Yang Hui杨辉, Qin Jiushao秦九韶 and Zhu Shijie朱世杰, the last three of whom have previously been mentioned) appeared who took Chinese mathematics to its highest peak ever with developments that pre-dated Europe in the same areas by several hundred years. During the Ming Dynasty itself, the mathematician Zhu Zaiyu (朱载堉 1536–1611), Xu's older contemporary by twenty-six years, had published several path-breaking works between 1583 and 1603 in musical theory, in which a mathematical solution for the calculation of equal temperament was provided for the first time in world mathematical history.[17] Had Xu been trained in Chinese mathematics beyond making elementary

[17] Zhu's work on equal temperament predated the same development in Europe made by Marin Mersenne (1588–1648) and Simon Stevin (1548–1620) (Chen Cheng-I 1996, pp. 66, 111). The argument that this constituted a case of diffusion (Joseph 2011, pp. 294, 447 n 24; Needham and Robinson 1962, pp. 214–8; Hart 2013, pp. 110–28) has, unsurprisingly, been disputed; that, however, is not our concern.

calculations, he would have been fully aware of all this. Yet, he claimed that 'Since the Han Dynasty', mathematics that was supposed to have existed in China's purported past golden age 'has been obliterated'. Furthermore, he claimed that mathematicians 'since the Han Dynasty…have arbitrarily fumbled about, like blind men shooting for a bull's-eye'. As seen, Ricci claimed that Chinese mathematics proceeded through the 'wildest imagination'. Thus, Xu made this comment on post-Han Chinese mathematics, not because he had been trained in it and found it worthless, but because he was told this by Ricci.

The above further shows that Xu had not come to eulogize Euclid because as someone trained in the Chinese mathematical tradition, he found Euclid, which is from a very different tradition, superior, but simply because he was told by Ricci that Euclid was the be all and end all of mathematics, something that he completely accepted from his teacher. That this is so is unsurprising from the perspective of our analysis of scholarly tradition. It is, hence, unsurprising that in the whole preface, Xu said not a single word about why Euclid was such a great work. 'In fact', at this time, as Hart (2013, p. 244) observes, 'Xu understood little of even the basics of Euclidean geometry'.

In the above, we remarked that Xu must have learned about the purpose of learning science as seen in the hand-maiden approach from Ricci. We now have proof of this from Xu's preface as well. In the preface, Xu distinguished between the 'greater use' and 'lesser use' of European learning brought to China by Ricci, 'the greater is self-cultivation to serve Heaven; the lesser is to investigate objects to fathom principles; one aspect of the principles of objects is also called the numerical arts [mathematics]'. The term translated as 'Heaven' is *tian* (天). As noted (note 9 above), this word had (still has) several meanings in Chinese, the Zhou people used this word to refer to the supreme anthropomorphic deity. This sense of *tian* had since remained a central notion in dynastic China as in the term Mandate of Heaven, the mandate given by the supreme deity to rule. Thus, according to Xu, the 'greater use' of European learning was to serve the supreme deity; as seen, Xu had previously said that the Christian God was none other than the 'Emperor on High', which was the term the Shang people gave to the supreme deity, but which the Zhou people renamed as *tian*; thus, when Xu said that 'the greater use' of European learning was to serve *tian*, this is equivalent to saying that it was to serve the Christian God. This is in accordance with the hand-maiden tradition, since being able to better understand God's design implies being able to better serve God.

Xu's mentioning of the 'lesser use' of European learning shows us how Xu was able to make sense of the hand-maiden approach from the Confucian point of view. The phrase 'to investigate objects to fathom principles' (*ge wu qiong li*格物穷理) comes from the Song Dynasty's neo-Confucians. The neo-Confucians' philosophy is called 'the study of *li*' (*lixue*理学).[18] The notion of *li* was understood in two senses by the neo-Confucians. In one sense, it refers to natural principles of objects of the natural world, with principles understood to incorporate operating mechanisms, natural patterns, innate nature, and so on.[19] This is the *li* in the phrase 'to investigate objects to fathom principles'. In modern terminology, this phrase simply means to engage in natural scientific investigation so as to understand the natural principles of nature. In its second sense for the neo-Confucians, *li* refers to metaphysical truth. In the neo-Confucians' view, 'investigating objects to fathom principles' (Xu's 'lesser use') in depth will ultimately sublimate into a leap of the mind whereby it will be able to intuit *li* as metaphysical truth (Xu's 'greater use'). Thus, while Xu accepted the purpose of learning European science as stated in the hand-maiden approach that he had been taught by Ricci, he was able to make sense of it from the Confucian perspective as well.

Finally, in the preface, as previously, Xu made sense, from the Confucian angle, of accepting the 'superiority' of European learning and the need for China to learn it in accordance with Ricci's teaching by interpreting European learning as recovering for China what it once possessed in its past golden age, but which it had since lost due to the corrupting influences of Buddhism and Daoist religion.

To conclude with regard to the second episode of the story of Euclid in China, in direct contrast to the cold reception that Euclid received in the twelfth century, Xu eulogized it some three centuries later. Conventional wisdom would have us believe that this was because Xu was a mathematician steeped in the Chinese mathematical tradition, and as such he came to recognize the 'superiority' of Euclid. If this were true, it would directly contradict our argument concerning tradition-specific intellectual resistance to a radically different scholarly tradition. Our above detailed

[18] On the following, see Lao Siguang (1983).

[19] Hence, the translation of *li* sometimes as 'pattern' is wrong, because natural patterns constitute merely one aspect of *li* as understood by the neo-Confucians.

analysis, however, has shown that this was decidedly not the case. Quite the contrary, all the historical facts concerned serve, instead, to confirm the validity of our argument.

Nonetheless, with little training in either science in general or mathematics in particular originally, Xu's conversion to Christianity and what all that implies as analysed above did make Xu turn to studying science and mathematics. As shown above, by the time of the translation of Euclid, his engagement was still rather shallow; even by 1613, when Xu and his fellow-converts to Christianity Li Zhizao (李之藻1565–1630) and Yang Tingyun (杨廷筠1557–1627) each wrote a preface to the *Guide to Calculation in the Unified Script* (*Tong wen suan zhi*《同文算指》), in which all three faulted Chinese mathematics for being 'inferior' to European mathematics, his and his fellow-converts' adeptness in mathematics was still very suspect.[20] But Xu's level of adeptness did improve over the years, which enabled him to lead an astronomical reform in the second half of the 1620s, to be discussed in the next section.

The third episode of the story of Euclid in China occurred under the Qing Dynasty, and its protagonist is the scholar-official Li Shanlan (李善兰1811–1882). In 1857, he and Alexander Wylie from the London Missionary Society published a translation of the last nine chapters of Euclid's *Elements*, to complement the translation of the first six chapters by Xu Guangqi and Matteo Ricci. If Xu, with no roots in the Chinese mathematical tradition, dismissed that tradition and eulogized Euclid without really knowing what he was talking about, what was Li's attitude towards Euclid?

Li Shanlan was born to a literary family whose lineage could be traced back five hundred years to the Song Dynasty.[21] He was educated by the renowned philologist Chen Huan (陈奂), but at the age of nine, he noticed a copy of *Nine Chapters on Methods of Calculation* on his father's bookshelf, and was able to master it through self-study within a year. As mentioned in the previous chapter, *Nine Chapters*, compiled during the -1st to 1st centuries, was, firstly, the text that founded what we call China's

[20] Authorship of the *Guide to Calculation* is attributed to Li Zhizao and Ricci. It is a translation of a 1583 work by Ricci's teacher in the Catholic Church, namely, Christoph Clavius, together with some problems taken from Chinese mathematical texts. Neither the authors, nor the writers of the prefaces were aware that these problems taken from Chinese mathematical texts were, in fact, 'arguably the most advanced and recognizably "modern" mathematics in the' whole work. For a detailed analysis of this, see Hart (2013, Chap. 5).

[21] For a summary biography of Li, see Martzloff (1997, pp. 173–6).

computational mathematical tradition; and, secondly, identical methods later referred to as Gaussian elimination in Europe, as well as determinantal-style calculations and solutions later found in Fibonacci in Europe, can already be found in *Nine Chapters.* Several years later, Li repeated the same feat with the first six chapters of Euclid as translated by Xu Guangqi and Matteo Ricci; shortly afterwards, he did so again with *Sea Mirror for Circle Measurements* by the Song mathematician Li Ye. It will be recalled from chapter Three that *Sea Mirror* took the Chinese mathematical tradition to its theoretical limits. There is a report that Li subsequently took lessons from a mathematician named Wu Zhaoqi (吴兆圻).

In 1845, at the age of thirty-four, Li became a tutor at the family of the well-known scholar Lufei Chi (陆费墀) in the city of Jiaxing in the Yangtze River delta area of Zhejiang province. By then, Li had already been writing original mathematical texts. During this period, he met and became associated with a number of mathematicians and scholars interested in mathematics, including the mathematicians Gu Guanguang (顾观光), Dai Xu (戴煦), Luo Shilin (罗士琳), and Xu Youren (徐有壬)—Gu was later to proof-read Li's and Wylie's translation of the last nine chapters of the *Elements.* In 1852, taking refuge from the Taiping Rebellion (1850–1864), Li moved to Shanghai, where he met Wylie. Wylie had come to China in 1846 and stayed until he left in 1877 for health reasons, having become fluent in Chinese, he read a work on logarithms that Li had published in 1846. Greatly impressed by that work, Wylie recruited Li to translate the last nine chapters of Euclid with him, as well as various scientific works in mechanics, botany, astronomy, and mathematics with a number of Wylie's fellow Protestant missionaries stationed in Shanghai. All these translated works were published during the 1850s.

Towards 1859–1860, Li joined the staff of the Governor of Jiangsu province, which is adjacent to Zhejiang province to its north, who was none other than the above-mentioned mathematician Xu Youren. Three years later, he was recruited by Zeng Guofan (曾国藩1811–1872), the famous scholar-statesman and general who was to put down the Taiping Rebellion in 1864. On Zeng's defeat of the Taiping rebels in Nanjing, Li followed Zeng to move to that city. On Zeng's sponsorship and at Li's urging, the complete *Elements* was published in 1865, and in 1867, Li's collected original mathematical works titled *The Mathematics of the Studio Taking the Ancients as Models* (*Zeguxizhai suanxue*《则古昔斋算学》) was published, in which eight out of thirteen sections were on European mathematics, while four sections were derived from the Chinese mathematical tradition. In 1869, Li was appointed as mathematics professor at the Tongwen Guan (同文馆) located in the capital Beijing, a post that he held until his death. This institute, supervised by William Alexander Parsons Martin of the Presbyterian mission, had been established by the Qing government; initially only with the objective of teaching European languages, it later extended into other disciplines.

From the above, it can be seen that in contrast to Xu Guangqi, who had no specialist mathematical training, Li Shanlin, China's premier mathematician of the nineteenth-century, was expertly versed in both the Chinese and the Greek-European mathematical traditions. Li had not first been deeply nurtured and worked in the Chinese mathematical tradition before he came across Euclid; on the contrary, he mastered both *Nine Chapters*, on the one hand, and, on the other, the first six chapters of the *Elements* within several years of one another *early in his mathematical life*. This enabled him to be *receptive to both traditions*, and, in his ensuing mathematical career, he engaged with *both* the Greek-European tradition through his translation work, and the Chinese tradition. *Contrary to Cohen, there existed no supposed 'cultural obstacles' to Li completely mastering the Greek-European tradition.* In contrast to Xu Guangqi and contrary to scholars who denigrate the Chinese tradition as being 'inferior' to the Greek-European tradition,[22] Li did not dismiss the Chinese tradition in favour of the Greek-European tradition. On the contrary, while holding the Greek-European tradition in the highest regard, Li was convinced that the Chinese tradition could do even better. When he took up the mathematics professorship at the Tongwen Guan, he did not use Euclid, but the above-mentioned Li Ye's *Sea Mirror*, which, to repeat, took the Chinese mathematical tradition to its theoretical limits, as his fundamental reference. Martzloff (1997, p. 176) observes:

> Li Shanlan's treatment of subjects derived from Chinese traditional mathematics (four sections out of 13 [of *The Mathematics of the Studio Taking the Ancients as Models*]) is frequently very original...as explicitly stated in Li Shanlan's preface [to the just mentioned work], its [referring to the search for originality] essential motive was the author's desire to beat Western mathematics at their own game, not by a slavish appropriation of 'barbarian' mathematical techniques but by highlighting traditional Chinese modes of

[22] The Graecophile Chen Fangzheng, who asserts that the third-century mathematician Liu Hui must have plagiarized from Archimedes (see Sect. 6.2 of Chap. 6), is one such scholar.

thought and explicitly taking his inspiration from his medieval predecessors, especially Zhu Shijie.[23]

In view of the high regard in which Li held the Greek-European tradition at the same time, his confidence in the value of the Chinese tradition cannot be put down to xenophobic nationalistic pride.[24] Nor was his confidence misplaced at all. In terms of complementarity to science, the Chinese tradition can in no way be considered as being 'inferior'. Thus, Kvasz (2015) points out that modern science is not based upon triangles, circles, and other geometric figures but upon functions and differential equations. Hart (2024) similarly observes: 'Linear algebra[25] is one of the mathematical foundations of modern science, with broad applications in physics, economics, statistics, and computer science; it is the essential mathematical framework for much of twentieth-century quantum mechanics along with twenty-first-century artificial intelligence, quantum information sciences, and the Second Quantum Revolution'.

To conclude with regard to the third episode of the reception of Euclid in China, Li Shanlan's mastering of and high regard for Euclid show claims of 'cultural obstacles' to be entirely unfounded and ill-conceived; it was because early in his mathematical life, Li was nurtured (through self-study)

[23] Zhu, mentioned earlier in the present section, was one of the quartet of thirteenth-century mathematicians who took the Chinese tradition to its highest peak, and regarded by many scholars to be the greatest of the four.

[24] It is worth noting that Li lived in an age of great externally generated political shock to China. In 1839, the First Opium War between Britain and China broke out, ending in 1842 with China ceding the island of Hong Kong to Britain. In 1856, Britain and France launched the Second Opium War against China, which resulted in China's cessation of Kowloon peninsula (across the harbour from the island of Hong Kong) to Britain as part of Hong Kong; it also led to the British and the French invasion of Beijing in 1860 and the sacking and burning down of the Imperial Summer Palace in the capital's suburb. From the path-dependency perspective, such immense external impact was prone to lead to the wholesale repudiation of the Chinese tradition on the part of many Chinese intellectuals, as indeed it did—to refer to just one instance from several decades later, the Beijing University professor Chen Duxiu (陈独秀1879–1942), founder of the Communist Party of China in 1921, precisely called for a wholesale repudiation of Chinese culture in favour of a wholesale adoption of Euro-American culture. On the other hand, such external impact could also have led in the opposite direction on the part of some other intellectuals, namely, a blind insistence on the 'superiority' of Chinese culture and repudiation of Euro-American culture. That Li Shanlan held the Greek-European mathematical tradition in the highest regard shows that he was not among this latter category of intellectuals.

[25] See Hart's analysis of the Chinese origins of linear algebra examined in Chap. 6.

in both the Chinese and the Greek traditions that he was able, throughout his life, to be expertly versed in both of them, this is in accordance with what can be expected from our perspective concerning the role of scholarly tradition; and, finally, Li provides an illuminating contrast to the case of Xu Guangqi, confirming our above analysis of how and why Xu came to dismiss the Chinese mathematical tradition and eulogized the Greek-European one.

7.3 Huff's Cultural-Essentialist Neo-Eurocentric Claim That China Suffered From a 'Deficit in Intellectual Curiosity'

Cohen's cultural-essentialist view, with which this chapter began, claims that culture in a general sense obstructed certain civilizations such as China from learning science from other societies; in the literature, there is another cultural-obstacle argument, which claims that one particular aspect of the cultures of civilizations such as China, namely, a purported 'deficit in intellectual curiosity', resulted in the failure of these civilizations, though equipped by the Jesuit missionaries with update knowledges and the associated scientific paraphernalia, to make any scientific advances such as that occurred in Europe in the seventeenth century and thereafter. This is the thesis advanced by the historical sociologist of modern science Toby Huff (2010), who refers to the telescope especially to make his point:

> I began this inquiry with the invention of the telescope in 1608 because it is linked to a burst of creativity and scientific curiosity that flowered across Europe in the seventeenth century. Second, the telescope was quickly taken around the world in the early seventeenth century. Yet, the arrival of the telescope in China, India, and the Ottoman Middle East did not produce a similar upsurge in scientific curiosity...the failure of the telescope to trigger an exciting new burst of scientific creativity, especially in astronomy, around the world serves best to highlight a deficit in scientific curiosity that seems to have prevailed outside Europe from before the seventeenth century all the way to the end of the twentieth century. That is an extraordinary record of *cultural disparity*. (Huff 2010, p. 300, our emphasis)

Huff's 'comparative approach' (p 300) from 'a global perspective' (subtitle of Huff's book) covers, as can be noted, China, India, and the Ottoman Middle East; since we are more familiar with China, the following will be confined to it with reference to Huff's thesis. Before addressing it, however, it is necessary to make a short detour to provide the necessary context. As noted at the end of the previous section, in the second half of the 1620s, Xu Guangqi led an astronomical reform. In this reform, the telescope figured prominently. To understand the dynamics of this reform, a brief account of the state of science in China before the Jesuits came would be useful. Chinese mathematics, as pointed out earlier, reached a new height of development in the thirteenth century. Of relevance to us in the present context are two mathematical astronomers who straddled the Song Dynasty and the succeeding Yuan Dynasty (1279–1368) established by the invading Mongols, namely, Wang Xun (王恂1235–1281) and Guo Shuojing (郭守敬1231–1361). Guo was the astronomer who invented equatorial mounting, some two-and-a-half centuries before it first appeared in Europe in Tycho Brahe's instruments in 1585 (Chen Cheng-I 1996, p. 235). It should be noted that while both Wang and Guo lived into the Yuan Dynasty, they were born some forty to fifty years before the fall of the Song and, hence, were brought up in the Song scholarly tradition—in fact, in terms of learning, scholars living in the relatively short-lived (98 years) Yuan Dynasty all traced their heritage to the Song period.

China's dominant astronomical model since the Eastern Han Dynasty (25 to 220), called the Celestial-Sphere model (*Huntian shuo*浑天说), utilized algebraic interpolation to track the celestial bodies and make predictions. Algebraic models can only provide approximations, hence, calendars devised on their basis needed to be revised periodically to keep up their relative reliability. In the early Yuan Dynasty, the above-mentioned Guo Shuojing and Wang Xun and others devised a calendar known as the *Shoushi* Calendar (授时历), which was implemented in 1281. The Yuan Dynasty fell in 1368, succeeded by the Ming Dynasty (1368–1644), which adopted the *Shoushi* Calendar; in 1385, some minor modifications were made to it, whereby it was renamed as the *Datong* Calendar (大统历). The *Datong* Calendar remained in use until the early seventeenth century. This means that the calendar first implemented in 1281 remained in use for over three hundred years with only very minor changes. This is a graphic indication of how learning in general and science in particular atrophied after the early Yuan Dynasty and throughout most of the Ming Dynasty (Chen Cheng-I 1996, p. 228; Chen Meidong 2008, p. 339). This is not to say that no progress was made at all in science under the Ming. The most significant is the mathematician Zhu Zaiyu mentioned in the previous section. Nonetheless, that there was a general trend of decline since the early Yuan Dynasty is the view of most scholars.

As is well-known, Galileo's telescope was invented in 1609, the first mention of it, referred to as a 'clever instrument' (*qiaoqi*巧器), in China was in 1615. Then in 1630, a book completed in 1626 by the Jesuit J. Adam Schall von Bell, who had arrived in China in 1619, with his Chinese collaborator Li Zubai (李祖白) titled *Treatise on the Telescope* (*Yuan jing shuo* 《远镜说》) was published, which gives details of the telescope's optical design. A year earlier, in 1629, under Emperor Chong Zhen (崇祯), Xu Guangqi was given the task of calendar reform. Xu put the Tycho Brahe model introduced to China by the Jesuits,[26] the abovementioned *Datong* Calendar, and the Muslim Calendar (*Huihui li* 回回历) which had been taken over from the Yuan Dynasty[27] to a comparative empirical test; unsurprisingly, with the *Datong* Calendar severely out of date, and the Muslim Calendar based upon outdated models in comparison to the Brahe model, the Brahe model proved to be superior in making predictions of eclipses. In 1632, a proposal to construct three telescopes was approved. Despite Xu's untimely death in 1633, a telescope was constructed the following year in 1634 under the leadership of Xu's successor Li Tianjing (李天径1579–1659). Thus, from the first mention of the telescope in China in 1615 and the publication of its optical design in China in 1630, to the construction of the first telescopes by the Chinese for observational purposes, respectively only nineteen and four years had elapsed. In addition, the Ming reformers compiled the 137-volume *Chong Zhen Treatise on Astronomical Science* (*Chong Zhen lishu* 《崇祯历书》), which represented a one-sided adoption of European astronomical views, despite opposition among some Chinese astronomers who questioned the solid-sphere worldview and maintained the notion of the universe as an infinite empty space (Chen Cheng-I 1996, pp. 228 ff.; Chen Meidong 2008, pp. 339–41).[28]

The keen and rapid positive response of the Ming reformers, with the support of the emperor, to European astronomy and the telescope puts

[26] At that time, Jesuit missionaries told the Chinese only of the Brahe model, without mentioning Copernicus's heliocentric model, which was due to the Church's interdiction; in consequence, they also did not tell the Chinese of Kepler's work, which was based upon the heliocentric model.

[27] As part of this inheritance, Ming's Imperial Observatory had a Muslim Division.

[28] It is worth noting that it is because Xu had not been brought up and trained as an astronomer in the Chinese tradition, that he did not find the solid-sphere view problematic.

paid, again, to any cultural-essentialist claim that the Chinese were incapable of absorbing foreign scientific ideas due to purported 'cultural obstacles'. Why such a response? Xu Guangqi's tilt towards European science as a result of his conversion to Christianity and coming under the dominance of Ricci is part of the reason. But over and above that, there are several other very good reasons that have little to do with Xu's conversion to Christianity. Firstly, ancient Chinese astronomy had had a long tradition from the very beginning of making and recording detailed celestial observations, the positive response of astronomers brought up in or familiar with such a tradition to the telescope is in line with this tradition; similarly, ancient Chinese astronomy had had a long tradition of putting competing models to the empirical test and opt for the model that proves to be superior, hence, the reformers' positive response to the Brahe model was also in line with this tradition. Furthermore, as an instrument, the telescope is non-tradition-specific, hence minimizing scholarly-tradition-derived resistance.

The positive response of the Ming reformers can also be appreciated from another angle. In Chap. 3, it was pointed out that according to PD theory, a self-reinforcing-sequence development does not change track except under huge external impact. According to Kuhn's notion of paradigm, when anomalies that can no longer be handled by means of ad hoc modifications continue to pile up, the paradigm falls into crisis. With reference to this, it was pointed out that such a crisis, which constitutes a huge external impact, is much more likely to occur in empirical sciences than in other disciplines such as philosophy. Astronomy is an empirical science in that astronomical models constantly face the test of empirical observations in competition with competing models. The predictive superiority of the Brahe model constituted a huge external impact on a self-reinforcing-sequence development. Hence, from the perspective of PD theory, the reformers' positive response is also readily explicable.

Ming's astronomical reforms, however, soon faded. This occurred under another huge external impact, which constituted a contingent event in relation to intellectual development, namely, intense political turmoil that was to lead within ten years to the dynasty's demise. Emperor Chong Zhen, who sponsored the astronomical reforms, came to the throne in 1927. Precisely in that same year, widespread starvation occurred in Shaanxi province, which led to a huge peasant uprising; similar uprisings subsequently occurred across the whole country, which increasingly posed a menace to the country's socio-political stability. In 1635, a year after reformers had constructed the first telescopes for observation in China, uprising leaders from all over the country convened a meeting in an attempt to unify their struggles against the Ming. Shortly thereafter, rebels led by one of them seized Fengyang county in Anhui province and burnt down the tomb built by the founder of the Ming Dynasty for his parents located there. In Chinese culture, this was a most disrespectful act, which signalled the rebels' objective of overthrowing the dynasty. At the same time, the nomadic Manchus to the northeast, under the name of the state of Great Jin (*Dajin* 大金, known in history as *Houjin*后金), had already begun an initially incremental invasion of China in 1618, and the Ming's internal turmoil provided them with a golden opportunity to accelerate their invasion. In 1644, as rebels converged upon the capital Beijing, with the assistance of a Ming military officer, the Manchus quickly did the same. Faced with a hopeless situation, Emperor Chong Zhen took his own life, thereby ending the dynasty's rule of almost three hundred years.

The Ming was succeeded, not by the rebels, but by the Manchus who established the Qing Dynasty in 1644, which was to last until 1911. As previously mentioned (note 2 above), many cultural-essentialist claims about China are made by making reference to what occurred under the Qing. This is most unfortunate and misleading. The Manchus originated from the amalgamation and reconstitution of various Jurchen (女真) tribes to the northeast of China in the early seventeenth-century, whereupon the Manchu leaders organized Manchurian society into administrative divisions known as the Eight Banners (八旗), which functioned as armies in wartime. As the Manchus progressively conquered the northeastern part of the Ming, significant numbers of Ming generals and soldiers surrendered to the Manchus. The Manchus organized them into the Han Eight Banners to correspond to the Manchu Eight Banners. The elites of the Han Eight Banners were further recruited into the Manchu Eight Banners and given Manchu names, who were required to learn and speak the Manchu language, and so on. At the time of the overthrow of the Ming in 1644, Manchu Eight-Banner soldiers numbered only 300,000; by the mid-Qing Dynasty well over a century later, the total number of Manchus living in China, including those belonging to the Han Eight Banners, still amounted to only three odd million (Fang Meimeng 2011, p. 140). This was in a country with a population of some 300 million. Hence, from beginning to end, the Manchu ruling elites were extremely wary of the possibility of rebellion. The following are a few illustrations of this.

The Manchus instituted an apartheid-like separation between the Eight-Banner population (inclusive of the Han Eight Banners) and the rest of the population, with the former under the administration of the Eight-Banner system instead of the geographically based governmental administrative system, though this separation progressively degenerated over the course of the dynasty (Sun Xue 2016, p. 51). A campaign of literary persecution (*wenzi yu*文字狱) began right from the beginning of the dynasty that lasted for some one-and-a-half centuries (fifty-six percent of the entire duration of the Qing), in which individuals implicated in real or suspected cases of the expression of pro-Ming or pro-Han and anti-Manchu views, thoughts or sentiments were imprisoned, exiled, or executed. Literally from the very first days of the overthrow of the Ming until the dynasty's demise, all non-Manchu Chinese males were compulsorily made to wear the Manchu pig-tail and Manchu-styled attire, not only to symbolize their submission, but also as a means of cultural assassination. This extreme wariness of the danger of rebellion made the ethos of the Manchu rulers and their courts exceptionally conservative, with considerations and objectives very different from those of a Han-Chinese dynasty. Furthermore, while some individual Manchu emperors were highly appreciative of Chinese culture, with one of them, Emperor Qianlong (乾隆 reigned 1735–1796), the one who met Britain's Lord McCartney on the latter's mission to China in 1793, being quite adept in Chinese poetry and calligraphy,[29] the Manchu leaders, nonetheless, hailed from a very different culture. The result of these and other factors can be seen by comparing how the Qing government treated astronomical reform to the reforms under Ming's last emperor.

The newly established Qing government did not appoint Chinese scholar-officials to head the Imperial Bureau of Astronomy, but appointed the above-mentioned Schall, on whom the Ming had in 1638 bestowed with the official title of Mandarin, to the bureau's directorship.[30] Schall shortened the *Chong Zhen Treatise on Astronomical Science* compiled by Xu Guangqi and his colleagues, and renamed it *Treatise on Astronomical Science According to the New Western Methods* (*Xiyang xinfa lishu* 《西洋新法历书》), on the basis of which the *Shi Xian* (时宪) Calendar was devised. For the entire twenty-year duration of his directorship, Schall

[29] It is thanks to his love of Chinese culture that many ancient Chinese cultural artefacts, which he voluptuously collected, survived and have since been passed down to the present day.

[30] The following account is based upon Chen Cheng-I (1996, pp. 231–41).

kept astronomical development in China at the level attained by Xu Guangqi and his colleagues thirty years ago.

The appointment of Schall was not well received by Chinese scholar-officials. In 1662, Emperor Kangxi (康熙reigned 1622–1722) succeeded to the throne at the age of ten, with ruling power held by four Manchu regents. In 1664, the Chinese court scholar-official Yang Guangxian (杨光先), who possessed no expertise in astronomy but was opposed to European astronomy,[31] with the support of the four regents, accused (justifiably) Schall of plagiarizing from the *Chong Zhen Treatise*, and succeeded to remove him from office. Yang was appointed to lead the Imperial Bureau of Astronomy, since he possessed no expertise in astronomy, he relied on the Muslim Division of the bureau and replaced the *Shi Xian* Calendar devised by Schall with the Muslim Calendar. In 1668, Kangxi assumed the duty of emperor at the age of fifteen. Schall's assistant Ferdinand Verbiest, who had taught Kangxi mathematics and astronomy, took this opportunity to complain to Kangxi that Yang's calendar was inaccurate. Kangxi decided to end the power struggle between Verbiest and Yang by ordering a test between the Muslim Calendar and the *Shi Xian* Calendar; since the latter was devised on the basis, ultimately, of the *Chong Zhen Treatise* compiled by Xu Guangqi and his colleagues, the outcome was left in no doubt. Kangxi removed Yang and put Verbiest in charge of the Imperial Bureau.

The following year, Kangxi commissioned Verbiest to refit the Imperial Observatory. This project took seven years to complete, whereupon Verbiest wrote the *Records of the Observatory Instruments* (*Ling tai yixiang zhi* 《灵台仪象志》, which includes a table of star measurements supposedly obtained with the help of the newly constructed instruments. As a physics professor teaching in America, Chen Cheng-I (1996, pp. 233, 234, our emphasis) comments on the new instruments and the star measurements as follows:

> In retrospect, one cannot but be puzzled by the questionable selections and designs of the instruments used in equipping the Imperial Observatory. All the instruments were essentially traditional instruments relying on naked-eye observations. Except for the sextant, all the instruments were previously

[31] Yang's opposition cannot be put down to any purported 'cultural obstacles', since there were Chinese scholar-officials who accepted European astronomy, an acceptance that, as shown, can be satisfactorily explained even without reference to Xu's conversion to Christianity.

> known in China...An obvious deficiency of the newly constructed instruments was the absence of optical aids, since none was equipped with an optical device and *the telescope was totally absent.*
>
> ...the measurements listed in the Verbiest's book were not new measurements, but based on those of the Ming astronomers first published in 1634 [thirty-nine years ago]...

Chen notes that while Verbiest had certainly received an education and instruction in science and astronomy while still in Europe, he was neither an astronomer nor instrument specialist. Verbiest was succeeded by another missionary Bernard Stumpf in 1715. Stumpf proposed to construct a bronze quadrant and destroyed the Yuan and Ming instruments stored at the Imperial Observatory. Chen comments that Stumpf's quadrant was 'clumsy in design' and 'less valuable than' the 'simplified instrument' (*Jian yi*简仪), the world's first known instrument with an equatorial mounting, constructed by the previously mentioned Guo Shoujing in the thirteenth century which Stumpf had destroyed. For 182 years from the founding of the Qing Dynasty until 1826 (which account for *sixty-eight percent of the entire duration of the Qing*), the Imperial Bureau of Astronomy was headed by Jesuit missionaries except for the non-astronomer Yang Guangxian; and it was only towards the end of his reign that Kangxi invited Chinese astronomers outside of the bureau to participate in an upgrade project. The attitude towards astronomical reform of the Qing government stood in stark contrast to that of the Ming under Emperor Chong Zhen.

Let us now consider Huff's argument. He begins with a discussion of the arrival in China of Ricci and other Jesuit missionaries, and the eulogization of European learning by Xu Guangqi and other Chinese scholars. He describes Xu as a '*brilliant* Chinese scholar' (p. 74, our emphasis)[32] and 'an *exceedingly able* astronomer and mathematician' (p. 90, our emphasis); he also mentions one Wang Zheng (王征whom he renders as Wang Cheng) in the following way: 'In the following year, 1627, the convert known as Dr. Philip (Wang Cheng, 1571-1644), working in collaboration with [Johannes] Schreck, published an impressive little book called *Diagrams and Explanation of the Marvelous Devices of the Far West.* Dr. Philip was another *exceptional* scholar who had passed his Metropolitan Examination in 1622...' (pp. 85–6, our emphasis). The descriptions of

[32] References giving only page number(s) hereinafter refer to Huff (2010).

'brilliant', 'exceedingly able', and 'exceptional' are, of course, intended to imply that the superiority of European learning could immediately be recognized by the crème de la crème of the Chinese intelligentsia. There are, however, some problems with this, to put it straightforwardly, little literary trick. As seen, it took Xu Guangqi seventeen years (from 1580 to 1597) to proceed from passing the first level of the civil service examinations to passing the provincial-level examination, and another seven years to achieve *jinshi* status, which qualified him to become a capital official; that hardly qualifies him as a 'brilliant' scholar. Similarly, Wang Zheng passed the provincial-level examination in 1594, but failed the metropolitan examination a total of nine times before finally passing it in 1622 at the ripe age of fifty-two (Kim 2020), which hardly qualifies him as an 'exceptional scholar'. Both Xu and Wang suffered prolonged frustration as also-rans in the civil service examination system.

Checking the index of Joseph (2011), which provides, inter alia, a lengthy discussion of the history of Chinese mathematics, one finds no entry for Xu Guangqi; similarly, checking the index of Martzloff's *A History of Chinese Mathematics* (1997), one finds that all references to Xu concern his translation in conjunction with Ricci of Euclid's *Elements*. As we have shown above, Xu is a non-entity in the history of Chinese mathematics. One wonders how Huff is able to describe him as 'an exceedingly able mathematician'. As to 'an exceedingly able astronomer', Xu's role in the history of Chinese astronomy is solely in relation to Ming's astronomical reform from 1629 until his passing away in 1633. As far as the development of China's indigenous astronomy is concerned, he is again a non-entity, *which is why*, unlike astronomers raised in the indigenous tradition, he did not, as already observed (see note 28), find the Jesuits' solid-sphere model problematic. Incidentally, there is also a problem with the English translation of the title (*Diagrams and Explanation of the Marvelous Devices of the Far West*) of the joint-work of Wang Zheng and Schrek as mentioned by Huff. The Chinese title of this book is 《远西奇器图说录最》 (*Yuanxi qiqi tushuo luzui*), which is more accurately translated as 'Selected Records of the Diagrams and Explanations of Strange Machines from the Far West' (Kim, *op.cit.*). The term translated as 'Marvelous Devices' by Huff is *qiqi*, which means strange devices.

In a section titled 'Platform for a Scientific Takeoff?', Huff states:

> the original accomplishment of 1631-34. Given all these innovations and aids to understanding the new astronomy, along with the telescopic

> discovery machine, one might have expected the presumed excellence of the Chinese in science[33] to have propelled them forward in the 17th century, making significant discoveries and laying the foundations for a unified celestial and terrestrial physics. One might also have expected them to make innovations with regard to the telescope such as the invention of longer and more powerful telescopes produced by Europeans from the 1640s onward. Some Chinese scholars [referring to scholars of the Qing period] took a great interest in the new Western mathematics and astronomy, as historians have reported. Some minor celestial discoveries were made, but no significant astronomical innovations were forthcoming, nor were any innovations made regarding the telescope. The main drift of the scholars close to the emperor was toward severely restricting the use of the new natural studies.... (pp. 97-8)

And in the concluding section titled 'Taking stock', he states:

> all studies to this date indicate that the Chinese made no improvements on the telescope: they did not build larger, more powerful telescopes comparable to those in Europe...Even the most gifted mathematician of the late seventeenth century, Mei Wending (1633-1721), had difficulty accepting the new system...Adopting revolutionary scientific ideas was too much to ask...Chinese astronomers in the seventeenth century did not undertake such a systematic study of the sky as was carried out by Flamsteed...There was what may be called a *curiosity deficit* [Huff's emphasis], with the result that nothing extraordinary was undertaken or fulfilled in astronomy by the Chinese in this period...The arrival of the telescope in China did not have the fructifying effect on scientific inquiry that it had on astronomy in Europe. (pp. 109–13)

The above statements provide an excellent basis on which to assess the merit of Huff's thesis. Firstly, it will be recalled from our quote from Huff with which we began the present section that he argues that the invention of the telescope is 'linked to a burst of creativity and scientific curiosity that flowered across Europe in the seventeenth century', and he contrasts this with 'the failure of the telescope to trigger an exciting new burst of scientific creativity, especially in astronomy, around the world'. He believes that this contrast proves his cultural-essentialist thesis, for it 'serves best to highlight a deficit in scientific curiosity that seems to have prevailed outside Europe...That is an extraordinary record of *cultural disparity*' (our

[33] This sarcastic remark seems to be aimed at Needham.

emphasis). With that in mind, let us return to Huff's above first quote, in which he (correctly) draws a distinction between, on the one hand, 'the original accomplishment of 1631-34', i.e. the astronomical reform originally led by Xu Guangqi ending with the construction of the telescope in 1634 by Xu's successor Li Tianjing, and, on the other, 'the scholars close to the emperor' in the subsequent period. Now, if it were a question of 'cultural disparity', then, China's 'deficit in scientific curiosity' should apply *generally* to all scholars, in which case how can one account for the positive response to the telescope on the part of the reformers of 1629 to 1634? This is especially the case were the conventional narrative, as exemplified by Jacques Gernet (see above), that Xu Guangqi and others were converted to Christianity *as a result of* their recognition of the superiority of European learning true.[34] In praising, for the reason pointed out above, these reformers with misleading superlatives, does Huff realize that he is cancelling out his own argument of 'cultural disparity'?

Further, as Huff himself is aware of, after 1634, Ming's astronomical reform, as we pointed out above, rapidly faded, hence, Huff's 'scholars close to the emperor' whose 'main drift' 'was toward severely restricting the use of the new natural studies' could only be referring to scholars of the Imperial Bureau of Astronomy under the Qing Dynasty. As discussed in the above, from 1644 with the appointment by the Qing government of Schall as director of the Imperial Bureau onwards, for 182 years, that bureau was headed by Jesuit missionaries, except for the period of four years from 1664 to 1668, when the non-astronomer Yang Guangxian succeeded to remove Schall from office to be replaced by himself. Hence, the 'scholars close to the emperor' whom Huff refers to can only, for 178 years, be the Jesuit directors of the Imperial Bureau, plus, for four brief years, Yang. Yang was a non-astronomer, hence, can his failure be construed as due to a 'deficit in scientific curiosity' on the part of Chinese scholars *in general*? Can the failure to 'make innovations with regard to the telescope' under the 178-year-long directorship of the Jesuits be construed as due to such a 'deficit' *on the part of Chinese scholars*? These questions are so obvious that only someone who begins with the a priori premise of 'cultural disparity' can fail to see.

According to Huff, given that the Chinese had been equipped with updated knowledges (with the exceptions mentioned in note 26 above) and the associated scientific paraphernalia, 'one would have expected'

[34] There is no indication in Huff's entire discussion that he objects to this narrative.

them to make the same progress as that achieved by Europe in the seventeenth century, and since they did not, it could only be due to a 'deficit in scientific curiosity'. Along the same line of thinking, Huff remarks that 'adopting revolutionary scientific ideas was too much to ask', even of the 'most gifted' Chinese scholars such as the mathematician Mei Wending mentioned by Huff. In accordance with the superlatives in which Huff praises Xu Guangqi and his colleagues of the 1629–1634 reform, they must also be among the 'most gifted' of Chinese scholars, was it, then, 'too much to ask of' them to be receptive to the 'revolutionary scientific ideas' of the Brahe model and so on? But besides cancelling out his own argument at every turn, what is even more important is that Huff's 'one would have expected' and 'too much to ask' arguments betray an extraordinarily simple-minded conception of intellectual development. In this conception, scholars everywhere are supposed to be like blank calculators, input 1 plus 1 into them, 'one would have expected' that they all give the sum of 2. This is supposed to constitute a 'comparative approach' from a 'global perspective', in which what 'one would have expected' is completely abstracted away from the concrete circumstances of how a specific development in a particular society is subject to multiple determinations by a variety of factors (both internal externa) and events, both synchronically and diachronically. Nor is there any recognition that from the sixteenth century onwards, Europe's pursuit of knowledge and the scholarly agency underpinning it, instead of being driven forward by any 'intellectual curiosity' in the abstract, was inextricably bound up with the imperial enterprise, as we show in detail in the next chapter.

Bibliography

Brook, Timonty. 2001. 'Xu Guangqi in his context'. Chapter 2 of Jami, Catherine, Peter Engelfriet, and Gregory Blue (eds.). *Statecraft and Intellectual Renewal in Late Ming China: The Cross-Cultural Synthesis of Xu Guangqi (1562-1633)*. Leiden: Brill.

Chen Cheng-I. 1996. *Early Chinese Work in Natural Science*. Hong Kong: University of Hong Kong Press.

Chen Meidong. 2008. *Ancient Chinese Astronomical Thoughts*. 陈美东, 《中国古代天文学思想》(北京:中国科学技术出版社, 2008).

Cohen, H. Floris. 1994. *The Scientific Revolution: A Historiographical Inquiry*. University of Chicago Press.

Cohen, H. Floris. 2012. 'From west to east, from east to west? Early science between civilizations', book review of Arun Bala, *The Dialogue of Civilizations in the Birth of Modern Science*, Avner Ben-Zaken, *Cross-Cultural Scientific Exchanges in the Eastern Mediterranean*, Sylvain Gouguenheim, *Aristote au Mont-Saint-Michel. Les racines grrcques de l'Europe Chretienne*, Toby Huff, *Intellectual Curiosity and the Scientific Revolution. A Global Perspective*, and Jonathan Lyons, *The House of Wisdom. How the Arabs Transformed Western Civilization*'. *Early Science and Medicine* 17, pp 339-350.

Cohen, H. Floris. 2024. (in press) 'Dialogue and comparison compared'. Chapter 18 of Arun Bala, Raymond W.K. Lau, and Mei Jianjun (eds.) *Multicivilizational Exchanges in the Making of Modern Science: Needham's Dialogical Vision.* Palgrave Macmillan.

Elman, Benjamin A. 2000. *A Cultural History of Civil Examinations in Late Imperial China*. Berkeley: University of California Press.

Etymology of Words. 1987. Combined edition compiled by the Editorial Department of the Commercial Press, China. 商务印书馆编辑部，《辞源(缩印合订本)》(香港:商务印书馆香港分馆，1987)。

Fang Meimeng. 2011. 'The phenomenon of "the Manchurization of the Hans" under the Eight Banner System'. 方妹孟，'八旗制度下的"汉人满化"现象'，《历史研究》，2011， 页139-140.

Gernet, Jacques. 1985. *China and the Christian Impact: A Conflict of Cultures*, trans. Janel Lloyd. New York: Cambridge University Press.

Hart, Roger. 2013. *Imagined Civilizations: China, the West, and Their First Encounter*. Baltimore: John Hopkins University Press.

Hart, Roger. 2024 (in press). 'Ancient Chinese origins of modern western science; or, the early history of linear algebra'. Chapter 8 of Arun Bala, Raymond W.K. Lau and Mei Jianjun (eds.), *Multicivilizational Exchanges in the Making of Modern Science. Needham's Dialogical Vision*. Palgrave Macmillan.

Hobson, John M. 2006. *The Eastern Origins of Western Civilisation*. 5th printing with corrections. Cambridge: Cambridge University Press.

Huff, Toby. 2010. *Intellectual Curiosity and the Scientific Revolution: A Global Perspective*. Cambridge University Press.

Joseph, George G. 2011. *The Crest of the Peacock Non-European Roots of Mathematics*, 3rd edition. Princeton: Princeton University Press.

Kim Yung Sik. 2020. 'A philosophy of machines and mechanics in seventeenth-century China: Wang Zheng's characterization and justification of the study of machines and mechanics in the Qiqi tushuo'. *East Asian Science, Technology, and Medicine* 31, pp 64-95.

Kvasz, Ladislav. 2015. 'Mathematical language as a bridge between conceptualization of motion and experimental practice'. Pp. 229-48 in Raffaele Pisano (ed.) *A Bridge Between Conceptual Frameworks. Sciences, Society and Technology Studies*. Dordrecht: Springer.

Lao Siguang. 1983. *New Edition History of Chinese Philosophy*, Volume 3(1). 勞思光, 《新編中國哲學史(三【上】)》(臺北: 三民書局, 1983).

Lau, Raymond W.K. 2020. *Intellectual Developments in Greece and China: Contingency, Institutionalization and Path-Dependency*. Newcastle upon Tyne: Cambridge Scholars Publishing.

Martzloff, Jean-Cluade. 1997. *A History of Chinese Mathematics*. Translated from the French original of 1987 by Stephen S. Wilson. Berlin: Springer.

Needham, Joseph, and Kenneth Robinson. 1962. *Science and Civilization in China, Volume 4. Physics and Physical Technology, Part 1. Physics*. Cambridge: Cambridge University Press.

Needham, Joseph. 1969. *The Grant Titration*. Taylor and Francis.

Sun Xue. 2016. 'A brief analysis of the changes in the relationship between the Manchus and the Hans with reference to the Qing Dynasty's policies'. 孙雪, '从清代政策简析满汉关系变化', 《边疆经济与文化》, 2017年二期, 页49-51.

CHAPTER 8

Late-Renaissance Scholarly Agency and the Breakthrough into Modern Science

In order to account for the breakthrough into modern science, it is first necessary to ascertain what that breakthrough consists of. Different scholars have given slightly varying answers to this question; as an illustration, the historian of seventeenth-century science Peter Dear (1988) has identified a total of six essential and innovative features that define the new science, including the methodical use of experimentation, the use of mathematics as a tool of disclosing nature (usually referred to as the mathematization of nature), and so on. Given the limited objectives of the present work as mentioned at the start of the Introduction, there is no need for us to consider each and every feature identified by various scholars. Instead, our purposes can be well served by making reference to Cohen (2011) on the issue.

As pointed out in Chap. 4, in his analysis of the rise of modern science, Cohen speaks, firstly, of three 'cultural transplantations', and then of 'three revolutionary transformations' that were responsible for the breakthrough of ca. 1600. In discussing the 'transplantations', he speaks of 'Alexandria' as the code for the mathematical sciences, and 'Athens' as the code for natural philosophy. We have previously (Chap. 3) pointed out that 'cultural transplantation' is a misnomer, but these transplantations do not concern us in the present context. Cohen examines the three 'revolutionary transformations' in chapters that are titled respectively 'The first transformation: Realist-mathematical science', 'The second transformation: A

R. W. K. Lau, *Rethinking the Needham Question*,
https://doi.org/10.1007/978-981-97-9472-0_8

kinetic-corpuscularian philosophy', and 'The third transformation: To find facts through experiment'. Cohen describes the first transformation as transforming Alexandria into 'Alexandria-plus', the second as Athens into 'Athens-plus', and concerning the third, he describes it variously as 'specifically European-colored mode of investigation' (p. xvi), 'culturally marked pieces of nature-knowledge in Renaissance Europe' (p. 114), 'Nature-knowledge colored the European way' (p. 134), and 'the culture-bound European mode' (p. 139). In short, of the three 'revolutionary transformations', the first two were transformations by late-Renaissance scholars of the Greek corpus, while the third is Europe's own 'culture-bound mode' of nature-knowledge. By realist-mathematical science, Cohen refers especially to the breakthroughs obtained by Kepler and Galileo; by kinetic-corpuscularian philosophy, he refers to Descartes' version of atomic theory; and by finding facts through experiment, he refers especially to Bacon. To add to Cohen's schema, it should be noted that there are two aspects to the rise of atomic theory, firstly, the rise of the mechanical philosophy, and, secondly, the mathematization of nature.

Summing up, for our present purposes, we analyse the how and why of the breakthrough into modern science in terms of the following issues: The breakthroughs achieved by Kepler and Galileo; the rise of the mechanical philosophy; the mathematization of nature; and the use of experimentation to test hypotheses about nature. The first issue is covered in the present chapter in the context of our analysis of the relationship between late-Renaissance scholarly agency and the breakthrough of ca. 1600. As part of this analysis, it will be pointed out that Cohen's presentation of the mathematical sciences becoming realist as the transformation of Alexandria into Alexandria-plus is a Eurocentric misrepresentation; it will also be pointed out why a critique of Cohen in this regard from the dialogical perspective is insufficient, and hence it is necessary to transcend dialogism in critiquing Eurocentrism. The last three will be covered in the next chapter, where they are discussed in the context of our overall critique of Cohen's Eurocentric-cum-neo-Eurocentric thesis concerning the rise of modern science.

8.1 Arabic Scholarship's Trend Towards the Realist Direction

Kepler's and Galileo's turning of the mathematical sciences realist (in the sense of placing equal importance on *both* the mathematical *and* physical aspects in modelling) did not occur out of the blue, there were various preconditions serving as its basis. One such precondition concerns a particular trend of Arabic scholarship which Latin Europe absorbed. The Greek tradition, as it was originally developed, is a purely theoretical tradition, best epitomized, as discussed in Chap. 3, not only in the works of the founders of that tradition, namely, Parmenides and the early Pythagoreans, but also in Eudoxus' founding of Greek astronomical modelling. The Greek tradition was unique, all other ancient civilizations (Babylonian, Indian, and Chinese) developed what may be called a pragmaticist tradition, which took the form of a computational tradition with regard to mathematics. The original Greek tradition was enriched during the Hellenistic era. As also mentioned in Chap. 3, Ptolemy lived in Egypt where there was a strong presence of Babylon's empirical-observational astronomical tradition, as a result of which, while he upheld Eudoxus' project of maintaining the a priori principle of 'perfect' circular celestial motion, he had to inject into his model empirical operational capabilities. The close proximity between Greece, Egypt, and Babylon was a contingent coincidence that formed the background of the above. When we consider the rise of Arabic scholarship, we find a similar contingent coincidence having a crucial effect.

The just-mentioned *contingent coincidence*, which is of the nature of an *external factor*, refers to the location of the Islamic world of the Middle Ages in between Byzantine and India, in close proximity to both. In consequence, the rise of secular learning in the Arabic world did not only involve the adoption of the Greek corpus, but also the diffusion of Hindu mathematics from India. Beginning tentatively under the Umayyads (Saliba 2007, Chap. 2), the taking up of the Greek corpus was at full steam from the late eighth century onwards under the Abbasids. In tandem with this, written evidence suggests that Indian numerals possibly spread to the Mesopotamian region as early as in the mid-seventh century; in any case, around 825, al-Khwarizmi wrote his famous *Book of Addition and Subtraction according to the Indian Calculation*, the first text to deal with the new (to Arabic mathematicians) numerals (Joseph 2011, p. 462). As is generally known, al-Khwarizmi's work was one of the main sources through which Indian numerals and Arabic algebra spread to Europe; and the terms 'algebra' and 'algorithm' in Europe's mathematical lexicon came from the titles of two of his works translated into Latin. The subsequent absorption of Hindu mathematics by Arabic scholars since al-Khwarizmi is well-known, and need not be retold here.

The importance that the Arabs imported not only the Greek corpus, embodying a theoretical tradition, but also Hindu mathematics, embodying a computational tradition, is that in inheriting the Greek theoretical tradition, from very early on, Arabic scholarship, thanks to the influence of India's pragmaticist tradition, showed a strong tendency towards the realist direction. In the following, we discuss this trend with reference to the optical theory of Ibn al-Haytham (Alhacen, ca. 965-ca. 1040), and the revision of Ptolemy's model by Arabic astronomers that culminated in the work of al-Shāṭir. Since, both cases are well known, our account will be brief.

In Greek optical theories, Aristotle and the atomists were concerned with the physical side of vision, and argued for an 'intromission' view, i.e. the agent of vision emanates from the object to the eye.[1] On the other hand, coming after Aristotle and the atomists, Euclid was concerned only with the mathematical side, developing a geometrical theory of perception based on the visual cone. According to this theory, radiation emanates from the eye in the form of a cone; perception occurs when the rays within the cone are intercepted by an opaque object. Euclid's 'extromission' view was shared by Ptolemy.

Taking up a development begun by al-Kindi in the ninth century, Ibn al-Haytham critiqued the extromission view as being physically implausible or even impossible. For instance, he pointed out that it is impossible for bright objects to injure the eye in accordance with the extromission view, and that that view assumed that in viewing the heavens, the rays emanating from the eye were able to fill up all space up to the fixed stars, which was implausible to the extreme. However, al-Haytham realized the power of the mathematical model of the visual cone. Hence, what he did was to integrate the physically much more satisfactory intromission view with the mathematical tool of the extromission view. Briefly, he argued that rays emanating from external objects form a visual cone sensed by the crystalline humour or lens of the eye (this physiological idea of the eye came from Galen), with the visual field as base and the centre of the eye as

[1] The following account is based upon Lindberg (2007, pp. 313–20).

vertex, thereby 'unit[ing] the mathematical and physical approaches to vision in a single theory' (Lindberg 2007, p. 316).

Al-Haytham's *Optics* was translated into Latin in the late 12th or early 13th centuries, exerting a powerful influence in Europe. Roger Bacon in the 1260s took up his theory, and it was through his influence 'that Alhacen's optical theories, including his combined physical-mathematical-physiological approach, became dominant in Western thought...When Johannes Kepler began to think about visual theory in the year 1600 (...), he took up the problem where Bacon, Pecham, Witelo [the latter two being influenced by al-Haytham through Bacon], and Alhacen had left it' (ibid.: 320). Thus, unlike Eudoxus, Euclid, and Ptolemy who, in accordance with Greece's classical theoretical tradition, were not concerned with whether or not their mathematical models were physically possible, al-Haytham, while accepting the power of mathematical models, stressed that they must at the same time be physically possible as well. This is the realist trend of Arabic scholarship that we refer to, a trend that can also be observed in its approach to astronomical modelling.

In maintaining the a priori premise of 'perfect' circular celestial motion, which was flatly contradicted by the observed irregularities of the movements of the visible planets, Ptolemy advanced three models to, in his phrase, 'save the phenomena', namely, the eccentric model, the equant model, and the epicycle-on-deferent model. All these models lack physical possibility, with the equant model being the most implausible. By the eleventh century, from Al-Andalus in the far west to Bukhara in the far east of the Islamic world, and from specialist astronomers to philosophers, 'the physical absurdity of proposing that a physical sphere could move uniformly, in place, around an axis which did not pass through its center' (the equant model) was generally known and discussed. 'This absurdity permeated almost all of the models, which were proposed in Ptolemy's *Almagest*', though the equant problem was the most notorious (Saliba 2007, p. 95).[2]

In this critique of Ptolemy's astronomical models, al-Haytham was, again, at the forefront. Critiquing Ptolemy's model of the moon, al-Haytham remarked: 'The epicycle diameter is an imaginary line, and an imaginary line does not move by itself in any perceptible fashion that produces an existing entity in this world...no motion exists in this world in any perceptible fashion except the motion of [real] bodies' (quoted on

[2] The following is based upon this source, references to which are given in page number(s) only for simplicity.

pp. 97–8). As is well known, the objective of Arabic astronomers upon realizing the physical impossibility of Ptolemy's models was to preserve the power of mathematical models but make them physically more plausible. In the course of this development, the astronomer Naṣīr al-Dīn al-Ṭūsī devised a new theorem known in English as the Ṭūsī Couple, which successfully combines two circular motions into a straight motion. Shortly afterwards, the Damascus astronomer Mu'ayyad al-Dīn al-'Urḍī established a new theorem known in English as Urḍī's Lemma with reference to Ptolemy's equant model. In the fourteenth-century, Ibn al-Shāṭir, another astronomer from Damascus, successfully converted the eccentric and equant models to the epicycle-on-deferent model on the basis of the Ṭūsī Couple and Urḍī's Lemma. Further, by means of modifying Aristotle's physical model, he successfully made the epicycle-on-deferent model physically plausible.[3] As a result, he was able to achieve the same level of predictive power as Ptolemy but without the latter's eccentric and equant models. This is a major turning point in the history of Arabic and European astronomy, because for the first time, a mathematical model that possesses substantial predictive power exists which is physically plausible. As is now well-known, the models of the sun, the moon, and the five visible planets that Copernicus put forth in the early sixteenth century were all plagiarized from al-Shāṭir. Of the four theorems used by Copernicus in these models, two are from Greece, the other two are none other than the Ṭūsī Couple and Urḍī's Lemma.

To sum up, for reasons already explained, there existed a strong trend towards the realist direction in Arabic scholarship in its development of the Greek corpus to new heights; *this realist trend, which was embodied in the Arabic works concerned*, was diffused to Latin Europe as the latter absorbed Arabic learning. What Latin Europe took up in the mathematical sciences was *not* the original, purely theoretical Greek tradition or even the Alexandrian tradition of Ptolemy and others, but *the Arabic tradition embodying this trend*. Kepler's and Galileo's breakthroughs cannot be understood without reference to this. In the next section, we examine the other preconditions underpinning Kepler's and Galileo's breakthroughs.

[3] Incidentally, Bala (2006, Chap. 7) points out that these Arabic innovations were enabled by Indian trigonometry that had diffused to the Arabic world.

8.2 Rivalry with Islam, Imperial Conquest, the Reformation, and Late-Renaissance Scholarly Agency

Latin Europe's inheritance of the realist trend embodied in Arabic scholarship constitutes an *internal factor* in accounting for the breakthrough of ca. 1600. There is another indispensable *internal factor*, namely, what we refer to as late-Renaissance scholarly agency among scholars such as Kepler and Galileo; but to account for this internal factor, it is necessary to consider *a particular synthesis of various external factors*, which, in relation to intellectual development, are all *contingent* in nature.

The fall of Constantinople in 1453 was a severe shock to Europe and a major turning point in history with immense *unforeseen consequences* in more than one way. Firstly, Aristotelianism was absolutely dominant and without rival in Latin Europe from 1200 to 1450 (Grant 1996, p. 162), but Greek scholars fleeing to western Europe upon the advance of the Ottomans brought with them Greek works on Neoplatonism, Hermeticism, atomism, Archimedean mathematics and so on, which led to a second translation movement and the end of Aristotelianism's monopoly, though it was not until the latter seventeenth century that Aristotelianism was completely dethroned (Grant ibid.: 203). Both Kuhn (1957, pp. 128–31) and Yates (1979 [1964], pp. 153–4) have suggested that Copernicus's heliocentrism was inspired by Hermetic sun worship; and Webster (1980, p. 2) notes 'The self-evident impact of Neoplatonism in England after 1660'. Aristotelianism was undermined not only because of the existence of newly introduced rivals, but also for other reasons to be discussed shortly below.

Since antiquity, Europe had had great demand for Asian products such as spices and silk. 'Though Europe had traded with East Asia for centuries, it had always been done through intermediaries' (Pomeranz and Topik 2006, p. 19). Since the rise of Islam, these intermediaries were none other than the Muslims, both in the northern route through west Asia, and in the southern route through Egypt, which, since the mid-thirteenth century, had been under the Islamic Mamluks. We have previously noted (see end of Chap. 3) Latin Europe's rivalry with and hatred of Islam ever since Islam expanded into Europe. After the fall of Constantinople, Athens fell in 1456, by 1481, the entire Balkans was under the Ottomans. It was this that made Latin Europe determined to find a way to bypass Muslim intermediaries in its trade with Asia, especially in spices, eventually leading to the so-called 'voyages of discovery' beginning in 1492 (Keay 2005).

We do not know whether or not any Foucauldian genealogical analysis has previously been done on how the narrative of 'explorers' such as da Gama, driven by 'curiosity', set out on these journeys of 'discovery' was constructed; if not, it would constitute an interesting project. In any case, Europe knew well about Asia as the source of spices, which did not need to be 'discovered'; the Arabs had rounded the Cape, sailed up Africa's west coast and entered the Mediterranean through Gibraltar in the mid-fifteenth century (Hobson 2006, p. 138), hence da Gama did not discover any new route; the indigenous peoples of the Americas, migrating from Asia, had discovered those continents long before Columbus; the only real discovery was the Atlantic sea-crossing which, however, was made by accident. Columbus and so on were not 'explorers' driven by 'curiosity', their journeys were commercial expeditions financed by private investors as well as royal houses and advocated by the Pope (see Hobson 2006, pp. 136–7 on the papal bulls concerned) with one sole objective: to grab the spice trade, by force if required, from the hated enemy, i.e. Muslims traders. They were, indeed, adventurers, but in the just-mentioned sense, not in the sense of curious explorers exploring for its own sake. This is starkly illustrated in one of the two charters that Charles II granted in 1660, to be referred to shortly below, one of which established the Company of Royal Adventurers Trading into Africa which principally traded in slaves—those who engaged in this violent and inglorious trade were called 'adventurers' in the company's name.

Before the Europeans came to Asia, trade along the maritime route from China to west Asia and Africa's east coast conducted by Islamic, Indian, and Chinese merchants had always been on strictly commercial terms, based on 'the principle—common throughout the region—that the sea belonged to no one' (Pomeranz and Topik ibid., p. 17). The Europeans overthrew that principle because they came to Asia with the above-mentioned predatory motive and mentality: 'The Portuguese government was the first to attack the [above] principle', 'Within twenty years of sailing into Asian waters, they created forts at two of the three places where major west-bound trade routes could be blocked…They claimed a monopoly in the pepper trade and the right to board or sink any ship in the hemisphere to which they had not issued a pass' (ibid., pp. 17–8). Whereas the Europeans came to Asia originally for spices, they soon discovered that they could colonize in order to produce other products that

Europe lacked but demanded, and for which they could not exchange on commercial terms since there was no Asian demand for what Europe produced (ibid., pp. 10–1), and in the mentality with which they came, they proceeded to do so forthwith. The above is not political correctness, but historical fact. Finally, it should be added that the motive and mentality of the Europeans did not constitute any 'essence' or 'uniquely European quality' on their part, because these were engendered by their long-standing rivalry with and hatred of Islam under *a synthesis of the contingent circumstances* of the Ottomans' advance into Europe, their great demand for Asian produce, and the lack of Asian demand for European products.

The 'voyages of discovery' and colonial conquests consequent upon them have tremendous bearing on our present topic in several ways. We consider one threefold issue here, and another issue in the following chapter. Grant (1996, p. 167) notes: 'The discovery [sic] of the New World by Columbus in 1492 destroyed the ancient and medieval view of the earth. It made European scholars realize that Aristotle's knowledge of the earth, and that of the ancients generally, was not only greatly limited but also often flatly wrong.' Similarly, Hooykaas points out that reports of Portuguese travellers[4] revealed many empirical facts that contradicted ancient Greek theory, such as Aristotle's denial that the tropics could be inhabited, and Ptolemy's mathematically derived conviction that all dry land was confined to part of the northern hemisphere (Cohen 1994, pp. 354–7). The following provides some details.

The Spaniard living in Florence Amerigo Vespucci claimed to have made four voyages to the Americas between 1497 and 1504,[5] whereas the veracity of some of his claims has been disputed, there is no dispute with regard to his trip of 1499, in which he joined an expedition licensed by Spain, led by Alonso de Ojeda as fleet commander and Juan de la Cosa as chief navigator, an expedition that was partly financed by Vespucci himself. In a letter to a friend on this trip, Vespucci reported seeing all kinds of incredible animals and birds previously unknown to Europe, 'He concluded with a damning criticism of Pliny's *Natural History*, the traditional authority on the subject. As Vespucci noted, "Pliny did not touch upon a thousandth part of the species of parrots and other birds and animals"' (Poskett 2023, pp. 16–7). In relation to Ptolemy's above-mentioned

[4] Hooykaas mentions Portuguese travellers, as is to be seen presently, the same applied, to a much greater extent, to Spanish travellers.

[5] The naming of the 'New World' as America was due to none other than him.

mathematically derived claim, Vespucci wrote after returning from Brazil in 1503: 'The opinion of the ancients was, that the greater part of the world beyond the equinoctial line to the south was not land, but only sea', 'But this opinion is false, and entirely opposed to the truth' (quoted in ibid., p. 38). In relation to Aristotle's claim that neither the 'torrid zone' nor the 'frigid zone' were habitable by humans, the Spanish Jesuit missionary Jose de Acosta, who went to South America in 1571 and spent fifteen years there in search of converts, wrote: 'I must confess I laughed and jeered at Aristotle's meteorological theories and his philosophy, seeing that in the very place where, according to his rules, everything must be burning or on fire, I and all my companions were cold' (quoted in ibid., p. 18).

The Aztecs had a rich culture of herbal medicine, which greatly impressed the Europeans.[6] The Spanish Franciscan friar Bernardino de Sahagun went to the 'New Spain' (modern-day Mexico) on a missionary journey in 1529, and spent the rest of his life of more than fifty years there, studying Aztec beliefs, culture, and history. With the help of locals and using Aztec taxonomy, he compiled the *General History of the Things of the New Spain* (1578), which contains detailed discussions of Aztec herbal medicine. For Sahagun, this work 'was primarily a religious work…[by means of it,] Sahagun hoped [to] convince Christians back in Europe that the Aztecs were a "civilized" race [there existed contemporaneous views that they were sub-human] capable of receiving the word of God'. The *General History* was bought by Ferdinando de' Medici, Grand Duke of Tuscany and head of the Medici family in 1580, hence the work has since been popularly known as the *Florentine Codex*. Sahagun's own intentions notwithstanding, 'For Ferdinando de' Medici, the *Florentine Codex* was essentially a commercial catalogue: a list of the most valuable natural resources that the New World had to offer'. The importance of the commercial value of 'New World' plants, especially as herbal medicine, for Europe is evident not only in Ferdinando de' Medici's buying of the *General History*, but in the following as well. 'In 1570, Philip II of Spain ordered a major survey of the natural history of the "New World"; at the head of the survey, Philip appointed his personal physician, Francisco Hernandez. Over the next seven years, Hernandez travelled across New Spain, collecting herbs and learning about Aztec medical practices'. In 1545, the University of Padua established the first botanical garden in

[6] This paragraph is based upon, and quotes are taken from, Poskett (2023, pp. 22–3).

Europe to grow 'New World' plants, soon followed in Pisa and Florence, and then all over Europe.

The realization, as a result of what Europeans going to the 'New World' found through first-hand experience, that not only were the ancients' knowledge severely limited, but that they were often flatly wrong, 'forced such scholars as Joao de Castro', in Hooykaas' words quoted by Cohen, 'to acknowledge the superiority of experience over and above rational speculation in the realm of the factual'. 'New World plants…soon came to feature in new works of natural history and medicine [appearing in the 1500s], many of which promoted the value of experience over ancient texts' (Poskett 2023, p. 29). Citing Vogel and Rankin (2016, p. 478), Poskett observes: 'Many began to question the authority of ancient texts, instead emphasizing the importance of collecting evidence and testing ideas against experience' (p. 38). This trend of prioritizing experience over rational speculation or abstract reasoning in European academia is perhaps best epitomized by the Spanish throne's appointment of Juan Lopez de Velasco as Chief Cosmographer at the Council of the Indies; in 1577, he organized a *questionnaire* to be sent out to every Spanish American province, which contained fifty questions on the geography of the various places (pp. 40–1). Noteworthily, Poskett (p. 99) observes that '[Francis] Bacon was reading Spanish accounts of travels to the New World', a point that Hooykaas also underlined.

The threefold effect of the above is: (a) it served to further undermine the authority of Aristotelianism, which was already losing its monopoly as discussed above; (b) by the same token, this served to enhance the *willingness to break with orthodoxy*; and (c) it engendered what can be called an *empirical turn* on the part of some science scholars in contrast to the tradition of abstract reasoning of Aristotelian scholasticism, i.e. the prioritizing of experience over abstract reasoning mentioned above. Brahe, who was more of an observer than a mathematician (Cohen 2011, p. 129), represents a striking embodiment of this turn, and his observations in the latter sixteenth century contributed to the eventual overthrow of the Ptolemaic-Aristotelian model. This empirical turn was of epoch-breaking significance, but before discussing that, let us consider (b) a little further.

We mention the Reformation in the title of the present section, which may puzzle some anti-Eurocentric readers: Do we mean to endorse Weber's popular Eurocentric Protestant-ethic thesis? Of course not. In our view, the Protestant ethic was itself one embodiment of a general shift in Latin European ethos in the post-1453 period, and did not constitute an originary causal factor, but that is outside our present scope—though we shall touch upon the issue of inner- versus other-worldly asceticism in the next chapter. For our present purposes, the Reformation's importance, the same as (b) above, lies in its effects on encouraging a willingness to break with orthodoxy. Whereas it may well be true that Copernicus was inspired by Hermetic sun worship, his readiness to go against both Aristotelianism and Christianity's endorsement of it occurred in an environment where such a shift in intellectual climate was taking place. While on the Reformation, we may perhaps note that its occurrence constitutes an internal European factor, but can hardly be interpreted as any 'uniquely European essence', since religious schisms occur elsewhere as well, and hence, our reference to it does not constitute any endorsement of Eurocentrism.

On this newly emerging willingness to break with orthodoxy, it may be recalled that at the start of Sect. 3.7 of Chap. 3, we quoted science historian Colin Ronan who says, inter alia, that in China, there was 'no aspiration to break the mould of existing orthodoxies as inspired men such as Galileo'. The way it is formulated by him, this constitutes a cultural-essentialist argument. To show why he is wrong concerning Europe,[7] let us consider Galileo mentioned by him. One of Galileo's important breaking-with-orthodoxy concepts concerns the relativity of motion, used against Aristotle's argument of a stationary earth, which constituted the orthodoxy supported by the Church. In putting forth his argument, Galileo illustrated it with a shipboard example. As a matter of fact, the issue of a rotating or stationary earth had previously been debated by the scholastics Buridan (d. ca. 1358) and Oresme (d. 1382), both at Paris University. Buridan devised an 'arrow' argument (which is equivalent to Galileo's later 'tower' argument) to support the stationary position. Oresme critiqued Buridan by giving a shipboard example similar to the one that Galileo was later to use. Oresme's argument was later to convince Copernicus, but he (Oresme) did not himself pursue it further and gave his reason for not doing so, which Lindberg (2007, p. 285) explains as follows:

[7]As to China, we have considered cultural-essentialist arguments concerning it at various points in previous chapters, such as Huff's 'intellectual curiosity' thesis discussed in Chap. 7. In the next chapter and in the appendix, there will be other occasions to do so.

> His [Oresme's] purpose [with the shipboard example], he stated, was to offer an object lesson for those who would impugn the faith by rational argument. His success at formulating persuasive philosophical arguments for an idea as 'opposed to natural reason' as the rotation of the earth demonstrated the unreliability of rational argument and, therefore, the caution to be used where rational argument touches the faith, as this one did.

In other words, Oresme did not want to break with orthodoxy so as to maintain Christian faith. In a different intellectual climate, the roots of which, as seen, cannot be accounted for in essentialist terms, Copernicus and Galileo were not restrained in the way that Oresme was. All three of them were Europeans, but their 'aspiration', to use Ronan's term, with regard to orthodoxy differed, in different intellectual climates. This is an illustration of the point about *path-dependent* analysis that we mentioned in Chap. 3, namely, that *timing* is a crucial variable. In this case, the same argument made at different times met with different fates.

The questioning of relying on abstract reasoning alone not only gave rise to an empirical turn, but also an inter-related turn, which can best be illustrated with reference to Bacon. When Bacon was only 16 (ca. 1578), he disliked Aristotle because his teachings were not conducive to practical benefits (Bajaj 1990, p. 65, n 17). Bacon said:

> Truth, therefore, and utility, are here perfectly identical...There is a most intimate connection between the ways of human power and human knowledge...and that which is most useful in practice is most correct in theory. (quoted in Bajaj *ibid.*, p. 44)

Hall (1962, p. 15) states in this relation:

> the arguments of the Baconian school: that true command—and therefore real if unwitting knowledge—of natural processes had been won by the arts [referring to craft] rather than by sciences, and that the scholar would often become more learned if he would consent to apprentice himself to the craftsman.

In a word, Bacon disliked scholasticism because he believed that real knowledge can only be gained by the scholar learning from the practical craftsman, and, hence, real knowledge and practical usefulness are identical. We may call this a *practical turn*, i.e. the double belief that knowledge cannot be attained through abstract reasoning alone, but through combining reasoning with practical application, and, hence, that real knowledge is necessarily practically useful knowledge. Petrus Ramus (1515–1572) also expressed this practical turn well in his critique of the scholastics: 'Get out of your armchairs, ye philosophers, and gain wisdom in the workshops' (quoted in Cohen 2011, p. 131). Practical orientation had always underpinned alchemy, an issue to be discussed in the next chapter, what was new about the practical turn that is being discussed here is that it occurred in areas where abstract reasoning had been dominant previously.

In principle, the empirical turn can be independent of the practical turn. For instance, in astronomy, one may, in an empirical turn, emphasize that observation is just as important as abstract theorization without thinking in terms of practical application. But in practice, the empirical and practical turns emerged together as unforeseen consequences of one and the same process of the unfolding of the imperial enterprise. In other words, the two turns were inseparable, and cannot be separated from one another as they can, from an abstract vantage point, conceptually be done in principle. This can clearly be seen from the discovery of plants and animals previously unknown to Europe which had great *practical* (commercial and medicinal) value, a discovery that engendered the drive to get to *empirically* know more about what lied in the 'New World' waiting to be known. As Bacon put it, 'The growth of the sciences [depended on] the exploration of the world'. This was so not only in relation to natural history, but also in relation to, for instance, astronomy, as the following instance illustrates, though it dates from the 17th instead of sixteenth century. In 1672, the French astronomer Jean Richer traveled to the French colony of Cayenne in South America, the voyage was sponsored by Louis XIV through the Royal Academy of Sciences in Paris, while the ship was provided by the French West India Company. Richer made various astronomical observations focusing on the movements of the planets as seen in the southern hemisphere, and catalogued the stars close to the equator. Such data were crucial to enhance navigational techniques for empire expansion. In fact, the Royal Academy of Sciences was founded in 1666 precisely to support such scientific voyages. (Poskett 2023, pp. 101–2)

With regard to Bacon's statement quoted in the previous paragraph, Poskett (2023, p. 99) comments: 'What Bacon had in mind was the Spanish Empire of the fifteenth and sixteenth centuries. His vision of science drew directly from the House of Trade in Seville [on which see shortly below]. That is exactly how the Royal Society was imagined'.

Thus, from the turn of the sixteenth-century onwards, there was established in European scholarly consciousness (other than diehard scholastics) an integral intermingling of knowledge and practical usefulness. Hence, instead of seeing the empirical and practical turns existing separately from one another, it is more appropriate to think in terms of a combined ***practical-empirical turn*** occurring among an increasing number of European intellectuals during that period.

Just as the practical-empirical turn was an ***unforeseen consequence*** of the imperial enterprise (on this point, see later on), that enterprise also constituted an indispensable ***pre-condition*** for putting that combined turn into practice. Hence, right from the beginning, European states were instrumental in promoting the pursuit of knowledge in the practical-empirical mode. In 1503 and 1524, the Spanish throne established the House of Trade and the Council of the Indies respectively.

> Together, these two organizations provided some of the first salaried positions for studying science outside of universities. Geographers, astronomers,[8] natural historians, and navigators were all employed by the Spanish Crown…This then was the first time that modern science became fully institutionalized in Europe. Not in universities[9]…but as part of a Spanish project ***to know, and to conquer***, the Americas. (Poskett 2023, p. 40, our emphases)

This direct link between 'to know' (the Council of the Indies) and 'to conquer' (the House of Trade) was subsequently replicated in Britain. In 1660, Charles II granted two charters, as is well-known, one established the Royal Society, while the other, which is less well-known, established the Company of Royal Adventurers Trading into Africa, later known as the Royal African Company, which principally traded in slaves. One-third of the founding members of the Royal African Company became fellows of the Royal Society, while the latter invested in the former (Poskett 2023, pp. 98 9).

[8] As just seen in relation to Jean Richer, astronomy was indispensable to navigation.

[9] Spain's Council of the Indies was, thus, the forerunner of the scientific societies that were founded later, such as the Academia dei Lincei, founded in 1603, and the Royal Society, founded in 1660. Like the Council of the Indies, these scientific societies worked outside the universities (though many of their members originated from none other than the universities) to bypass the resistance of university-based scholastics who were opposed to the practical-empirical mode.

On the part of some European intellectuals advocating or working in the mode of the practical-empirical turn, the direct link between 'to know' and 'to conquer' existed on a *subjective* level, i.e. for these intellectuals, to know was in order to conquer, and to conquer was in order to know. Bacon was the most well-known instance of this. Bajaj (1990) notes that all of Bacon's books were addressed to the king, and:

> [Bacon] often recommended to King James I various ways of expanding his empire. Such expansion would be useful, he urged [in a piece presented to the king in 1606], both for acquiring material benefits and for the honour of civilizing barbarians through spreading truth and casting out superstition. (p. 63)

Bacon himself stated:

> let mankind regain their rights over nature, assigned to them by the gift of God, and obtain that power, whose exercise will be governed by right reason and true religion. (quoted in Bajaj *ibid.*, p. 50)

For Bacon, as seen previously, truth and practical utility are identical, and knowledge should serve to exercise power. While Bacon often spoke of humankind exercising power over nature, he also spoke, as in the second quote above, of that power being exercised in accordance with right reason (which the 'barbarians' did not possess) and true religion (Christianity). In other words, for Bacon, true knowledge is practically useful knowledge that enables, inter alia, the exercise of power over the 'barbarians', hence, his frequent exhortations to the king to expand his empire. That was why Bacon held the inventions of printing, gunpowder, and the magnet, which he did not know originate from China, in such high regard:

> It is well to observe the force and virtue and consequences of discoveries. They are to be seen nowhere more conspicuously than in those three which were unknown to the ancients...: namely, printing, gunpowder, and the magnet. For these have changed the whole face and state of things throughout the world, the first in literature, the second in *warfare*, the third in *navigation* [crucial to the imperial enterprise]...no empire, no sect, no star, seems to have exerted greater power and influence in human affairs than these three mechanical discoveries. (quoted in Needham 1954, p. 1, our emphases)

It would, of course, be unwise to believe that every scientist driven by the practical-empirical turn was, like Bacon, an advocate of the imperial enterprise on a subjective level—whether or not that is so on the part of any particular scientist requires empirical research to ascertain. But, as shown, the direct link between 'to know' and 'to conquer' did *objectively* exist for all scientists working in the practical-empirical mode. Without political patronage that was directly or indirectly part and parcel of the imperial enterprise, 'to know' in the practical-empirical mode lacked the required preconditions. For instance, with reference to Galileo, the Venetian Arsenal with which he collaborated constituted the Republic's state-owned naval power, and Galileo presented his telescope to none other than the Doge of Venice. Saliba (2007, p. 252) further points out: 'the Academia de [dei] Lincei soon enjoyed the membership of none other than Galileo...one of the earliest projects of the Academia de Lincei was the re-publication of the survey of the medical plants of the new colonies in Mexico'. With regard to some other scholars, the existence of the above link went beyond the objective level: Newton invested in the slave trade, Locke (who was elected to the Royal Society in 1668) owned shares in the Royal African Society, while Boyle was a director of the British East India Company (Poskett 2023, p. 97). Even were Newton, Locke, and Boyle, unlike Bacon, not vocal advocates of the imperial enterprise (concerning which we are not aware whether or not any research has been done), they nonetheless actively participated in it.

In the above, we remarked that just as the practical-empirical turn was an unforeseen consequence of the imperial enterprise, that enterprise constituted an indispensable precondition for putting that combined turn into practice. This, let us stress, should not be confused with the Hessen thesis (see Chap. 2) or other theses of a similar nature, such as that suggested by Needham. Our argument concerns solely the emergence of the practical-empirical turn as a consequence of the imperial enterprise[10] and the latter being a precondition for putting that combined turn into practice, it does not say that the imperial enterprise produced the breakthrough into modern science, though, as we show shortly, the practical-empirical turn was instrumental in enabling that breakthrough. As explained in Chaps. 3 and 5 with regard to what social explanation can and cannot do, and to

[10] Note that we speak of the imperial enterprise, not, as in Needham's argument, the rise of capitalism or the bourgeoisie coming to power. The imperial enterprise of the sixteenth-century was, to begin with, an enterprise of merchant and financial-speculative capital, not of productive capital, though it soon developed into slavery-based plantation production.

internalist and externalist explanations, intellectual development is an intellectual phenomenon, hence the *substance* of intellectual development (ideas and so on) can *only* be explained by means of *internal scholarly factors*. In our argument, without the *mediation* of an *internal* factor (the practical-empirical turn), on the condition of the presence of other *internal* factors (such as the absorption of Arabic scholarship's realist trend), an *external* factor (the imperial enterprise) can never explain why late-Renaissance scholars were *intellectually able* to make the *scholarly* advances that they did. In short, as an external factor, the imperial enterprise did not directly produce the breakthrough into modern science, what it did was to engender a change in an internal factor, and it was this change that contributed to the making of the breakthrough.

At this point, it may, perhaps, be useful to briefly comment on a related issue, namely, the question of why Latin Europe was able to forge ahead and breakthrough into modern science, while Arabic scholarship went into decline. Saliba (2007—a view that, we understand, he continues to hold up to the present) has suggested a social explanation, namely, the relatively change in the wealth of Latin Europe and the Arabic world, which much affected political patronage of scholarship, after the former had succeeded to seize the lucrative spice trade from the latter. That such a social explanation is inadequate with regard to Latin Europe has just been explained in the previous paragraph; as to the case of the Arabic world, we hazard to speculate that our process approach may be able to suggest some leads for investigation. In Chap. 3, we made a contrast between the development of the institution of the theologian-natural philosopher in Latin Europe and the strict distinction between theology and secular knowledge in the Arabic world, arguing that they were the contingent outcomes of two different double contingencies. It seems possible that it was this contingently produced strict distinction between theology and secular knowledge in the Arabic world that the relative decline of that world's wealth which affected political patronage and hence protection of secular scholarship could have had such consequential effects on the development of Arabic secular scholarship.

Returning to our present argument, Galileo is an embodiment of the combined practical-empirical turn. Although Galileo subjected the scholastics to merciless attack, there was, in fact, significant continuity between him and the scholastics on the subject of motion. The shipboard example has been pointed out above. But the continuity went beyond that. Thus, Lindberg (2007, p. 306) remarks:

> out of this purely intellectual labor [of the scholastics] came a new conceptual framework for kinematics and a variety of theorems (...) that figured prominently in the kinematics developed in the seventeenth century by Galileo...It is implausible to suppose that Galileo was ignorant of his fourteenth-century forerunners.

Similarly, Grant (1996, p.169) observes:

> Their [the scholastics'] contributions on various aspects of motion, both kinematic and dynamic, were striking; Galileo himself could not actually enhance them...

However, if there existed such a continuity, there was, on the other hand, a critical difference between the scholastics and Galileo. Crombie (1959, pp. 92–3) observes:

> [I]t was the clerics of the universities who were responsible for the main theoretical conceptions of science, and their interest in them was in general epistemological or metaphysical or logical [only]...They went through the whole discussion without making a single measurement. The direction of interest is quite different when we turn to Galileo. Instead of methodological principles and mathematical possibilities for their own sake, he firmly took the motions actually found in nature as the object of his inquiries...[Galileo's work] illustrate[s] a general change of interest between the fourteenth and the seventeenth centuries from scientific method as such to scientific method applied to the facts of nature.

Similarly, Grant (1996, p. 151) notes:

> Galileo...sought to apply mathematics to real problems in the physical world. By contrast, medieval applications of mathematics to natural philosophy were usually of a hypothetical character divorced from empirical investigation.

Thus, what distinguished Galileo from the scholastics was his practical-empirical turn (what Crombie in the above quote from him calls 'a general change of interest'), which was instrumental to the making of his breakthroughs. In the 1580s, Galileo attempted to reform Aristotle by means of Archimedean mathematics, but as his abandonment of his draft titled 'On motion', written in 1590 while he was at the University of Pisa, shows, his efforts along that line failed (Machamer and Miller 2021). Most of Galileo's breakthroughs were achieved during the time from 1592 to 1610 when, having left Pisa, he was in Padua. There he collaborated closely with the Venetian Arsenal (navy). Renn and Valleriani (2001) argue that of the two new sciences presented in Galileo's *Discourse*, his theory of motion 'is largely rooted in the practical knowledge on ballistics as it was accumulated by contemporary experts on artillery', and his analysis of the strength of materials is rooted in 'the practical challenges of the construction of large galleys as they were faced by the foremen of the Venetian Arsenal'. It should, however, be pointed out that arguments such as that of Renn and Valleriani are not reducing Galileo's theorization to practical engineering knowledge, Galileo brought with him a theoretical awareness and sensitivity from his scholarly training to the practical problems faced by the Arsenal's engineers, which enabled him to draw theoretical conclusions from tackling these practical problems. Neither armchair scientists such as the scholastics nor the engineers without prior theoretical training could have done the same. As Hall (1962, pp. 18–9) notes: 'Science indeed owes much to technology; but we must remember that the debt was itself created by natural philosophers and other men of learning...even if the pivotal invention of the telescope were a craft invention, its scientific potentialities were perceived by scholars'. Relatedly, it is as a result of the practical-empirical turn *among scholars* that engineers, mariners, surveyors, and physicians played a significant role in the development of seventeenth-century science.[11]

To sum up, in the post-1453 period up to the eve of the breakthrough of 1600, there occurred *a fundamental threefold shift in intellectual climate* in Latin Europe, characterized by a practical-empirical turn which represented a qualitative change in the mode of enquiry compared to the previous mode, a greater readiness to break with orthodoxy, and the progressive undermining of Aristotelianism. The former two components of this threefold shift pertain to the *agency* of late-Renaissance European scholars, bringing about changes in their *values* (the practical-empirical turn embodied a relative change in the values placed, on the one hand, on abstract reasoning and, on the other, on the practical and the empirical) as well as their *drives* (or propensity or tendency or orientation or aspiration, as one may like to describe it), namely, the enhanced readiness to break

[11] Hall (1962, p. 3) points out that the role of instrument makers became significant only in the eighteenth century.

with orthodoxy. The case of Galileo shows the key role played by these *agency* variables.

From the angle of historical explanation, Aristotelianism's loss of its monopoly was the result of a *contingent* (contingent in relation to intellectual development, ditto below) *external factor*, namely, the Ottoman's westward expansion. The colonial enterprise was a *contingent unforeseen consequence* of the Ottoman's advance. The further undermining of Aristotelian orthodoxy by reports of Portuguese and Spanish travellers was also of the nature of a *contingent unforeseen consequence*. From the point of view of intellectual development, the Reformation was a *contingent external factor*, hence, its encouragement of a willingness to break with orthodoxy is similarly of the nature of a *contingent unforeseen consequence*. The undermining of the reliance on abstract reasoning alone and the concomitant emergence of the practical-empirical turn were similarly of the nature of a *contingent unforeseen consequence*.

8.3 The Need to Transcend Dialogism to Resolve the Positive NQ on a Non-essentialist Basis and Rebut Eurocentrism

In Sect. 8.1, it was pointed out that there existed a trend towards the realist direction in Arabic scholarship. With regard to astronomy, on which we focus in the following for the sake of simplicity, this consisted in reforming Ptolemy (whose models lack physical possibility), by mathematical means devised on the basis of Hindu mathematics, so as to retain the operational power of his mathematical models but at the same time make them physically more plausible. This development culminated in the work of al-Shāṭir, from whom Copernicus plagiarized except for his replacement of geocentrism by heliocentrism. However, in retaining the epicycle-on-deferent model in reformed form, al-Shāṭir remained within Ptolemy's 'saving the phenomena' framework, in the sense that his model remained a mathematical device, instead of being a physical model. The same applied to Copernicus.

In speaking of mathematical science becoming realist, Cohen refers to Kepler's and Galileo's efforts to make heliocentrism a physical model in addition to being a mathematical model. Thus, Kepler posited the sun, which was God-the-father to him, to be a magnet to explain why planets revolve round it. Kepler's model of Mars was contradicted by observation, he revised it, but the revised model still failed the empirical test; in the climate of the empirical turn, he abandoned the dogma of 'perfect' circular motion and replaced it with elliptical motion which stood up to observation. This represented a *qualitative breakthrough* in comparison to both al-Shāṭir and Copernicus, for *by making changes to theory on the basis of empirical observations, Kepler broke with the 'saving the phenomenon' tradition.* Copernicus's heliocentric model met with various physical objections,[12] on the basis of which it was rejected by most. Galileo's new theory of motion, achieved in consequence to his practical-empirical turn, was able to overcome these objections; further, observations made by him with his telescope led to the falsification of the Aristotelian-Ptolemaic model and confirmation of the heliocentric model. Both developments led Galileo to support heliocentrism as a physical model. It can be seen that Cohen's realist turn in mathematical science was enabled by scientists such as Kepler and Galileo, who *consummated Arabic scholarship's realist trend*, and who were working *in a scholarly climate of the practical-empirical turn* that we have discussed above on a non-essentialist basis.

Cohen describes Kepler's and Galileo's works as the transformation of Alexandria into Alexandria-plus. It is now clear that this constitutes a *Eurocentric* misrepresentation because it completely bypasses the trend towards the realist direction of Arabic scholarship that Latin Europe inherited[13]: Kepler's inheritance, in his work on optical theory, of the tradition founded by al-Haytham and popularized in Latin Europe by Roger Bacon and his followers provides direct testimony to this inheritance. In his critique of Cohen (2015) in relation to this *from the dialogical perspective*, Bala (2024) also points this out with reference to al-Haytham's optical theory.[14] However, Kepler and Galileo *did go beyond* Arabic scholars in breaking away from the 'saving the phenomenon' tradition, and *the dialogical perspective is unable to explain, within its parameters, this advance.* This recalls our comment in Chap. 2 in relation to the rise of neo-Eurocentrism, that for dialogism, how and why late-Renaissance scholars epitomized by Galileo were able to breakthrough into modern science on

[12] This contrasts sharply with the cases of Eudoxus and Ptolemy. And this contrast provides testimony to the influence of Arabic scholarship's realist trend absorbed by Latin European scholars.

[13] It also ignores the extent to which Galileo was indebted to the scholastics Buridan and Oresme.

[14] Our discussion in Sect. 8.1 has partly been inspired by Bala's critique, for which we are indebted to him.

the basis of what they had learnt from non-European inputs, remains, in Bala's wording, a 'black box'. It can now be seen that it is only by going beyond dialogism and take into account the variable of late-Renaissance *scholarly agency*, which can be fully analysed on a non-essentialist basis, that it is possible to crack open this 'black box'. To put it in the theoretical-methodological terminology of the Introduction, it is because dialogism does not engage in any analysis of the phenomenon of intellectual development itself, that variables such as scholarly agency can nowhere be found within its parameters,[15] hence the need to transcend it.

Bibliography

Bajaj, Jatinder K. 1990. 'Francis Bacon, the first philosopher of modern science: a non-Western view'. Pp 24-67 in Ashis Nandy (ed.) *Science, Hegemony and Violence: A Requiem for Modernity* (1990[1988]). Oxford University Press.

Bala, Arun. 2006. *The Dialogue of Civilizations in the Birth of Modern Science.* Palgrave Macmillan.

Bala, Arun. 2024 (in press). 'Explaining the rise of modern science: A dialogical perspective'. Chapter 16 of Arun Bala, Raymond W.K. Lau and Mei Jianjun (eds.), *Multicivilizational Exchanges in the Making of Modern Science. Needham's Dialogical Vision.* Palgrave Macmillan.

Cohen, H. Floris. 1994. *The Scientific Revolution: A Historiographical Inquiry.* University of Chicago Press.

Cohen, H. Floris. 2011. *How Modern Science Came into the World: Four Civilizations, One 17th-Century Breakthrough.* Amsterdam University Press.

Cohen, H. Floris. 2015. *The Rise of Modern Science Explained. A Comparative History.* Cambridge: Cambridge University Press.

Crombie, A.C. 1959. 'The significance of medieval discussion of scientific method for the scientific revolution'. Pp 79-101 in Marshall Clagett (ed.) *Critical Problems in the History of Science.* University of Wisconsin Press.

Grant, Edward. 1996. *The Foundations of Modern Science in the Middle Ages.* New York: Cambridge University Press

Hall, A. Rupert. 1962. 'General Introduction'. In Marie Boas, *The Scientific Renaissance 1450-1630.* London: Collins.

Hobson, John M. 2006. *The Eastern Origins of Western Civilisation.* 5th printing with corrections. Cambridge: Cambridge University Press.

[15] There is, of course, the further point that supporters of dialogism share the common mistake of seeing certain types of explanation (such as the consideration of values and drives) as necessarily essentialist.

Joseph, George G. 2011. *The Crest of the Peacock Non-European Roots of Mathematics*, 3rd edition. Princeton: Princeton University Press.
Keay, John. 2005. *The Spice Route.* London: John Murray Publishers.
Kuhn, Thomas. 1957. *The Copernican Revolution.* Harvard University Press.
Lindberg, David. 2007. *The Beginnings of Western Science*, 2nd edition. Chicago: University of Chicago Press.
Machamer, P. and D.M. Miller. 2021. 'Galileo Galilei'. *Stanford Encyclopedia of Philosophy*, available online, accessed 1 Jun 2022.
Needham, Joseph. 1954. *Science and Civilization in China, Vol. 1: Introductory Orientations.* Cambridge: Cambridge University Press.
Poskett, James. 2023[2022]. *Horizons: A Global History of Science.* UK: Penguin Books.
Pomeranz, Kenneth, and Steven Topik. 2006. *The World that Trade Created*, 2nd edition. Armond: M.E. Sharpe.
Renn, J. and M. Valleriani. 2001. 'Galileo and the challenge of the Arsenal', lecture delivered in Florence on 21 March 2001, available online, accessed 23 May 2022.
Saliba, George. 2007. *Islamic Science and the Making of the European Renaissance.* Cambridge, Mass: MIT Press.
Vogel, Klaus and Alisha Rankin. 2016. 'Cosmography'. Pp. 469-96 in *The Cambridge History of Science, vol 3 Early Modern Science.* Cambridge: Cambridge University Press.
Webster, Charles. 1980. *From Paracelsus to Newton, Magic and the Making of Modern Science.* Cambridge University Press.
Yates, Frances. 1979[1957]. *Giordano Bruno and the Hermetic Tradition.* University of Chicago Press.

CHAPTER 9

Cultural Essentialism and Cohen's Eurocentric–cum–Neo-Eurocentric Argument Concerning the Breakthrough into Modern Science

As mentioned at the start of the previous chapter, we analyse the breakthrough into modern science in terms of four issues, namely, the breakthroughs achieved by Kepler and Galileo; the rise of the mechanical philosophy; the mathematization of nature; and the use of experimentation to test hypotheses about nature. This roughly parallels Cohen's argument concerning the rise of modern science, with the first issue corresponding to his first 'revolutionary transformation', the second and third issues to his second 'revolutionary transformation' (what he calls the rise of 'kinetic corpuscularianism'), and the fourth issue to his third 'revolutionary transformation' (which he argues to be 'a specifically European-colored mode of investigation'). The first issue was discussed in the previous chapter together with a critique of Cohen's argument concerning his first 'revolutionary transformation'. The present chapter examines the second, third and fourth issues, in the course of which Cohen's arguments concerning his second and third 'revolutionary transformations' are critiqued.

Concerning corpuscular theory, there actually existed two different varieties in Latin Europe. On the one hand, there was a chymical[1] version

[1] The old term 'chymistry' refers to early modern alchemy-chemistry, 'a discipline that still viewed the transmutation of base metals into gold (*chrysopoeia*) as viable and yet contained much in addition that is identifiable to us moderns as chemistry' (Newman 2006, p. xi). In

R. W. K. Lau, *Rethinking the Needham Question*,
https://doi.org/10.1007/978-981-97-9472-0_9

that traced back to the alchemical tradition beginning in the late thirteenth century, and which continued into the 'chemical atomism' of the nineteenth century, this version had little to do with Democritean atomism—hereinafter, for convenience, we refer to the atomic view tracing its lineage to Democritus and other ancient Greek atomists as Democritean atomism; on the other, there was a (Democritean) physical version that arose in the early seventeenth century. When Cohen speaks of the rise of 'kinetic corpuscularianism' as his second 'revolutionary transformation', what he is referring to is the physical version.

Unlike Hermeticism and Neoplatonism which quickly gained influence in the post-1453 period, Democritean atomism failed to do so for a long time. This was probably because, despite the greater readiness to break with orthodoxy in the sixteenth century, Democritean atomism's determinism was directly contrary to the notion of ethical responsibility. This situation changed in the early seventeenth century when Descartes picked it up. For the Democritean atomists of Greece, atoms are unbreakable, and their focus was on speculating about the shape, size, and position of different kinds of atoms in order to account for different sensual properties of wholes composed of them such as sweetness and so on. Descartes' Democritean atomism differs decidedly from the ancient Greek version. Firstly, his atoms are not unbreakable, but infinitely divisible. More crucially, he focused on the motion of atoms by interpreting the determinism posited in atomic doctrine as being governed by the laws of motion, hence Cohen's term '*kinetic* corpuscularianism'. This represented a conceptual innovation on Descartes' part.

Cohen (2011, p. 521) states that 'If it were possible to mark what was distinctively novel about the Scientific Revolution in just one word, that word would be "motion"...as a fundamental unit of analysis and/or explanation'. Galileo is customarily and justifiably regarded as the central figure in the breakthrough of ca. 1600 precisely because of his work on motion. Many historians are of the view that Descartes' kinetic turn was inspired by Galileo, despite Descartes' denial of ever having been influenced by Galileo, a denial that, as Cohen (2011, p. 241) notes, historians dismiss as implausible in view of Descartes' travels through Italy in the

the following, we speak of alchemy when reference is made to earlier time periods, and of chymistry when reference is made to the period immediately preceding and including the seventeenth century; when reference is made in a general way, as in the present context, we speak of either chymistry or alchemy as appropriate to the context.

early 1620s. In making his kinetic turn, Descartes developed the mathematical tools for analysing the laws of motion governing the interactions between atoms, thereby setting the trend of the mathematization of nature. This issue will be discussed in Sect. 9.3 below, where Cohen's representation of his second 'revolutionary transformation' as a direct transformation of Athens into Athens-plus is shown to be a Eurocentric misrepresentation.

The rise of the mechanical philosophy is usually linked to the rise of Democritean atomism in the seventeenth century. Thus, in speaking of the 'acceptance of the mechanical model of reality', Needham referred to his so-called 'Western predilection for [Democritean] atomism' (Needham 1969, p. 14; Needham et alia. 2001, p. 227). The mechanical philosophy or worldview is in contrast to the holistic-organic worldview. In today's understanding, the latter sees wholes being structures possessing its own properties which cannot be reduced to the properties of its component parts. Thus, wholes are like organisms, hence the term 'organic'. The oft-repeated classic example is the water molecule: Water can put out fire, whereas both hydrogen and oxygen are highly inflammable. The mechanical worldview, sometimes also referred to as the atomistic worldview and is associated with the term 'analytic' or 'analytical', on the contrary, reduces the properties of wholes to the properties of their component parts, as the ancient Greek atomists did; in other words, it sees wholes as nothing more than arithmetic sums of their parts. Thus, whereas the organic worldview focuses on the whole, the mechanical worldview focuses on the parts. The mechanical philosophy understood in this general sense arose in the seventeenth century from developments in chymical corpuscular theory which, as noted, had little to do with Democritean atomism. This will be discussed in Sect. 9.2. On the other hand, by virtue of his analysis of the kinetic relations between the parts (Democritean atoms), Descartes' approach also embodied the mechanical worldview. This will be discussed in Sect. 9.3 as part and parcel of the discussion of Descartes' mathematization of nature.

With regard to the rise of experimentation, Cohen (2011, pp. 114, 134) speaks firstly of the emergence of what he calls 'Europe's coercive empiricism' in the 15th and 16th centuries, and that investigations in the 'coercive empiricist' mode constituted 'culturally marked pieces of nature-knowledge in Renaissance Europe', 'Nature-knowledge colored the European way'. Such investigations included the work of da Vinci. Cohen argues that his third 'revolutionary transformation' was a 'transition from "coercive empiricism" to "fact-finding experimental science"' (p. xxx), 'a specifically European-colored mode of investigation intent upon *accurate description and practical applications* [as in some of da Vinci's pursuits] that had arisen by the mid-15th century began to consolidate around 1600, under the aegis of Francis Bacon's calls...into a fact-finding, practice-oriented mode of experimental science' (p. xvi). Cohen's view of the origins of the experimental method and of the role of Bacon are inconsistent with those of other science historians who point out that the use of experimentation to test hypotheses about nature actually emerged in alchemical practices borrowed from the Arabs, whose origins traced back to the late thirteenth century, and which became part of the new science with the work of Robert Boyle in the 1650s. The development from alchemy to Boyle involved no 'revolutionary transformation', it was Bacon's rhetoric that has misled some scholars into thinking that there was. This will be discussed in Sect. 9.2.

Upon completing his discussion of his three 'revolutionary transformations', Cohen asks why they all occurred at the same time in the early seventeenth century. In answering his own question, Cohen turns the outdated ideological 'voyages of discovery' narrative into a cultural-essentialist neo-Eurocentric pseudo-explanation of this timing. Just as Cohen is able to find support for his cultural-obstacle argument with regard to knowledge transmission across different civilizations (see Sect. 7.1 of Chap. 7) from none other than Needham, Cohen is able to repeat the same feat in relation to his pseudo-explanation. A critique of Cohen's pseudo-explanation is provided in Sect. 9.4, where a number of claims made by Needham concerning China (implicitly conceptualized in terms of 'inhibitory' factors that prevented China from developing along European lines—see Sect. 2.2 of Chap. 2 on the conceptualization of 'inhibitory' factors) cited by Cohen for support are shown to be contrary to the simplest historical facts about China.

9.1 Is Cohen's 'Coercive Empiricism' a 'Specifically European-Colored Mode of Investigation'?

We would like to begin with some comments on Cohen's claim concerning 'Europe's coercive empiricism' of the 15th and 16th centuries as 'a specifically European-colored mode of investigation intent upon *accurate description and practical applications*'. Although his argument that it was from 'coercive empiricism' that the seventeenth century's experimental method had been 'consolidated' is untenable, this is nonetheless worth doing because of the cultural-essentialist nature of Cohen's argument. In speaking of 'specifically European-colored', it is clear that for Cohen, 'coercive empiricism' had arisen in Europe from purely internal European dynamics, i.e. it constituted Europe's 'essence'. Da Vinci, regarded by Cohen as a Weberian ideal-typical figure of 'coercive empiricism', grew out of the tradition known as naturalism that had arisen in Italian art since the early fourteenth century, hence to understand the origins of 'coercive empiricism', it is necessary to take a look at the emergence of naturalism.

This takes us to the history of the relationship between Latin Europe and Byzantine. By the late eighth century, the Barbarian kingdoms in central and western Europe including the British Isles had all sworn allegiance to the Vatican, at the same time, they had all adopted Latin as the official language. In contrast, Byzantine not only followed what later came to be called Orthodox Christianity, it also adopted Greek as the official language in 620. Starting even earlier from the late fourth century onwards, the church in Rome (city) and the church in Constantinople had been in conflict with one another. Initially, the conflict was mainly over jurisdiction (over other churches), it subsequently extended to conflicts over rituals, practices, and doctrine. These conflicts eventually led to the two churches' mutual ex-communication of one another in 1054. Thus, Latin Europe and the Byzantine world constituted two very different societies, hostile to one another in political, cultural, and religious terms. This mutual hostility was worsened by the Crusades that began in 1095, which led to the massacre of Latin Catholic inhabitants of Constantinople in 1182; in retaliation, Normans of the Kingdom of Sicily sacked Thessalonica in 1185; this was followed by the capture and pillage of Constantinople during the Fourth Crusade in 1204, and the imposition of Latin patriarchs in the Crusader states.

Ironically, it was under these circumstances that during the thirteenth century, Italian art came under the heavy influence of Byzantine art: 'thirteenth-century Italian art was influenced by Byzantine art...Greek [Byzantine] mosaicists had been called to the court of King Roger II of Sicily in the twelfth century (...),[2] where they founded a new school of Italo-Byzantine art. Byzantine influence in thirteenth-century Europe is also in part explained by the Sack of Constantinople in 1204 by the

[2] The king and his family enjoyed good relationships with the Byzantine court.

Crusaders, who devastated the churches and the Great Palace. The artistic works taken by the Crusaders—...—were scattered throughout Europe, where their *refined style and impressive craftsmanship* inspired local artists. Byzantine art impressed through its sophisticated style...The earliest Italo-Byzantine paintings demonstrate Italian artists' reliance on Byzantine models' (Hartt and Wilkins 2010, pp. 41–2, our emphasis). In Byzantine art, 'Every shape is treated as an *abstracted* form, severe and clear-cut' (p. 47, our emphasis). The painter Cenni di Pepi (fl. ca. 1272–1302), better known by the nickname of Cimabue, was the last of Italo-Byzantine painters. In sum, Byzantine art was abstract in form, it spread to Italy and, then, elsewhere in Latin Europe due to the good relationships that King Roger II of Sicily enjoyed with the Byzantine court and in consequence of the Crusades, and Italian art came under its dominance due to its sophistication.

Then, 'In the early Trecento [1300s], a new style of painting emerged that revolutionized the art of Florence, Tuscany, and eventually that of the entire Western world. The man who initiated this new style is Giotto di Bondone (c. 1277-1337)...For his contemporaries and successors the virtue of Giotto's style seems to have been based in its fidelity to the human, natural, Italian world they knew, as against the artificial manner from the Byzantine East' (Hartt and Wilkins ibid., p. 73). The style that Bondone initiated is known as naturalism. To understand the background to the rise of naturalism, it would be useful to take a look at a new literary trend that arose in Italy contemporaneously with naturalism known as humanism, which was characterized by seeking inspiration from antiquity. To repeat a quote from Chap. 3:

> Indeed the return to the classical world was seen as the answer to the threat from Islam to European culture. George Sarton commented that the humanists' love of Greece and Rome was nourished by their hatred of Islam. To counter the culture of the Arabic world, the revival of antiquity was carried forth on many fronts. He wrote that the Renaissance was 'partly a reaction against influences (especially those represented by Avicenna and Averroes). The anti-Arabic drive was in full swing in Petrarch's [1304-1374] time'. (Goody 2004, p. 48)

In other words, as a result of the hatred of Islam and under the increasing dominance of Arabic learning, there occurred a reaction in Italy's literary circles that took the form of looking to antiquity for inspiration. As noted, in initiating the style of naturalism, Bondone was reacting against the dominance of Byzantine art in Italy; while Byzantine was not the same as Islam, there existed mutual hostility between the Latin and the Byzantine worlds, much like there existed mutual hostility between Latin Europe and Islam. Hence, the rise of humanism and the rise of naturalism which accompanied one another were both of the nature of a reaction against the dominance of a hostile target. In relation to this, it is instructive to note that both Petrarch, a core figure of the humanist movement, and Bondone hailed from Tuscany; it was in Florence where Bondone established the naturalist movement, while Petrarch spent his childhood in a village near the city.[3] In view of the above circumstances, it is easy to understand why in reacting against the dominance of Byzantine art, Bondone, the same as the humanists, looked to antiquity for inspiration. Hartt and Wilkins (ibid., p. 20) note that 'This ancient material' from antiquity that Bondone looked to, 'had been available throughout the Middle Ages, but during that period it had little effect'. Clearly, it was only under the above circumstances that it now became Bondone's inspiration.

In ancient Greek art, Bondone found the perfect *antithesis* to Byzantine's abstract style, for the former stressed lifelikeness in representation. Lifelikeness refers to the ideal of capturing the world as it is seen by the human eye: 'fourteenth- and fifteenth-century painting in Italy can be understood as an attempt by artists to capture naturalism...how to represent on a two-dimensional surface what is seen by the human eye' (ibid., pp. 17–8). Which is why the style is called naturalism. The naturalists' pursuit of lifelikeness led to a merging of art and an empirical pursuit, to a certain level, of nature-knowledge, such as first-hand study of anatomy, a point much stressed by Cohen. In the early fifteenth century, this led to the development of linear perspective in painting, which combined art with mathematics and optics, especially the optical theory of al-Haytham discussed in the previous chapter.

To sum up, with regard to its origins and its tradition of pursuing lifelikeness, naturalism had not arisen from any purely internal European dynamics, but as an outcome of the dominance of Byzantine art in Italy in the thirteenth century in an environment of mutual hostility between Latin Europe and Byzantine, which resonated with the former's

[3] Incidentally, Dante, another core figure in the humanist movement who consigned Muhammed to the eighth circle of the inferno, also hailed from Tuscany and was a friend of Petrarch's father.

contemporaneous hatred of Islam. Lifelikeness itself was adopted from Greece (which, as shown in Chap. 4, constituted a different civilization from Latin Europe) instead of being an original Latin European aesthetic, while the development of linear perspective in the course of naturalism's combining of art and nature-knowledge relied heavily on Arabic learning. None of the above constitutes any 'specifically European-colored mode' as claimed by Cohen. Given that da Vinci, born like the humanist Petrarch and the naturalist Bondone in Tuscany and raised in the Florentine milieu, grew out of this tradition, the same conclusion applies to him. Thus, even were it true, which is not the case, that the experimental method of the seventeenth century had developed from a 'consolidation' of Europe's previous 'coercive empiricism', as claimed by Cohen, this does not lend any support to his Eurocentric claim that the experimental method was 'specifically European-colored'.

9.2 Testing Hypotheses About Nature by Experimentation and the Rise of the Mechanical Philosophy

The use of experimentation in physics in Arabic scholarship goes back to al-Haytham, who 'performed experiments designed to prove or disprove the truth of optical theories' (Lindberg 2007, pp. 362–3).[4] But before the rise of modern science, the experimental proto-science par excellence was not in physics, but in alchemy, not only in the Arabic world, but also in China and elsewhere. In Cohen's discussion of his third 'revolutionary transformation', no mention is made of alchemy at all. However, contrary to Cohen, it has been well-established by scholars such as William Newman and Lawrence Principe that the seventeenth century's use of experiments to test hypotheses about nature originated not from any 'consolidation' of some previously existing 'coercive empiricism', but from alchemy. There was very substantial continuity in developments between the pre-Boyle era and Boyle, without any 'revolutionary transformation', though in inheriting this lineage of development, Boyle did not only take it further, but also raised developments to a new level of *conceptualization* by speaking in terms of the mechanical philosophy. Boyle's mechanical philosophy had little to do with *physical* Democritean atomism; instead, it was derived from *chymical* corpuscular theory, originating not from the Greek Democritean atomists, but, it may come as a surprise to most, Aristotle.

[4] Incidentally, in China, artisans-cum-scholars of the Later Mohist (*Houqi mojia*后期墨家) school also performed optical experiments in the -4th and -3rd centuries.

The confusion concerning the origins of the mechanical philosophy is partly due to Boyle himself. In his early manifesto titled *Of the Atomicall Philosophy*, he began by celebrating the recent resurgence of atomism 'by Gassendus [Gassendi], Magneus, Des Cartes & his disciples' (quoted in Newman 2006, p. 160).[5] It is important to note that by 'atoms', Boyle did not mean Democritean atoms, but simply corpuscles that resist further breaking down in laboratory processes, an issue to be discussed below. Boyle's verbal association of himself with Descartes and others subsequently became the standard historiographical tradition concerned: 'Older accounts of the mechanical philosophy, with a few important exceptions, routinely picture Boyle and his peers as drawing mainly on the self-styled opponents of Aristotle such as Gassendi and Descartes and on the ancient atomists for their corpuscular theory' (p. 6). As a matter of fact, however, Boyle's corpuscular theory, hence also his mechanical philosophy, had a very different origin, a chymical one. In his eagerness to associate himself with the 'new philosophy' of Descartes and others, Boyle deliberately hid his lineage from alchemy. In this, Boyle was followed 'by eighteenth-century chemists and philosophers eager to detach themselves from a dark and "irrational" past [i.e. alchemy]' (p. 8). In this way, the above-mentioned standard historiographical tradition was established.

From the late thirteenth century onwards, Latin European alchemy was based upon the conviction that alchemical *laboratory operations* can *reveal* the fundamental components of matter by means of *analysis* (breaking down to these components). Latin Europe's alchemy was borrowed from the Arabs. At around the end of the thirteenth century, a work titled *Summa perfectionis* (Sum of Perfection) was published under the name of Geber. 'Geber' is a partial Latin transliteration of Jabir ibn Hayyan, a semi-mythical figure of the eighth century, to whom various works (numbering in the hundreds, only some of which have survived) written in Latin were attributed. According to Newman, *Summa perfectionis* is not a Latin translation of an Arabic text, but written, on the basis of Arabic models, by a European author, whom he identifies as Paul of Taranto, a Franciscan alchemist from southern Italy (p. 26). However, since this identification remains hypothetical only, we refer to *Summa*'s author as Geber below.

[5] The following discussion is based upon Newman (2006), except indicated otherwise. References giving only page number(s) hereinafter are to this work. To give credit to where it is due, while Newman is our source, it should be mentioned that in developing his analysis, he collaborated closely with Lawrence Principe.

Paul of Taranto was, indeed, the author of another alchemical text titled *Theorica et practica*, to be mentioned in due course. By the time Latin Europe borrowed alchemy from the Arabs, the discipline had had a long tradition of development in the Arabic world. Both Ibn Sīnā (Avicenna, ca. 980–1037) and Ibn Rushd (Averroes, 1126–1198) had intervened in the debates concerning alchemy, and they were regularly referenced by Latin European scholars in similar debates subsequently appearing in Latin Europe (pp. 108, 110). Hereinafter, our focus will be on alchemy as it developed in Latin Europe.

To understand what alchemy involved, it is necessary to begin with Aristotle. According to him, each (species of) thing that exists of itself constitutes a 'substance' which is defined by its 'essence', thus, it is the 'essence' of, say, birds ('birdness') which makes birds birds as a distinctive 'substance'; birds may be of different colours, shapes, and so on, but these qualities merely constitute 'accidents', which are inessential and can change without affecting the 'birdness' of birds. Following Empedocles, Aristotle also hypothesized that all empirical things consist of four root elements, namely, fire, air, water, and earth,[6] which are associated with the primary qualities of hot, dry, wet, and cold (hot and dry in fire, and so on). The scholastics took this doctrine further when they considered how the four elements become the individual things of the phenomenal world. The common view was to think in terms of hylomorphism, the interaction of form and matter. To simplify, we focus on the doctrine advanced by Thomas Aquinas in the 1270s. According to him, it is the 'substantial form' of each particular 'substance' that acts on matter (the four elements which are constituted by an undifferentiated 'prime matter'[7]) to make that 'substance' what it is. This process of generation involves the primary qualities of the four elements interacting, as a result of their mutual passion for one another,[8] to produce a single medial quality which provides the disposition necessary for the induction of the 'substantial form'

[6] Aristotle also posited a fifth element, namely, aether, the purest fire; for our purposes, we can bypass this.

[7] 'Prime matter' is pure potentiality without actual existence separately from the primary qualities of the four elements.

[8] The hypothesis of mutual action resulting from mutual passion was inherited from Aristotle. In his analysis of the cause of motion, Aristotle posited an 'unmoved mover' as the final cause of motion. Each celestial sphere in his astronomical model possesses its own 'unmoved mover', and it is out of each sphere's 'love' or 'desire' for the 'perfection' of its 'unmoved mover' that it rotates at the speed and in the orientation that it does.

concerned. The four antecedent elements are 'corrupted' (destroyed) in the process in which they are combined to form the particular mix concerned, hence, the 'substantial form' that is imposed on matter to produce the particular 'substance' concerned is called the 'form of the mixture'.[9] In the 'substance' so generated, the four elements, which have been 'corrupted', are present within the mixture only 'virtually' by virtue of the primary qualities.

Aquinas' Aristotelian holistic view is not identical to that of today's holism. In today's understanding of holism, a whole possesses its own particular properties that cannot be reduced to the properties of its component parts by virtue of the particular structure in which those parts are combined. In combining to form the whole, the component parts *remain* in their original form, as atoms remain in the molecule. But in Aquinas' doctrine, in a 'substance' (whole), there no longer exist distinct component parts, and the 'substance' consists merely of 'prime matter' upon which is imposed its particular 'form of the mixture' to make it the 'substance' that it is.

As is clear, 'the Thomistic theory of mixture [see note 9 on the term 'mixture'] left no place, then, for such entities as mercury and sulfur, the immediate ingredients of the metals according to traditional alchemical theory' (p. 38). It is in order to rebut Aquinas that Paul of Taranto wrote his *Theorica et practica*. In this work, besides general arguments, he relied on examples. For instance, he said that when the unit is subtracted from a quaternary, the ternary emerges, the smaller number always existed within the larger one in its own species. Most interestingly, he argued from alchemical laboratory findings, such as: 'mercury and sulfur persist beneath the substantial form of a complete metal, and the sulfur, at least, can be revealed by the metal's dissolution. This principle appears from the sulfurous stench and smoke when metals are calcined at an intensely hot fire, as also by the color of their calces' (p. 41). Relying on the early technology of dilute acids and bases used in alchemical experiments, 'the metals', Paul argued, 'could not be calcined, their calces dissolved, sublimed, and then finally reduced, if the metals had been corrupted all the way down to the level of the four elements' (p. 42), i.e. if the immediate ingredients of the

[9] In Aristotelian theory, the meanings of 'mixture' and 'compound' are exactly the reverse of today's meanings, i.e. the Aristotelian 'mixture' refers to what we would call a compound today, and 'compound' to what we would call a mixture today (see Newman's note on terminology on pp. xi–xiii).

metals did not retain their existence as distinct ingredients. This argument foreshadowed Daniel Sennert's much later use of 'reduction to the pristine state', which was copied verbatim by Boyle to advance his mechanical philosophy. In short, relying on alchemical laboratory findings involving the processes of sublimation, calcination, and dissolution, Paul was able to argue convincingly against the Thomistic theory of mixture.

In *Summa perfectionis*, Geber advanced a corpuscular theory of matter; his inspiration did not come from the Greek atomists, but was grounded upon alchemical experimental findings, and his theorization was based, ironically, on none other than Aristotle. In Greece, in reaction to Parmenides' doctrine of reality (Being) being one and indivisible, Empedocles was the first to advance what are collectively referred to as element theories. Democritean atomism posited that elements consist of indivisible particles that are able to move around because of the existence of void in between these particles. It has often been taken that all versions of Greek element theories think in terms of particles, but that is not the case. Aristotle, living after the first generation of element theorists (Empedocles, Anaxagoras, Leucippus, and Democritus), stated that Empedocles's elements are particulate, but the fact is 'it is not immediately clear whether Empedocles was imagining matter as continuous and flowing or as made up of minute discrete particles' (Guthrie 1965, p. 149). Aristotle himself argued against the existence of void, which implies that, logically, he could only be thinking, like the Stoics after him, in terms of flow instead of particles. Yet, as noted, Geber formulated his corpuscular theory on the authority of Aristotle. How was that possible?

Geber's reference is to the fourth book of Aristotle's *Meteorology*, which speaks of 'pores' and 'corpuscles', and proposed various corpuscular explanations of material change. 'The fourth book of Aristotle's *Meteorology* has long puzzled Aristotelian scholars' (p. 67), for the obvious reason that it is so contrary to the general thrust of Aristotle's doctrines concerned. Hence, many scholars have dismissed it as spurious, but, Newman adds, more recent scholarship has been inclined to accept its authenticity. In any case, it is worth noting that the long road from the thirteenth century to the complete overthrow of Aristotelianism in favour of atomism (both the chymical and physical versions) in the seventeenth century began with Geber relying doctrinally on none other than Aristotle himself (*Meteorology IV*) to advance his corpuscular theory.

Briefly, Geber posited that the four elements of fire, air, water, and earth consist of minute particles, these particles combine to form 'very strong compositions' of the larger corpuscles of sulfur and mercury (quicksilver),[10] which, in turn, combine to form the corpuscles of other metals. As can be noted, the positing of sulfur and mercury remaining as distinct entities in other metals is directly contrary to the Thomistic theory of mixture. If Geber resorted to *Meteorology IV* as the doctrinal basis of his corpuscular theory, he supported the latter by means of alchemical experimental findings. First, Geber pointed to the observed fact that the sublimed mercury and sulfur collect in the sublimatory vessel as tiny droplets (mercury) or minutely divided powder (sulfur), which seems to reveal their particulate structure. Secondly, both mercury and sulfur sublime without leaving any residue, which, Geber argued, shows that, because they constitute 'very strong compositions', they do not decompose further and remain as distinct entities.

In the *Summa*, Geber also performed 'a battery of assaying tests to arrive at the composition of the metals in terms of varying proportions of mercury and sulfur in volatile and fixed varieties...Geber then explains cupellation, cementation, and the other assaying tests available to him in terms of his corpuscular theory...' (pp. 33–4). 'Geber's corpuscular theory...would prove immensely appealing to the self-styled atomists and corpuscularians of the seventeenth century....Underlying this belief [of theirs] was the clear conviction, inherited from medieval alchemy, that processes such as sublimation, calcination, and dissolution in corrosives provided ocular testimony to the analysis of matter into its more fundamental corpuscular constituents' (p 44). To sum up, 'One thing that appears very clearly from the work of Geber is his insistence on the fact that laboratory operations can *reveal* the fundamental components of matter by means of analysis [breaking down to these components]...' (p. 45, our emphasis). This is the first statement, made around the end of the thirteenth century, concerning, in modern terminology, the use of experiments to test hypotheses about nature.

Geber's corpuscular theory grounded on alchemical experimentation was taken up by the Swiss lay preacher and medical writer Paracelsus von Honhenheim (1493–1541), and, after him, Andreas Lebavius (ca. 1550 to 1616). It is, however, unnecessary to discuss their views for our purposes. Instead, we come immediately to Boyle's immediate predecessor, namely, Daniel Sennert (1572–1637), professor of medicine at Wittenberg.

[10] Mercury is a metal though it exists in liquid form at room temperature; it is called quicksilver in metallic form. Geber's reference to mercury is to quicksilver.

To begin with, it will be recalled that the particles that Geber spoke of refer to that which resists further division by means of laboratory techniques, not to Democritean atoms. Sennert, in the post-1453 era when works of Greek atomism, Neoplatonism, and so on had become available to Latin Europe, spoke of atoms, but in the sense of Geber's particles, and not in the Democritean sense of either physical or mathematical indivisibility. As a matter of fact, Boyle, Lavoisier (a century after Boyle), as well as the 'chemical atomists' of the nineteenth century, all spoke of atoms in this same sense. This point should be borne in mind whenever the term 'atom' is used in the following; when atom is referred to in the Democritean sense, it will be specified as such.

In 1600, Sennert published a work in which he presented himself as a staunch defender of Aristotelian philosophy and accepted Aristotle's refutation of Democritus. Eleven years later, however, in a work titled 'On the operations necessary to pharmacy', he endorsed the corpuscular view, citing Geber eighteen times in the space of fifty-six pages, and giving example after example of how such microscopic activity as it is effected by laboratory processes supported the corpuscular view (p. 89). In 1619, in a text titled, for short, *De chymicorum*, he openly committed himself to the atomic doctrine (in Geber's sense). It is, Newman points out, 'his gradually increasing appreciation of chymistry' and 'increasing immersion in chymistry' that was the 'prime mover' of his transition from Aristotelianism to atomism (p. 124).

In 1619's *De chymicorum*, Sennert presented his famous proof for the existence of atoms by means of the laboratory process of *reduction to the pristine state*, that traces its origins, as previously mentioned, to Paul of Taranto's *Theorica et practica*. Sennert's experiment involves the dissolution and recapturing of metals. Briefly, silver and gold are fused together to produce a seemingly homogenous alloy; the silver (but not the gold) is then dissolved by adding aqua fortis (nitric acid), whereupon it is no longer visible in the solution, then the silver is precipitated into a 'heap of innumerable atoms' by means of salt of tartar (potassium carbonate, a base). The solution of silver and aqua fortis was poured through filter paper, leaving no residue behind. On this experiment, Newman (pp. 121, 123) comments as follows:

> In summary, one can see how Sennert's use of dissolution and precipitation left little alternative to the conclusion that silver alloyed with gold and then dissolved in nitric acid retained its substance intact...[various problems

> arising from trying to explain this within the Aristotelian framework] all led to the same conclusion—that 'reversible reactions' [as revealed by the experiment] involved corpuscular interaction rather than Aristotle's perfect mixture...the silver was composed of minute, cohering particles, each of which retained the substantial form of silver within itself...pouring the solution of silver and aqua fortis through filter paper. The fact that no residue was left behind demonstrated the extreme minuteness of the silver particles...At this point, then, Sennert had full experimental evidence for the two theses that he had wanted to prove—that a metal seemingly mixed with an acid or another metal retained its own nature intact, and that it was composed of extremely tiny corpuscles.

The expression 'reduction to the pristine state' refers to the fact that after being fused with gold, silver is reduced from the alloy back to its original state. Laboratory operations showed that reduction to the pristine state was a general phenomenon: 'an increasing number of chemical reactions were revealing themselves to be reversible and the prevailing scholastic theories of mixture could not adequately account for the reacquisition of ingredients from a mixt [mixture in the Aristotelian sense]' (p. 159). In the above quote, it is stated that Sennert argued that the retrieved silver 'retained the substantial form of silver'. 'Substantial form', it will be recalled, is Aquinas' Aristotelian notion; it will also be recalled that Sennert's atoms are meant in Geber's and not the Democritean sense. Thus, while abandoning most of the Aristotelian framework, Sennert retained the understanding of the corpuscles of silver as Aristotelian mixtures.

Silver loses its colour when it is dissolved in a clear solution of nitric acid, how should such a change in quality ('accident' in the Aristotelian conception) be accounted for? To explain this, Sennert introduced his notion of immutation, 'the vaguely described rearrangement of atoms' (p. 131) during the experimental process. The introduction of this notion brought about an *unforeseen* effect[11]:

> His reliance on the *diakrisis* [separation of atoms as in dissolving silver in nitric acid], *synkrisis* [combination of atoms], and immutation of minute

[11] *Contingent* unforeseen consequences do not only occur as a result of external factors or events, but also in the very substance of the development of ideas as well. In Lau (2020, chap. 5), we show how this played a crucial role in the process leading to Leucippus' and Democritus' arriving at their doctrine.

> atoms to explain a host of visible changes shifted the burden of explanation away from immaterial form to the structure of the matter in which it inhered...The elimination of the silvery color of the metallic mass resulted from the mere fact that the silver had been divided into particles that were too small and dispersed to see, not from the loss of an old form and the imposition of a new one. (p. 153)

In other words, Sennert's move of introducing the notion of immutation to explain qualitative change (silver losing its colour when dissolved in nitric acid) rendered his retention of the notion of 'substantial form' superfluous. This is because qualitative changes were now seen as resulting from changes in the relations between component parts ('rearrangement of atoms') or what Newman calls 'the structure of the matter' in the above quote. In this understanding, there is no role at all for the Aristotelian notion of 'substantial form'. This paved the way for Boyle's abandonment of this last vestige of the Aristotelian framework retained by Sennert.

Robert Boyle was originally a writer of romances and moral epistles. In the early 1650s, he was tutored by the New England èmigrè alchemist and medical practitioner George Starkey (1627–1665) in chymistry; at the same time, he also developed a great interest in Francis Bacon's empiricism from the circle around the German expatriate scientist Samuel Hartlib (1600–1662). In the mid-1650s, he wrote *Of the Atomicall Philosophy* (only a few leaves of which survived), in which he pledged allegiance to a corpuscular theory of matter. As previously mentioned, in this early manifesto, Boyle began by celebrating the 'new philosophy' of Gassendi, Descartes, and others, giving the impression that his atomism was inspired by their 'new philosophy'. However, Boyle's atomism was actually *grounded on Sennert's chymical experimentation, but without acknowledgment*:

> In this early manifesto of atomism, Boyle borrowed heavily, and without acknowledgement, from the works of Daniel Sennert...Boyle's two examples of reduction to the pristine state—they are taken almost verbatim from Sennert's Hypomnemata physica, *where they appear, juxtaposed, in the same order in which Boyle gives them*...it we consider [Boyle's] Atomicall Philosophy as a whole, it is striking that Sennert's experiment provides the *only* direct evidence for the existence of atoms to be found in the text. (pp. 160, 166–7, 169, our emphases)

In the above quote, we have emboldened the facts that prove irrefutably Boyle's unacknowledged copying from Sennert. Over his career, Boyle consistently kept Sennert away from his own theory:

> In the many places where Boyle actually does cite Sennert, he typically presents him as scholastic, chymist, and physician, not as a corpuscularian. In only one out of fifty-two references does Boyle even intimate that Sennert upheld a corpuscular theory of matter, and there Boyle misrepresents Sennert as an opponent of chrysopoeia. (p. 160, n 13)[12]

Boyle's unacknowledged reliance on Sennert continued in *Sceptical Chymist* printed in 1661. Boyle did, of course, go beyond Sennert. The goal of the mature Boyle was to explain qualitative changes—the phenomena of the sensible world—in non-Aristotelian terms. As noted above, in introducing his notion of immutation to account for qualitative mutations, Sennert had unawares rendered his retention of the Aristotelian notion of 'substantial form' superfluous. Boyle developed this to its logical limit by completely dispensing with that notion:

> [Boyle] would rely heavily on the experimental strategies pioneered by Sennert, though now revised to reveal not only the particulate structure of matter but the ways in which that structure could be altered to engender qualitative change. (p. 175)

To explain this, it is appropriate, at this point, to examine the ways in, and extent to, which Gassendi *did* influence Boyle. Firstly, it is worth noting that Gassendi was a bit of a rare breed in France: In Woolhouse's (1988) examination of eight empiricist philosophers of the 17th and 18th centuries, seven were British, with Gassendi, a stern critic of Descartes, as the sole Frenchman among them. Secondly, and relatedly, it is interesting to note that *Gassendi himself was much influenced by chymistry*: 'It is well known that Gassendi had a strong appreciation of the analytical capabilities of chymistry' and he 'eagerly employed chymistry to show that new *qualities* could be induced where they had not existed before' (pp. 191–2, our emphasis). For instance, by means of hammering, iron becomes hot and a long, thin piece of silver becomes springy; in both cases, a mechanical action (hammering) induces a qualitative change (becoming hot or

[12] For a detailed discussion of this, see Newman and Principe (2002, pp. 20–1).

becoming springy), in which the Aristotelian notion of 'substantial form' plays no role at all. Gassendi's chymical experiments impressed Boyle.

Gassendi explained qualitative changes resulting from a mechanical action by means of the notion of 'textual alteration', with 'texture' meaning the structural attributes of a group of corpuscles, which is said to be altered by mechanical means such as hammering—it will be noted that Gassendi's 'textual alteration' is conceptually akin to Sennert's immutation. Boyle developed this further and combined it with Sennert's reduction to pristine state to arrive at a complete abandonment of the Aristotelian conception of 'substantial form'. In his work titled *Colours*, Boyle gives an experiment: White mercury sublimate (mercuric chloride) is dissolved in water to form a colourless solution; then salt of tartar is added to produce an orange colour; then some oil of vitriol (sulfuric acid) is dropped and the colour disappears. This is the classic Sennertian reduction to pristine state (clear solution—orange solution—clear solution). What this proves, Boyle argues, is that changes in colour can be induced and removed 'by a bare Mechanical change of Texture in the Minute parts of Bodies'. The qualitative change in colour is *mechanical* because it is induced by a change of texture (how a group of corpuscles are arranged), which concerns the relations between component parts.

At this point, a note of clarification would be useful. It will be noted from the above quotes from Newman that when he refers to Sennert's notion of immutation ('rearrangement of atoms'), he speaks of 'the structure of the matter'; similarly, Gassendi's notion of texture is explained as the structural attributes of a group of corpuscles. In today's parlance, the term structure is associated with holism. From the contemporary holistic view, it is the specific relations between component parts that give the whole its *structure*, and it is as a result of this structure that the whole possesses properties of its own that cannot be reduced to the properties of the component parts. Both holism and the mechanical view speak of relations, but relations in holism are understood as *necessary* and *invariant*—it is such relations in which, for instance, hydrogen and oxygen atoms are combined to form a specific structure that is the water molecule; in contrast, the mechanical view regards relations as *accidental* and *variable*. It is clear that the relations between atoms that are conceived of in Sennert's notion of immutation and Gassendi's notion of texture, which Boyle adopted, are accidental and variable, *which can vary as a result of mechanical causes* such as hammering. Hence, the 'structure' spoken of by Newman with reference to Sennert and Gassendi must not be confused with structure as understood in contemporary holism.

Thus, by combining Sennert with Gassendi, Boyle was able to prove the mechanical philosophy experimentally, firstly, by showing the corpuscular nature of matter, and, secondly, by showing that qualitative changes are due to mechanical causes. In other words, it can be seen that *the use of experimentation to test hypotheses about nature (the corpuscular theory and the mechanical cause of qualitative change) and the mechanical philosophy were established in one and the same process, whose lineage was from alchemy.*

As noted, Bacon did have an influence on Boyle. Concerning Bacon, it should first be pointed out that scholars such as Paolo Rossi in the 1950s, Yates (1967, 1972), Rees (1975) and Newman (2004) have established that, like Gassendi, Bacon, and his followers 'found inspiration in the work of the alchemists' and were 'significantly influenced by' them (pp. 12, 190).[13] Newman (p. 223) states that Bacon's role in Boyle's thoughts was that he 'provided Boyle with the general method of composing experimental histories, inculcated a rigorous experimental methodology'. In other words, Bacon's influence on Boyle was in respect of the formal aspect of the formulation of his theory. *It was not Bacon, but Boyle's chymical heritage, which inspired Boyle to engage in experimental testing of hypotheses about nature, while Bacon himself was also inspired by alchemy.* Contrary to Cohen's crediting of Bacon for the seventeenth-century's experimental method, after commenting on Newman's research and analysis, Lindberg (2007, p. 364) observes: 'what credit is left for Francis Bacon (1561-1626), popularly celebrated as the founder (or a founder) of experimental science?...what he and the Baconian tradition of the seventeenth century gave us was not a new method of experiment, but a new *rhetoric* of experiment'. Further, Bajaj (1990) also points out that not only was Bacon's method not new at all, the detailed steps that he laid down were in fact not followed by actual practitioners of laboratory experimentation, both before and after him.

The influence of Descartes' physical (Democritean) atomism on Boyle was similarly peripheral, in his microstructural explanations, Boyle speculated on what he calls the 'mechanical affections' of matter, which include size, shape, texture, and, as a nod to Descartes, motion or the lack of it.

[13] This, in itself, casts serious doubts on Cohen's claim that Bacon 'consolidated' the 'specifically European-colored' 'coercive empiricism' of Renaissance art into the 'fact-finding, practice-oriented mode of experimental science' of the seventeenth century.

Boyle did want to merge his theory with physical atomism, and dreamed of pushing his explanation down to the lowest threshold of what he called *prima naturalia* or primordial particles, but his dream was unrealizable in the laboratory of his time or even a couple of centuries later. As Newman (p. 221) points out, even in the nineteenth century, the chemical atomic tradition admitted nescience about, and was unconcerned with, the atoms of the physical tradition.

9.3 Descartes' Kinetic Mechanical Mathematization of Nature

Descartes was a rationalist, with a frame of mind worlds apart from that of the chymical experimenters.[14] As noted, with the centre stage of the newly emerging science being occupied by the analysis of motion, Descartes' interest in Democritean atomism took a kinetic turn. To analyse the kinetic relations between Democritean atoms, conceptualized as being accidental and variable in accordance with the mechanical view, mathematics is required. Geometry is ill-suited for this purpose, because it deals with *continuous* magnitudes, whereas the mechanical worldview embodied in Democritean atomism sees wholes as simply arithmetic sums of *discrete* parts. Descartes resolved this through his discovery of co-ordinate geometry in which, in a synthesis of geometry and algebra, geometric straight lines, circles, and other figures can be represented by algebraic equations. In the view of Kvasz (2015), modern science is not based upon geometric figures, which cannot deal with either universal laws or interactions between bodies, but upon functions and differential equations, and the first move towards this mathematical approach was precisely made by Descartes.

As is commonly known, algebra from Hindu mathematics was diffused to Europe through Arabic mathematicians. Bala argues that the number-centric approach in Hindu algebra is ideally suited to the mechanical mode of analysis, because it fits the latter's assumption that the whole is nothing but the *sum* of its parts. In Bala's words:

> The notion that a whole can be treated as the sum of its parts is crucial for the development of seventeenth century physics and is the distinctive

[14] The present section is indebted to Bala (2024). No pagination is given for the quote from Bala because, as of the time of writing, the work has not yet been published.

> contribution of the Cartesian analytic method...Thus, the mathematical system inherited from the Indians provided a powerful instrument for the atomic mode of analysis...

Thus, Descartes' kinetic mechanical mathematization of nature cannot be conceptualized in a Eurocentric fashion, as is done by Cohen, simply as the transformation of Athens into Athens-plus. From his dialogical standpoint, Bala concludes that what was needed was 'the combination of kinetic corpuscularism [sic] with the articulated algebraic developments of the Indian number system'.

9.4 Cohen's Turning of an Outdated Ideological Narrative into a Cultural-Essentialist Neo-Eurocentric Argument

Having discussed his three 'revolutionary transformations' constituting the breakthrough into modern science, Cohen's next question is why these three transformations all occurred in the early seventeenth century. If it is true, as it is according to many historians, that Descartes' kinetic rendering of Democritean atomism was due to Galileo's influence, then the concurrence in the timing of Descartes' conceptual innovation and Galileo's breakthrough requires no further explanation. As said, the development of the experimental method involved no 'revolutionary transformation' at all; by the same token, neither the experimental proof of the corpuscular theory, nor the development towards the experimentally based mechanical philosophy can be seen as sudden, revolutionary breaks. In short, from the perspective of our analysis in the previous and present chapters, Cohen's question is, to a considerable extent, a non-question altogether.

Nonetheless, what is interesting about Cohen's question is not the question itself, but the answer that he provides to it:

> The only viable answer resides in the unusually extraverted *propensities* of European civilization—its 'roving' *qualities* and its outer-directed *religious orientation* (...)....Europe had by 1600 reached a point where the *will* toward innovation was furthered by an overall climate in which *daring* novelty stood at a certain premium. (Cohen 2011, pp. 273, 277)

Concerning 'religious orientation', Cohen has explained shortly earlier:

> *As a rule* in the world history of religions, then, abstention from earthly goods was directed away from the world; in the *exceptional* case of European Christianity, it was directed toward it. (p. 264)

In the first-quoted passage, Cohen advances a neo-Eurocentric argument of the cultural-essentialist variety. It is neo-Eurocentric because he accepts the debts owed by Latin Europe to non-European inputs (though, as analysed in Chap. 4, the structure of his book is contrary to this acceptance), but argues that it was certain 'uniquely European qualities' which enabled Europe to make the breakthrough of ca. 1600. But let us begin with the second quoted passage instead of the first. This is, of course, none other than Weber's Protestant-ethic thesis of inner-worldly asceticism, but with a double twist. Firstly, Weber meant his thesis to apply to Protestantism, especially Calvinism, only. Cohen seems unhappy with this—perhaps because doing so would make the inner-worldly ascetic ethic less *intrinsically* European?—hence, contrary to Weber, he claims, in summary fashion, that this ethic is actually characteristic of Christianity except in its earliest phases, by tracing it to 'St. Benedict's rules for the monastic life as promulgated in the 6th century', through 'the Franciscan Order in the 13th century' to the Reformation. To our knowledge, no supporter of Weber's thesis has done this before.

Secondly, Cohen claims that European Christianity's inner-worldly asceticism is 'exceptional', all other religions in the world, 'as a rule' are other-worldly. Again, this was not Weber's position. Whereas Weber did not state in so many words that Confucianism is also characterized by inner-worldly asceticism, it is generally agreed among Weber's followers that based upon his discussions on the issue in *The Religion of China* and *The Sociology of Religion*, this was, indeed Weber's understanding (Choi 2012). Cohen, in fact, may be aware of this. When he discusses Weber's thesis on the economic ethos of different religions in his historiographical inquiry of 1994, he mentions 'Confucianism and Taoism, Hinduism and Buddhism, and Judaism' in the same breath (Cohen 1994, p. 225). However, when he argues that inner-worldly asceticism is the 'exceptional' property of European Christianity in the above quote in his own analysis of the rise of modern science, he states: 'In the great religions of the world, be it Taoism, Hinduism, Buddhism, Islam, or early Christianity, the former variety [other-worldly asceticism] has been predominant as a rule' (p. 263). Confucianism has mysteriously disappeared on this occasion.

While Cohen's double twist with Weber's thesis is significant in itself, we are more concerned with his talk of European civilization's 'extraverted propensities', '"roving" qualities', 'will towards innovation' and placing a premium on 'daring novelty', which will certainly remind the reader of Ronan's 'impulse to exploration' and 'aspiration to break from orthodoxy'. We have already given a non-essentialist account of the last (Ronan's 'aspiration to break with orthodoxy') in the previous chapter, a similar account can be provided for the rest. Let us begin by quoting Cohen (2011, p. 135) again, to give a fuller picture of his argument:

> European society and culture [have been endowed] with an extraordinary *dynamism...restlessness* and *curiosity* instigated wide-ranging *explorations* culminating in the *Voyages of Discovery*.

In our analysis of the role of late-Renaissance scholarly agency in the rise of modern science, by this agency, we refer to the practical-empirical turn and the greater readiness to break with orthodoxy, and the newly emerging values and drives embodied in them, in a fundamental shift in the intellectual climate *among intellectuals*. The relationship between this agency among scholars such as Galileo and the breakthroughs achieved by them is *concretely* shown. In contrast, it can be seen that the 'propensities' and so on that Cohen speaks of pertain to, to quote from him, 'European society and culture' in general. How these *society-wide* 'qualities' posited by Cohen are related in *concrete* terms to the occurrence of his three 'revolutionary transformations' *among scholars* is left unexplained. In other words, this does not amount to any explanation at all. Rather, it is simply a statement of a pre-conceived cultural-essentialist view misconceived as an explanation, hence we refer to it as a pseudo-explanation below.

Secondly, Cohen speaks of 'restlessness and curiosity instigat[ing] wide-ranging explorations culminating in the Voyages of Discovery'. As shown, to point out that the 'voyages of discovery' narrative constitutes an imperial ideology is not political correctness, but historical fact. That this ideological narrative could have successfully been constructed and that it is still able to live on today are made possible *only* because Europe's imperial enterprise succeeded. To see it repeated in the twenty-first century in serious scholarship is both surprising and unexpected. The 'extraverted propensities', '"roving" qualities', 'will toward innovation', putting a premium on 'daring novelty' and 'dynamism' mentioned by Cohen are part and parcel of this ideological narrative. To give these 'qualities' their proper description, they are all part and parcel of a ***predatory will to conquer, control, and dominate***, engendered by circumstances that we have previously explained. Cohen does make a token admission of the violence of the European colonizers, but stated: 'Their aggression, however, did not stand in the way of a genuine interest in the rules and habits and the natural surroundings of the peoples they sought to subject' (p. 136). Two comments on this statement are in order. Firstly, as seen, on the issue of evidence of knowledge transmission from non-European sources to Europe, Cohen demands strict observance of the principles governing the criminal law court in historical investigation; in contrast, here, he asserts 'genuine interest', which is supposed to be divorced from the predatory will to conquer and so on, without providing a single shred of evidence. If Cohen provides no evidence to back up his assertion, let us look at some evidence relevant to it, going beyond the cases of Columbus, da Gama, and others like them lionized as 'curious' 'discoverers'.

In Chap. 8, we mentioned Amerigo Vespucci who castigated the ancients such as Pliny for being plainly wrong from his own first-hand experience on his trips to the 'New World'. Did he go there with, in Cohen's wording, a 'genuine interest in the rules and habits and the natural surroundings of the peoples [there]'? Vespucci was a Spaniard living in Florence looking after the business of Lorenzo di Pierfrancesco de' Medici, head of a junior branch of the Medici family; Lorenzo's business in Seville, Spain was handled by one Berardi, who also had his own business in African slavery. In 1492, Vespucci returned to settle in Seville, still looking after Lorenzo's business, but became increasingly involved in Berardi's business as well. Lorenzo, incidentally, was none other than Columbus' patron, while Berardi was also a sponsor of Columbus' first voyage. When Berardi passed away, Vespucci was the executor of his will, and he continued to provision ships going to the West Indies. The trip of 1499 mentioned in the previous chapter that Vespucci joined was for the purpose of investigating a rich source of pearls that Columbus had reported, this trip was partly financed by none other than Vespucci himself (Poskett 2023, p. 16).

Hans Sloane was an Irish physician and naturalist who went to Jamaica in 1687 and returned to London in 1689. Upon his return he published two large illustrated volumes of *The Natural History of Jamaica*. Was this a product of Sloane's 'genuine [scientific] interest'? Jamaica was, in Sloane's words, 'The Largest and Most Considerable of Her Majesty's Plantations in the Americas'. 'Sloane himself profited from this world [of British plantations in the 'New World']...he had access to one third of the profits of a large plantation on Jamaica. He also invested in a number of financial schemes in the Americas, including one to sell "Jamaican bark", another possible alternative to cinchona' (Poskett 2023, p. 140)—cinchona was the plant that proved to be highly effective in treating malaria found in Peru, which was monopolized by the Spaniards.

As mentioned in the previous chapter, the Spaniard Jose de Acosta went to the 'New World' in 1571 and spent fifteen years there; after returning to Europe, he published a book titled *The Natural and Moral History of the Indies*. De Acosta was a Jesuit missionary, and he went on his trip in order to seek converts. Similarly, Bernardino de Sahagun, also mentioned in the previous chapter, was a missionary. He spent more than fifty years in what is today Mexico studying the beliefs, culture, and history of the Aztecs, publishing the *General History of the Things of the New Spain*. As Poskett (2023, p. 22) observes, for Sahagun, this 'was primarily a religious work' with missionary objectives, despite its great intellectual value.

To sum up, the interest that Vespucci, Sloane, de Costa and Sahagun showed in 'the rules and habits and the natural surroundings of the peoples' of the 'New World' cannot, contrary to Cohen's above assertion, be separated from either imperial commercial interests or evangelical objectives. To assert the existence of a supposedly 'culturally unique' 'genuine interest' in or 'curiosity' about foreign things purportedly existing independently of these interests and objectives cannot stand up to scrutiny. As Poskett (2023, pp. 142, 146) observes with reference to natural history: 'Across the Americas, European naturalists quizzed enslaved people, hoping to discover new plants that might turn a profit, particularly if they had medicinal qualities'; 'wherever we look, the development of natural history in this period cannot be separated from the commercial world of trade and empire'.

The second comment on Cohen's 'genuine-interest' assertion that needs to be made is the following. It will be recalled from our discussion on scholarly tradition with reference to the example of Parmenides and Marx that though the former's 'reality versus appearance' distinction was, in part, mystically inspired, that did not prevent the anti-mysticism Marx from adopting it. Likewise, it may very well be that the success of Europe's predatory will to conquer in time did engender a genuine interest as a general cultural disposition; that, however, does not make this disposition, *given its origins*, an essentialist European quality. Further, it must not be forgotten that up to the present day, this interest, assuming that it does exist, is still very much *enabled* by Euro-American wealth and dominance over the rest of the world—very few Africans, for instance, would have any opportunity of developing and pursuing such an interest. Finally, other peoples at different historical periods have shown such an interest as well, *without* it being part and parcel of a predatory will to conquer and so on, as we show shortly below.

In Chap. 7, it was seen that Cohen was able to find support for his cultural-obstacle argument from none other than Needham. Here, again, he is able to find support for his cultural-essentialist pseudo-explanation from Needham:

> the exceptional poverty of Europe in the precious metals, the continual desire of Western peoples for commodities which they themselves could not produce (…)…By contrast China was a coherent agrarian land-mass, *a unified empire* since the third century B.C. [sic] with an administrative tradition unmatched elsewhere till modern times, *endowed with vast riches*…Europe a culture of rovers [cf. Cohen's '"roving" qualities'], was always uneasy within her boundaries [cf. Cohen's 'restlessness'], nervously sending out probes in all directions [cf. Cohen's 'curiosity']…The greater population of China was *self-sufficient*…generally *content* with only occasional exploration, essentially *incurious* about those far parts of the world…It may well be that here, at this point of tension, lies some of the secret of the specific European creativeness when the time was ripe. (Needham 1969, quoted in Cohen 2011, p. 135, our emphases)

This is a caricature that one still regularly comes across in popular Euro-American discourse (which, unfortunately, is also often taken up in non-Euro-American discourse due to Euro-American cultural dominance) as well as uninformed scholarly argument, and it is unfortunate that Needham completely shared it. Firstly, the point about Europe's purported 'specific creativeness' does not square with Needham's overall view at all. This is because since the socio-geographical explanation given by Needham applies trans-historically, one wonders how, lacking Europe's 'specific creativeness', China was able, as Needham believed, to be ahead of Europe not only in technology but also in science up to 1500. One also wonders why Europe's 'specific creativeness' had been lying dormant until then—this remains a mystery even in the context of Needham's untenable argument about modern science being the product of the 'rise of the merchant class to power' (see Chap. 2), which is what 'when the time is ripe' refers to.

More significantly, Needham made a number of contrasts between China and Europe, which we have italicized in the above quote with regard to China. These contrasts add up to the following narrative: China was a unified (implying and connoting stable) empire, rich and self-sufficient in resources, hence, it was content with itself, and incurious about the outside world, whereas the opposite applied to Europe. In this narrative, a simplistic socio-geographical explanation is given for a *cultural-essentialist* argument, and the claims concerning China's alleged contentment and incuriosity, which supposedly failed it with regard to the development of modern science, constitute the classic *'inhibitory'-factors* approach. Further, Europe's predatory will to conquer and so on is nowhere to be found in this narrative. Since the 'voyages of discovery' narrative, under the ideological influence of which Needham had clearly fallen, has already been deconstructed, we focus on Needham's claims with regard to China in the following. To keep matters simple, we shall restrict our discussion to the following for the moment: While Needham's knowledge of China with regard to technology and science is certainly formidable, the same cannot be said of his knowledge of China in other respects, and *each and every element of his just-mentioned narrative with regard to China is at variance with the simplest historical facts about China.* This is no exaggeration, and it is unimaginable that Needham was not aware of at least some of the historical facts that we are going to discuss, at the end of our discussion, we shall explain why, despite that, Needham could still have advanced his narrative.

Firstly, Needham claimed that China was 'a unified [implying or connoting stable] empire since the third century B.C. [sic]'. He counted from −221 with the founding of the Qin Dynasty, but China as a state based upon a particular ethic (after Confucius, it came to be called the Confucian ethic) began in −1046 with the founding of the Western Zhou Dynasty, which is why historians usually begin with −1046 in discussing China as a civilization, which is certainly the more appropriate starting point in our present context as well. But it should be noted that counting from −1046 instead of −221 does not have any substantive effects on our argument. From −1046 until 1911 when dynastic China ended spans 2957 years. We have calculated that out of this total, a unified state under Han-Chinese rule over the whole of China[15] existed only in 44.1% of the time (the

[15] China's borders varied over time, but for simplicity, we bypass this detail, which does not affect our argument.

Western Zhou, Qin, Han, Western Jin, Sui, Tang, and Ming dynasties); in 43.6% of the time, China was either fractured, or had large parts of the country occupied by invading nomads, or both; and in 12.3% of the time, the whole of China was ruled by invading nomads (the Mongolian Yuan and the Manchurian Qing dynasties). Even when the whole country was unified under Han-Chinese rule, nomadic invasions were a *perennial* threat—to illustrate with the three most powerful Han-Chinese dynasties in China's history, the Marshall Emperor of Han was engaged in a fifty-year war with the invading Xiongnu (see next paragraph); the invading Tibetans once occupied the capital city of the Tang Dynasty when the latter was fighting an internal rebellion; the Ming overthrew the Mongolian Yuan Dynasty in 1368 and drove the Mongols back to the north, but the latter kept preparing to invade China again, which they eventually did in 1449, in the course of which they managed to take the Ming emperor captive for four years, almost causing the Ming to collapse, and after the Mongols came the Manchus who were eventually to overthrow the Ming in 1644. The nomadic peoples who invaded and occupied (in part or in whole) China over the centuries numbered well over a dozen. China was, to borrow Needham's expression, 'uneasy within her boundaries' *throughout the ages.* This factor is of fundamental importance for understanding China's historical development, and those who fail to realize it are ill-equipped to understand that development, as we shall have various occasions to highlight below and in the Appendix.

Needham claimed that because Europe was 'uneasy within her boundaries', it 'nervously sen[t] out probes in all directions', whereas the opposite applied to China. However, the historical facts tell a diametrically opposite story. As every elementary pupil of China's history well knows, the Silk Road was established by the Marshall Emperor of Han (reigned −141 to −87), none other than because he was looking for allies in central Asia against the invading Xiongnu (匈奴 one branch of which was to become the Huns that invaded Europe later). His official historian Sima Qian described the mission of his envoy who made two journeys to central Asia as 'probing into the unknown' (凿空 literally chiselling into the void). In addition, as chronicled in the *Book of Han: Treatise on Geography* (《汉书·地理志》),[16] the Marshall Emperor also sent envoys by sea to the south,

[16] Historical chronicles with the title *Book of [dynasty name]*, in some cases with further specifications, compiled on the basis of official documents, are authoritative official histories (*Zhengshi*正史). Besides these, there are unofficial historical chronicles that fall into two types, known respectively as *Bieshi* (别史) and *Yeshi* (野史) (the distinction between these

who reached today's SE Asia and Sri Lanka. According to the *Book of Liang* (《梁书》),[17] after the fall of the Han Dynasty, diplomatic exchanges with various SE Asian countries and Simhala (in today's Sri Lanka) continued under the aegis of the Kingdom of Wu (东吴), which occupied the eastern and southeastern coastal regions of China, in the third century.

China was indeed rich in resources for an agrarian society, but it did not follow, as Needham claimed, that it did not rely on many kinds of imports. Whereas China's largest import category consisted of spices, it also imported a variety of other things, including military supplies such as hides, sinews, horns, sulfur, wax and oils, steel, steel swords, and a weapon known as *huoyou* or *meng huoyou* (火油、猛火油) from Arab merchants; timber from Japan probably for shipbuilding; large amounts of metals such as gold, tin, lead, and copper (Lo 2012, pp. 133, 217); manufacturing raw materials such as a dye from Persia for making the renowned 'blue and white porcelain' (青花瓷); silver for its currency; and so on. Thus, as early as during the reign of the Marshall Emperor of Han, maritime trade links between southeastern China and SE Asia and beyond had already come into existence; as is well-known, India became the mid-way station at which trading between Roman and Chinese merchandise occurred (Li Jinming 2015, p. 11)—the Romans had great demand for Chinese silk, while upper-class Han ladies were greatly fond of Roman glass jewellery.

It is true that as an agrarian society, from early on, the Chinese state was wary of the de-stabilizing effects on an agrarian society of a merchant class growing too wealthy and powerful, hence merchants were subjected to various restrictions most of the time, which also applied to maritime merchants. But that did not stop the maritime trade route from prospering. According to the *Book of Liang*, by the early sixth century, Chinese maritime merchants were aware that sailing further west from Sri Lanka, one could reach the same central Asian states relations with which China had previously established through the Silk Road. According to one Arabic source, during the Tang Dynasty (618 to 907), Chinese merchant ships were regularly sailing to the Persian Gulf (Lo 2012: 126). The top Tang official Jia Dan (贾耽 730–805), who was especially interested in geography and cartography, compiled two books and a map, respectively titled

two types need not concern us). While the authenticity of unofficial histories varies from case to case, historians do not lightly dismiss them but, instead, pay great though cautious attention to them as being supplementary to official histories. This is because official histories sometimes leave out materials through self-censorship of the officials compiling them.

[17] This is the official history of the last two of the southern dynasties of the period of the Northern and Southern Dynasties (南北朝 420–589).

Records of the Far Reaches of the Emperor's Splendour (《皇华四达记》), *Descriptions of Past and Present Regions, States, Counties, Routes and Foreigners in All Four Directions* (《古今郡国县道四夷述》), and *Map of Chinese and Foreign Territories of the World* (《海内华夷图》), which were presented to the throne in 798 and 801 (Lo 2012, pp. 117–121).[18] Jia compiled his books and map on the basis of detailed questioning of both Chinese and overseas individuals, including maritime merchants, who had direct knowledge of the matters concerned. In his books, Jia listed seven different routes to west Asia and eastern Africa, two of which were sea routes, one called the East Sea Route, along the coast from India all the way to Baghdad, the other called the West Sea Route, which, making use of the monsoon winds, crossed the Indian Ocean from the south of India to today's Tanzania, then up Africa's east coast to the Gulf of Eden and then to west Asia. Not only did the Marshall Emperor of Han send out 'probes in all directions', Chinese maritime merchants had since the -second century been engaging in long-distance maritime trade.[19] The difference between China and Europe in this respect is not that China lacked, to quote from Needham's above quote, 'the continual desire…for

[18] These works have not survived, but the books are listed, with their contents summarized, in the official *New Book of Tang: Accounts of Books: Geographical Titles* (《新唐书·艺文志·地理类》); while later maps that have survived are believed to have been drawn on the basis of Jia's map. Incidentally, cartography of foreign lands began systematically from the time of the Tang Dynasty onwards, becoming the duty of a special government unit (Lo 2012, pp. 120–1).

[19] In contrast, 'Though Europe had traded with East Asia for centuries, it had always been done through intermediaries…From the tenth century on, Venice had obtained a virtual monopoly on the transshipment of spices from Alexandria to Europe…It was only with the consolidation of Mongol power in Central Asia that the northern trade routes reopened, bringing Polo and other Europeans back into Central Asia for the first time, and into direct contact with China for the first time ever' (Pomeranz and Topik 2006, p. 19). While this is generally true, two caveats are, however, required, both relating to Rome. Firstly, Roman merchant ships did operate in the Arabian Sea from Egypt (which was under its rule) in trading with the ports in south India. Secondly, direct contact between Rome and Han did occur once. According to the *Book of Later Han* (《后汉书》), in the year 97, Han sent an envoy with the mission of visiting Rome, but, taking the land route of the Silk Road, he only managed to go as far as the Persian Gulf, at which point, failing to recruit assistance from the Arabs to go further west, he returned home; over seven decades later, a Roman mission taking the sea route (most probably on board Indian and/or SE Asian merchant ships in the journey east of India—see Glover 1989 on early trade relations between SE Asia and India—as there is no evidence that Roman ships ever rounded Cape Comorin on the southern tip of the Indian sub-continent) successfully reached China in the year 169.

commodities which they themselves could not produce', but that what China produced, such as silk, porcelain, and tea, were in great demand overseas, hence, it could trade them on strictly commercial terms for the just-mentioned products; in contrast, there was little demand in Asia for what Europe produced in the post-Roman era, which was why when Europeans came to Asian waters in the late fifteenth-century, they came with a predatory will.

Needham said that the Chinese, being allegedly self-sufficient, were 'content', 'essentially incurious about those far parts of the world', which Cohen (2011, p. 136) spices up as 'contemptuous indifference'. Why such a purportedly 'contented', 'incurious' and 'contemptuously indifferent' people were so quick in establishing long-distance land and sea trade routes seems inexplicable. Other historical facts also tell an entirely different story. Buddhism diffused to China during the Han Dynasty, precisely as a result of the establishment of the Silk Road, but Chinese converts were *not content with nor contemptuously indifferent about*[20] what they had learnt through this diffusion, hence, soon afterwards, *curious* Chinese Buddhist monks began making their way to central Asia and eventually India in search of Buddhist canons in the Sanskrit original so as to *deepen their knowledge* of the doctrine. The first to do so was Zhu Shixing (朱士行 203–282), China's first *samana* (Buddhist ascetic), who went to the Buddhist kingdom of Yutian (于阗) in today's southern Xinjiang in 260. Zhu pioneered a systematic trend in which 'hundreds of Chinese [Buddhist] monks who made pilgrimages to India during the first millennium CE' (Sen 2006, p. 24). In this trend, the Chinese pilgrims showed an intense *aspiration* to deepen their learning of Buddhist doctrine from *another civilization* by going back to the canons in original Sanskrit; and, as we show presently, their *curiosity* about and *interest* in other civilizations were not limited to religious matters. To illustrate this, we briefly discuss two of the most well-known of these pilgrims in the following.

In 399, the Chinese monk Fa Xian (法显 337–422) made a journey that took thirteen years, taking him to over thirty states in today's Xinjiang, central Asia, India, and Sri Lanka, eventually returning to China by sea. During the Tang Dynasty, the Chinese monk Xuan Zang (玄奘602–664) went on a similar trip in 627 that lasted seventeen years and took him to 138 states altogether. Both Fa Xian and Xuan Zang did not only return to

[20] In the present account, we italicize words so as to contrast with Needham's and Cohen's claims.

China with original Buddhist canons, once back in China, they wrote travelogues of their journeys, respectively *The Biography of Fa Xian* (《法显传》), known as *A Record of the Buddhist Kingdoms* (《佛国记》) in its first edition,[21] and *A Record of the Western Regions* (《西域记》).[22] These travelogues recorded not only matters relating to religion, such as the nature of Buddhist doctrines, rituals, and monastic institutions, but also meticulous accounts of 'the social and political conditions in South Asia and kingdoms situated on the routes between China and India', and, in relation to India specifically, 'the geography and climate, the measurement system, and the concept of time…urban life and architecture…the existing caste system, the educational requirements for the Brahmins…legal and economic practices, social and cultural norms, and the eating habits of the natives, and lists the natural and manufactured products of India' (Sen 2006, pp. 24, 29).

Travelogues were not only written by Buddhist monks travelling overseas. Previously, we mentioned that during the third century, envoys were sent by the Kingdom of Wu to southeast and south Asia to upkeep diplomatic relations with the states there, upon returning to China, one of these envoys, named Zhu Ying (朱应), wrote two travelogues titled *Records of Alien Things in the Kingdom of Funan*[23] (《扶南异物志》) and *Records of Foreign Countries in the Time of Wu* (《吴时外国传》).[24] During the Tang Dynasty, a soldier named Du Huan (杜环) was captured by the Arabs in a battle in today's Kazakhstan in 751, during his time in west Asia, he travelled extensively throughout the region as well as in north Africa. Upon his return to China in 762, he wrote a travelogue titled *Travel Records* (《经行记》); though the book itself has not survived, excerpts from it can be found in various authoritative sources. The most significant part of Du's book is probably its report on Islamic medicine. Travelogues dating from the Song Dynasty include *Response to Enquiries Concerning Places Beyond the Range*[25] (《岭外代答》) by Zhou Qufei (周去非 1134–89) and *Records of the Various Vassals* (《诸藩志》)

[21] An English translation of this work is available online at: http://onlinebooks.library.upenn.edu/webbin/gutbook/lookup?num+2124

[22] A recent English translation of this work is Li Rongxi (1996). 'Western regions' was the term used to refer to today's Xinjiang and beyond.

[23] This kingdom was in present-day Cambodia.

[24] Both works have not survived, but they were mentioned with summaries of their contents in official histories (see note 16) of subsequent eras.

[25] The Range refers to a series of mountains in southern China.

by Zhao Yushi (赵汝适 1170–1231). These works included detailed information on sea routes and states along the maritime trade route. Summarizing what these works show, Lo (2012, p. 119) observes: 'Thus the Chinese were well informed about the countries bordering the East and South China Seas and the Indian Ocean, and they had heard of Spain, North Africa, Sicily, and Asia Minor. Chinese ships may have visited the east coast of Africa down to Zanzibar and the island of Madagascar'.[26] Shortly after Polo, the Yuan Dynasty traveller Wang Dayuan (汪大渊) travelled all over the Indian Ocean twice by sea, the first time for five years from 1330 to 1334, and the second time for three years from 1337 to 1339. His travelogue, covering over two hundred places, is titled *Records of Islands and Barbarian Lands* (《岛夷志略》). There are also post-Yuan records of alien lands, but we shall stop here.

It is clear that the writing of records of overseas travels on the geography, people, indigenous produce, customs, and so on of the places visited was a consistent tradition from the fourth century onwards, if not earlier. Like Greece and Rome, China regarded most of its neighbours as 'barbarians',[27] but this did not prevent *an interest and curiosity to get to know about* these peoples and their societies. As Lo (2012, p. 120, our emphases) observes: 'Many accounts were written of these [overseas] expeditions to satisfy the *desire* of the Chinese at home who wished to *learn about* the strange peoples and exotic things across the sea'. Significantly, this interest and curiosity were *not* part and parcel of a predatory will to conquer and so on. *The Travels of Marco Polo* is known all over the world, whereas knowledge of the above records, all dating from pre-Polo times except for Wang Dayuan's *Records of Islands and Barbarian Lands*, is limited to a tiny number of people who do pay attention to China's history. This contrast is simply the result of Euro-American cultural dominance since the imperial age.

We remarked earlier that it was only because Europe's imperial enterprise succeeded that it was able to construct its predatory will to conquer and so on into the cultural-essentialist ideological narrative of Europe's 'dynamism', 'curiosity', 'exploratory spirit', and so on that remain, as illustrated by Cohen's pseudo-explanation, influential up to the present day. It

[26] On the possibility that Chinese merchant ships had visited the east coast of Africa down to Zanzibar and Madagascar during this period, see Schwarz (1936).

[27] There was one significant exception to this, namely, the high regard that the Chinese held for the Indians, from whom they imported Buddhism.

should now be obvious that it is because of their being dominated by this narrative that, regrettably, even scholars who should have known better about China such as Needham—as already remarked upon, Needham could not possibly be ignorant of at least some of the facts mentioned above, such as why and how the Silk Road was established—can overlook, ignore, or otherwise forget the simplest historical facts about China in culturally essentializing it.

Let us finish by summarizing the previous and present chapters. Cohen's cultural-essentialist neo-Eurocentric argument turns an outdated ideological narrative into a pseudo-explanation. His argument of the transformation of Alexandria into Alexandria-plus is Eurocentric because it completely ignores Latin Europe's absorption of the realist trend in Arabic mathematical sciences. A dialogical critique of this argument of Cohen's is, however, inadequate, because it cannot explain how and why Kepler and Galileo did advance beyond the Arabs. Such an explanation can only be found by analysing the variable of late-Renaissance scholarly agency, which can be done on a non-essentialist basis in terms of our theoretical-methodological framework.

Cohen's argument concerning his third 'revolutionary transformation' is erroneous in every respect: His argument of 'coercive empiricism' is Eurocentric; the experimental method of the seventeenth century had nothing to do with 'coercive empiricism'; his crediting of Bacon as being responsible for his third 'revolutionary transformation' mistakes Bacon's rhetoric for reality; there was, in fact, no 'revolutionary transformation' in this regard at all, for the development from Geber and Paul of Taranto to Sennert and Boyle was a continuous process. As the popularizer of the notion of the mechanical philosopher, Boyle's conception of that philosophy was derived from chymical experimentation, not physical Democritean atomism. Descartes' physical kinetic atomism owed considerably to Galileo's work on motion. In Descartes' contribution to the mathematization of nature in accordance with the mechanical worldview, the number-centric algebra of Hindu mathematics, diffused to Europe through Arabic mathematicians, played a crucial role; hence, Cohen's representation of Descartes' kinetic atomism as the direct transformation of Athens into Athens-plus constitutes another Eurocentric misrepresentation.

Finally, in stressing, as does dialogism, the role of knowledge transmissions from non-European sources to Europe in the rise of modern science, we, however, unlike dialogism, pay equal attention to the role played by European scholars. This can be seen not only in our call to transcend dialogism by taking into account the variable of late-Renaissance scholars' scholarly agency; but also in our highlighting of the contributions made by the scholastic scholars Buridan and Oresme, to whom Galileo owed a great debt, the contributions made by Latin European alchemists on the basis of what they had learnt from Arabic alchemy, and, of course, the achievements of the scholars directly responsible for making the breakthroughs. To see modern science as arising from the cross-fertilization of various civilizations, one must see Latin Europe itself as one of these civilizations.

Cohen's pseudo-explanation, with an illustrious precedent laid down by none other than Needham, as well as Huff's argument (see Chap. 7), is of the cultural-essentialist variety, which is probably the most important form of essentialist arguments today with the rise of neo-Eurocentrism. In Chap. 2, with regard to neo-Eurocentrism, we commented as follows:

> although it is useful to contrast neo-Eurocentric arguments with conventional Eurocentric ones, in terms of substance, there is actually very little that is really 'new' in the former; the only 'newness' consists in their application of conventional cultural-essentialist arguments but restricting that application to the period of the breakthrough into modern science and thereafter.

Whether in conventional or neo form, cultural essentialism claims that Europe possessed certain cultural qualities that were 'unique' to it. Of these cultural qualities, most are directly derived from the 'voyages of discovery' narrative: Ronan (in a conventional Eurocentric argument) speaks of Europe's 'impulse to exploration' (see Chap. 3); Cohen speaks of Europe in terms of '"roving" qualities', 'outer-directed', 'daring', and 'restlessness and curiosity instigating wide-ranging explorations'; Huff speaks of Europe's 'curiosity' by further specifying it as an 'intellectual curiosity'. Since, like Needham, Eurocentric and neo-Eurocentric scholars invariably conceptualize the NQ in dual form, while the 'voyages of discovery' narrative refers only to Europe, a more encompassing narrative is required to contrast Europe, which supposedly possessed the above 'unique' qualities, with non-Europe, which supposedly lacked those qualities. With reference to China as non-Europe, this more encompassing narrative takes the form of a contrast between 'maritime civilization' (Europe) and 'continental civilization' (China), hence, we refer to it as the 'maritime vs continental civilizations' (MvC) narrative.

In the MvC narrative, it is claimed that Europe, constituting a 'maritime civilization', was 'outwardly oriented' and 'curious' about the outside world, hence, 'lured' by the sea and what lay in its far reaches, 'fearless' of its hazards; in contrast, dynastic China, constituting a 'continental civilization', is said to be 'bound' in 'self-sufficiency' to the land, 'inwardly oriented' and 'incurious' about what lied beyond this 'self-sufficient' land, 'shying away' from the hazards of the sea. It is clear that the 'M' component of the MvC narrative is none other than the 'voyages of discovery' narrative, which has been critiqued in Chap. 8 as well as in the above. It is also clear that Needham's claims about China as being 'a coherent agrarian land-mass, a unified empire since the third century B.C.', 'endowed with vast riches' in 'self-sufficien[cy]', 'content' and 'essentially incurious about those far parts of the world' are part and parcel of the 'C' component of the MvC narrative, and while Needham did not use the term 'continental', his 'a coherent agrarian land mass' is synonymous to it. Our above critique of Needham's claims, however, constitutes only a partial critique of the 'C' component. In view of the special prominence of cultural essentialism today, a more comprehensive and systematic critique of this component is clearly required. To our knowledge, this has never been done before in the literature, hence, we include an appendix for this purpose.

Bibliography

Bajaj, Jatinder K. 1990. 'Francis Bacon, the first philosopher of modern science: a non-Western view'. Pp 24-67 in Ashis Nandy (ed.) *Science, Hegemony and Violence: A Requiem for Modernity* (1990[1988]). Oxford University Press.

Bala, Arun. 2024 (in press). 'Explaining the rise of modern science: A dialogical perspective'. Chapter 16 of Arun Bala, Raymond W.K. Lau and Mei Jianjun (eds.), *Multicivilizational Exchanges in the Making of Modern Science. Needham's Dialogical Vision.* Palgrave Macmillan.

Choi, Mihwa. 2012. 'Extreme asceticism: Confucian practice and Riesebrodt's religious virtuoso', *Journal of the Scientific Study of Religion* 51(3), pp 456-67.

Cohen, H. Floris. 1994. *The Scientific Revolution: A Historiographical Inquiry.* University of Chicago Press.

Cohen, H. Floris. 2011. *How Modern Science Came into the World: Four Civilizations, One 17th-Century Breakthrough.* Amsterdam University Press.

Glover, Ian C. 1989. *Early Trade Relations Between India and South-East Asia*. University of Hull Centre for South-East Asian Studies occasional paper.

Goody, Jack. 2004. *Islam in Europe*. Polity.

Guthrie, W.K.C. 1965. *A History of Greek Philosophy, Vol. 2, The Presocratic Tradition from Parmenides to Democritus*. Cambridge University Press.

Hartt, Frederick and David Wilkins. 2010. *A History of Italian Renaissance Art*, 7th edition. Pearson.

Kvasz, Ladislav. 2015. 'Mathematical language as a bridge between conceptualization of motion and experimental practice'. Pp. 229-48 in Raffaele Pisano (ed.) *A Bridge Between Conceptual Frameworks. Sciences, Society and Technology Studies*. Dordrecht: Springer.

Lau, Raymond W.K. 2020. *Intellectual Developments in Greece and China: Contingency, Institutionalization and Path-Dependency*. Newcastle upon Tyne: Cambridge Scholars Publishing.

Li Jinming. 2015. 'The development and changes in China's ancient maritime silk road'. 李金明，'中国古代海上丝绸之路的发展与变迁'，《新东方》，2015年第一期，页10-15.

Li Rongxi. 1996. *The Great Tang Dynasty Record of the Western Regions*. Berkeley: Numata Center for Buddhist Translation and Research.

Lindberg, David. 2007. *The Beginnings of Western Science*, 2nd edition. Chicago: University of Chicago Press.

Lo, Jung-Pang. 2012. *China as a Seapower, 1127-1368*. National University of Singapore Press.

Needham, Joseph. 1969. *The Grant Titration*. Taylor and Francis.

Newman, William and Lawrence Principe. 2002. *Alchemy Tried in the Fire: Starkey, Boyle, and the Fate of Helmontian Chymistry*. Chicago: University of Chicago Press.

Newman, William. 2004. *Promethean Ambitions*. Chicago and London: University of Chicago Press.

Newman, William. 2006. *Atoms and Alchemy. Chymistry and the Experimental Origins of the Scientific Revolution*. Chicago and London: University of Chicago Press.

Poskett, James. 2023[2022]. *Horizons: A Global History of Science*. UK: Penguin Books.

Pomeranz, Kenneth, and Steven Topik. 2006. *The World that Trade Created*, 2nd edition. Armond: M.E. Sharpe.

Rees, Graham. 1975. 'Francis Bacon's semi-Paracelsian cosmology'. *Ambix* 22, pp 81-101.

Schwarz, E.H.L. 1936. 'The Chinese connection with Africa'. *Journal of the Royal Asiatic Society Bengal Branch*, No. 4 of 1936.

Sen, Tansen. 2006. 'The travel records of Chinese pilgrims Faxian, Xuanzang, and Yijing'. *Education about Asia* 11(3), pp 24-33.
Woolhouse, R.S. 1988. *The Empiricists.* Oxford and London: Oxford University Press.
Yates, Frances. 1967. 'The Hermetic tradition in Renaissance science'. Pp. 255-74 in Charles S. Singleton (ed.), *Art, Science, and History in the Renaissance.* Baltimore: John Hopkins University Press.
Yates, Frances. 1972. *The Rosicrucian Enlightenment.* London: Routledge and Kegan Paul.

CHAPTER 10

Conclusion

The dialogical perspective argues that modern science is a multi-civilizational outcome, in that it had arisen on the basis of inputs or knowledge transmissions from various non-European civilizations into Latin Europe, either direct or indirect. Some of the developments arising from the process of such multi-civilizational exchanges are incontrovertible and firmly established by ample direct evidence, and their effects are well-established, such as Arabic scholarship's debts to Hindu mathematics and the Greek corpus; Arabic scholarship's taking of Greek learning to new heights; and Europe's debts to Arabic scholarship. Our arguments that Greece was a separate and distinct civilization from Latin Europe, and that Greece's development owed considerable debts to west Asia and Egypt, serve to strengthen the dialogical perspective. We accept dialogism as a valid and important starting point for resolving the positive NQ on a non-essentialist basis and for rebutting conventional and neo-Eurocentric accounts. However, we argue that it is necessary to transcend dialogism. This is because, by itself, dialogism is, firstly, unable to resolve the question of why Europe, based upon what it had learnt from non-European inputs, had been able to make the breakthrough into modern science itself, and, secondly and relatedly, it is unable to decisively rebut conventional Eurocentric accounts and to deal with neo-Eurocentric accounts at all. Both inabilities stem directly from dialogism's fatal limitations as a perspective in theoretical and methodological terms because it does not

R. W. K. Lau, *Rethinking the Needham Question*,
https://doi.org/10.1007/978-981-97-9472-0_10

engage in any analysis of the phenomenon of intellectual development itself. Additionally, from our analysis in Chaps. 8 and 9, it can be seen that dialogism has also failed to take into account important developments within Latin Europe itself that prepared the ground for the breakthrough into modern science, namely, the work of Buridan and Oresme on motion, and the work of the alchemists on corpuscular theory and the use of experimentation.[1] It is only by transcending dialogism in theoretical and methodological terms that its above-mentioned failures can be overcome, and a satisfactory resolution of the positive NQ on a non-essentialist basis can be achieved, and both conventional and neo Eurocentric accounts decisively rebutted.

In Fig. 10.1,[2] we try to provide a graphical representation of the genealogy of the rise of modern science from our non-Eurocentric perspective that transcends dialogism. This chart has considerable family resemblance

Fig. 10.1 (continued) the source of Brahe's model (Joseph 2011, pp. 418–9). (18) Transmission of the method of solving simultaneous linear equations and determinantal calculations and solutions between the late 12th to late 15th centuries; and transmission of the notion of the universe being an infinite empty space through Matteo Ricci in 1595. (19) Emergence of late-Renaissance scholarly agency: the practical-empirical turn and greater readiness to break with orthodoxy. (20) Latin Europe's rivalry with and hatred of Islam beginning with its counter-offensive against Islam in the eleventh century. (21) The fall of Constantinople. (22) The so-called 'voyages of discovery' and subsequent imperial conquests. (23) The Reformation. (24) Inheritance of pre-1600 Latin European developments by scientists involved in the breakthrough into modern science (specifically in two areas: Buridan and Oresme on motion; and alchemy in respect of corpuscular theory and the use of experimentation). (25) Possible transmission of Kerala School mathematics that might have paved the way for the development of calculus (Bala 2006, chap. 7). (26) The breakthroughs of Kepler, Galileo, Boyle, Descartes, and others

[1] Bala's (2024) argument that the use of experimentation was inspired by the import of several key Chinese technologies (gunpowder, the navigational compass, and printing) is unconvincing because it completely fails to show in concrete terms how such technologies were supposed to inspire the use of experimentation, except for mentioning that Bacon (whom Bala, following the erroneous mainstream view, regards as the so-called father of the experimental method) stressed the importance of the above three technologies.

[2] This figure originates from a series of earlier, simpler drafts exchanged between Arun Bala and us in the course of an extensive dialogue on the Needham Question from 2021 to 2022. It goes without saying that we bear sole responsibility for the present chart.

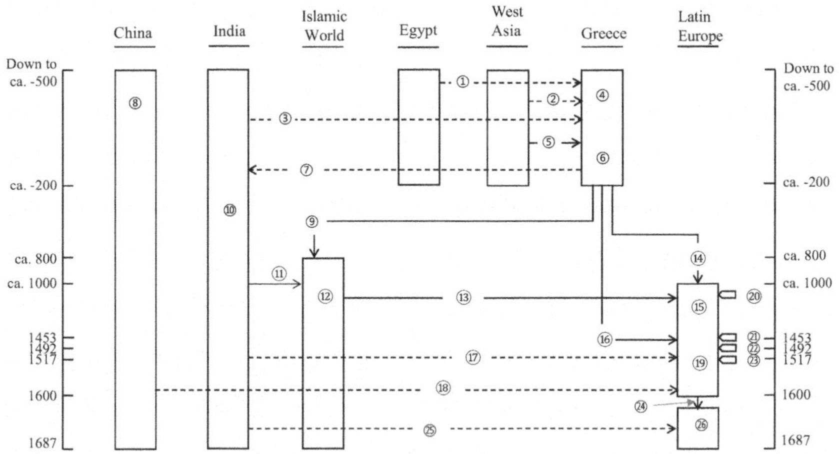

Fig. 10.1 Genealogy of the rise of modern science from a non-Eurocentric perspective transcending dialogism. Notes: (1) The chart is not to scale. (2) Arrows with solid lines represent transmission of scholarly traditions; arrows with broken lines represent transmission of individual substantial knowledges or ideas only; thick arrows on the far right represent the intervention of key external factors. (3) On the artificial separation of pre-1600 and post-1600 Latin Europe, see the explanation in the text. Key: (1) Transmission of Egyptian mathematics and philosophy to Greece. (2) Transmission of Babylonian science to Greece. (3) Diffusion of the belief in the transmigration of the soul. (4) The development of the classical Greek tradition on the basis of 1 and 2, plus the influence of 3 on the early Pythagoreans. (5) Infusion of the Babylonian pragmaticist tradition in astronomy and mathematics during the Hellenistic era. (6) The modification of Greece's classical tradition as a result of 5 into the tradition of Ptolemy and Archimedes. (7) Transmission of Greek astronomical and mathematical ideas to India. (8) The Infinite-Universe School of Astronomy appeared ca. -300s; the method of solving simultaneous linear equations and determinantal calculations and solutions appeared in *Nine Chapters on Methods of Calculation*, which was compiled between 1st to 1st centuries. (9) Graeco—>Arabic transmission beginning on a large scale in the late eighth century. (10) Indian developments in mathematics and astronomy. (11) Transmission of Hindu mathematics to the Arabic world. (12) Arabic developments on the basis of 9 and 11. (13) Arabic→Latin European transmission beginning ca. 1000–1100. (14) As subsidiary to 13, translation of some Greek works into Latin from the Greek original. (15) Latin Europe's development on the basis of 13 and 14 up to and including the sixteenth century. (16) Second translation movement consequent upon the Ottomans' advance into Europe. (17) Possible transmission of Nilakantha's geo-heliocentric model that could have been

to charts drawn by scholars arguing from the dialogical perspective (for example, see Fig. 1.4 on pp. 14–5 in Joseph [2011]). This is because, as mentioned, we accept dialogism as a starting point. We speak of genealogy to highlight the fact that our theoretical-methodological framework adopts a *process approach*. Greece and Latin Europe are in two separate columns because they constituted two separate and distinct civilizations. We distinguish between *two types of knowledge transmission* between different civilizations, namely, (a) transmission of scholarly tradition and (b) transmission of individual substantive knowledges or ideas only, where we are uncertain as to whether it is (a) or (b) due to limitations of our knowledge, we prefer to be cautious and regard it as (b). We stress this point because some of the transmissions that we designate as (b) in Fig. 1 may well actually be (a) instead. Hence, some of our designations should be understood as provisional only. In this regard, the point of Fig. 10.1 is to lay down the basic framework of our approach which makes a distinction between the two types of transmissions, this will not be affected by any substantive errors made by us in designating a transmission as (a) or (b). Further, (a) and (b) may not constitute a strict dichotomy, but a continuum, such that a transmission that we provisionally designate as (b), while not actually constituting (a), may gravitate more towards (a) than (b). This is an issue that requires further investigation, both conceptually (dichotomy or continuum?) and empirically. Finally, in making the distinction between (a) and (b), there is no suggestion that (b) is less important than (a), for insofar as transmissions of type (b) constitute indispensable inputs for the breakthrough of ca. 1600, they are no less important or impactful than transmissions of type (a). In sum, while we do think that our distinction between (a) and (b) points towards the right direction, we do not pretend to have said the final word on it.

In addition to incontrovertible transmissions, we also include transmission claims that we regard as worthy of attention and that, in our view, reach a minimum threshold of plausibility. Where necessary, it is up to the scholars concerned to further substantiate those claims in accordance with the principle of the balance of probabilities and plausibility, if incontrovertible direct evidence is absent. There are probably other such claims that we have, for simplicity, not considered but which should be included in Fig. 1 as well, but comprehensiveness is not necessary for our present purposes, which, as mentioned in the Introduction, are methodological. For instance, the reader will note that we mention only two transmissions from China, one indisputable, the other more than probable. This is because, for our purposes, there is no need to invite unnecessary controversy by mentioning other, disputable transmission claims, such as some of those made by Needham. This is, of course, not to say that efforts to fill in more details by considering other worthy transmission claims are not needed so that a more comprehensive chart can be drawn.

In the chart, we try to incorporate various components of our theoretical-methodological framework into it. The incorporation of the process approach, and the distinction between the transmission of scholarly tradition and the transmission of individual substantive pieces of knowledge alone, have been mentioned above. Besides them, item (19) in the chart refers to late-Renaissance *scholarly agency*, a variable that is absent in similar charts drawn from the dialogical perspective. Further, we also incorporate our *synthesis of externalist and internalist explanations* in the chart—whereas most items in the chart pertain to internal factors, items (20), (21), (22), and (23) pertain to external ones. In the previous chapters, we have had many occasions to discuss how various external factors had played a role, obviously, it is impossible to incorporate them all or even a majority of them; we include items (20) to (23) because they were key to the transformation of scholarly agency in the post-1453 era. Unfortunately, there are other components of our framework that cannot be graphically represented, such as the role of contingency and the effects of path dependency, but there is a limit to what graphical representation can achieve. Finally, we artificially separate pre-1600 and post-1600 Latin Europe, this is in order to stress the contribution of the work of Buridan and Oresme on motion, and that of alchemy to the breakthrough of modern science. This point is not only absent in analyses from the dialogical perspective, it is also not found in Cohen's thesis.

The critic of Eurocentrism Jack Goody (2006, p. 4) explains the argument of his *The Theft of History* in the following way:

> My argument is that while most [Eurocentric] historians aim to avoid ethnocentricity (like teleology), they rarely succeed in doing so because of their limited knowledge of the other (including their own beginnings). That limitation often leads them to make unsustainable claims, implicitly or explicitly, about the uniqueness of the west.

With regard to 'their own beginnings', chap. 4 has shown that the limited knowledge that Goody speaks of is the result of a particular process of the construction of the fiction of a pristine 'trans-historical West'. But this is not the point that we want to make here, which concerns the limited knowledge of non-European civilizations on the part of Eurocentric scholars. Since we are more familiar with China, the following is confined to its case. In the previous chapters, we have discussed four Eurocentric scholars in particular, namely, Cohen, Huff, Lloyd, and Sivin. We would like to consider them in the light of Goody's comment.

In commenting on his sources concerning China, Cohen states that his sources are Needham and Sivin (including Lloyd and Sivin 2002), and complains that 'a chronological survey of Chinese nature-knowledge still has not been written, certainly not in any language accessible to me' (2011, p. 50; cf. 2015, pp. 282–3). However, even had such a survey been written, as far as secondary sources are concerned, can one ever rely on any one or a limited number of such sources in the humanities and the social sciences? For instance, concerning Greek philosophy, Guthrie's six-volume *A History of Greek Philosophy* is the standard reference, but, disregarding the availability of post-Guthrie studies, as Guthrie himself makes clear at various places in his account, there exist different views on many key issues, hence making it necessary to consult other sources holding contrasting views as well in order to arrive at a more balanced understanding.

The point is, in order to get to know about a foreign civilization, it is necessary to dig into the literature instead of waiting for a 'definitive' general survey to appear. Doing so, one will find that in Chinese mathematics, there are several good general surveys in the English language, namely, Joseph (2011), Martzloff (1997), and Li and Du (1987), the accounts of which one can compare and contrast; while the works of the Singaporean-Chinese mathematics historian Lam Lay Yong published in English on various individual issues on the history of Chinese mathematics are highly stimulating. On Chinese astronomy, while the most authoritative, comprehensive, and detailed general survey that we are aware of (Chen Meidong 2008) is in Chinese, there are some good sources in the English language, such as Chen Cheng-I (1996). And so on.

Cohen relies mainly on Needham, whom he regards as a great scholar, supplemented by some individual works by Sivin. Sivin will be mentioned shortly below, as to Needham, on the one hand, his knowledge of China especially with regard to technological development is, indeed, formidable; however, it requires no great insight to see that most of the volumes of the SCC series, written decades ago, have been much overtaken by more recent studies. To refer to just one instance, in a work published just one year after Needham's passing away, Chen Cheng-I (1996, chap. 2) provides a comprehensive demolition of the arguments advanced by Needham and his collaborator Kenneth Robinson in volume 4 of SCC (Needham and Robinson 1962) concerning Chinese acoustics.

On the other hand, Needham's knowledge of China in other respects, from its history to Chinese philosophy, leaves much to be desired, as shown on various occasions in previous chapters. Yet, it is on the basis of just these two sources that Cohen makes various weighty comments, and passes various summary judgments, all of a sweeping nature, on China, and sometimes in unflattering language (such as speaking of 'contemptuous indifference' to spice up Needham's claim that the Chinese were 'incurious' about the outside world). Incidentally, Needham's unfamiliarity with China's history highlights another crucial point. Since in understanding a civilization's intellectual development, both internal and external factors are relevant, to be able to adequately comment on a particular civilization's intellectual development, some general understanding of that civilization's history, socio-economic system, and so on is indispensable. Besides the instances concerning Needham discussed in the previous chapters, another good illustration of this is provided in Chap. 7. In that chapter, it can be seen that to be able to correctly interpret Xu Guangqi's preface to the translation of Euclid in which he eulogized European learning, a knowledge of such things as the meaning to Confucians of referring to Yao, Shun, and the 'Three Ages', the distinction and relationship between the two senses of *li* of the Song neo-Confucians, and so on, is essential. Finally, Cohen probably thinks that Needham has exhausted whatever that can be said in favour of the non-Eurocentric case with regard to China. This, however, cannot be more wrong, for, as we have shown time and again, Needham, especially in his more general comments, remained very much under the domination of Eurocentrism.

Sivin was a sinologist who wrote on some aspects of Chinese science and medicine. It is safe to assume that, except for Huanglao-cum-Yin-Yang thoughts mentioned in Chap. 5, the mysticism-infused doctrines of which were used by the dominant school of Chinese medicine to conceptualize its clinical observations,[3] he was not familiar with other aspects of ancient Chinese thoughts. Lloyd is a formidable classical scholar, who subsequently became interested in China and the comparative study of intellectual developments in general. At least in the early stages, he relied

[3] We think this was why Sivin became acquainted with the Huanglao-cum-Yin-Yang school.

heavily on Sivin with regard to China. Yet, in the 1990s and early 2000s, when the main works of their collaborative project on comparing Greece and China were published, they proceeded to conduct a general comparison of the two civilizations with respect to a whole range of issues, from philosophy to the nature of the various schools of thought, in which they made a whole list of monumental claims concerning China. We have commented on some of these claims in Chap. 5. The extent to which their claims are well- or ill-founded is, perhaps, best epitomized by their astounding and incredible claim that China did not have a concept of nature because it 'did not need' such a concept.

Huff is a historical sociologist of modern science; it is safe to assume that he does not read Chinese, and hence he has to rely on secondary sources for his knowledge of China. In a general study of the negative NQ, with regard to China, other than the obligatory Needham, he relies heavily on two sources, namely, Sivin, and, especially, Derk Bodde.[4] As anyone familiar with the sinology of ancient Chinese intellectual developments well knows, Bodde and Sivin were like-minded sinologists with similar, strongly held views concerning such developments.[5] This should also be obvious to readers of their works unfamiliar with sinology who are at all aware of the fact that in the humanities and the social sciences, there exists a divergence of views. For a scholar whose objective is to acquire a balanced understanding of her subject matter, the need to consult other sources is evident. But Huff does not do so. As an illustration of the consequences of not doing so, one telling claim made by Huff on the authority of his main source is the following: 'Derk Bodde points out, "Throughout its history Confucianism has deprecated the use of debate as a means of advancing knowledge"' (Huff 2003, p. 279).[6]

[4] Sivin and Bodde respectively have ten and eleven references to them in the index, much more than any other author with the exception of the obligatory Needham. Huff's dependence on Bodde can be gauged from the fact that some authors are quoted in the text 'as cited in Bodde' (for example, see p. 301 n 44 and p. 304 n 59). Checking through the book's bibliography and index with regard to China, one finds that a significant number of the authors whose works are listed in the bibliography do not appear in the index at all. Even for those who do, one notices that several works by the same author may be listed in the bibliography while only one is actually cited in the text; and that most authors who do appear in the index have only one or two references to them in the text.

[5] Sivin has been discussed. Concerning Bodde, his position can be gauged from an episode related by Needham's collaborator Kenneth Robinson in Needham et al. (2001, pp. xiv–xx).

[6] The quote is taken from Bodde (1991, p. 178).

The present author was educated in colonial Hong Kong, which, as a matter of policy, suppressed the teaching of Chinese history and culture in general, but even in such an environment, we learned in junior high school in the 1960s that in the mid-Warring States period (the latter -300 s), there was a huge debate between the Confucians Mencius and Gaozi (告子) concerning the subject of *xing* (性, usually translated as 'human nature' in this context), which was continued by Xunzi (the Confucian mentioned in Chap. 5) in the late Warring States period. This is common knowledge among school kids all over the world who have received even the most minimal of instruction on ancient Chinese thoughts.

The period of the Qin and Han dynasties (−221 to 220) was a disaster for China's intellectual developments, except in astronomy which was of crucial political importance, as both dynasties engaged in a wholesale destruction of learning except for the state orthodoxy. From the Qin-Han period onwards, except for the brief renaissance in the mid-300s consequent upon the fall of Han, indigenous Chinese philosophy atrophied for a millennium. Nearly four centuries of intense political instability and disintegration followed after the fall of Han, during which period some regimes executed scholars for political reasons. As a result, Buddhist philosophy (along with Buddhist religion) was able to quickly gain ascendence until it achieved dominance. With the rise of the Song Dynasty (960–1279), under its stated twin policies of valuing the literati and guaranteeing that no one would suffer bodily harm for their views, indigenous Chinese philosophy revived, and debate among Confucians soon re-appeared, culminating in the *famous* face-to-face debate that lasted over three days in 1175 between leading Confucians Zhu Xi (朱熹) and Lu Jiuyuan (陆九渊), who represented two different Confucian schools of thought,[7] known in history as the Meeting at E'hu (鹅湖之会). If one reads Zhu Xi's proliferous works, one will find him often engaged in dialogue and debate with various correspondents. After the Song, as previously mentioned, except for the Ming period (1368–1644), China was ruled by the invader-dynasties of the Mongols and the Manchus until the end of the dynastic era. Under the Mongols and the Manchus, the environment simply did not exist for active intellectual developments; while during the Ming period, although the school founded by Lu Jiuyuan during the Song period became dominant,

[7] The first to appear (Zhu's school) tried to synthesize Confucian ethics with natural philosophy, which was challenged by Lu's school which argued for a return to the original spirit of Mencius' doctrine.

in the late Ming period, it was challenged by the Donglin School (东林学派); even under the Qing period, it continued to be challenged by scholars such as Gu Yanwu (顾炎武). It is, hence, clear that not only are the above-mentioned well-known debates not mentioned by Bodde, as a result of which Huff knows nothing about them, if one possesses a certain familiarity with China's history and considers the development of Confucianism in the context of the changing political environment over time (an external factor), it can be seen that instead of 'deprecating debate throughout Confucianism's history', when the external environment allowed, debate among Confucians never stopped.

From Weber in the early twentieth century to today, the readiness of Eurocentric scholars possessing only limited knowledge of China to comment and pass judgement on China is repeated case after case—besides the four scholars mentioned above, in the Appendix, we shall come across a fifth case, namely, the Harvard scholar David Landes. At the subjective level, the reason why this occurs probably varies from one scholar to another. However, in view of the consistent pattern of such occurrences, it is difficult to resist thinking that some underlying common factor is involved, and the most likely candidate for such a common factor would seem to be some unconscious mental assumption or mentality.

One most consequential notion that has arisen from the Enlightenment is the unilinear view of history. According to this view, all societies sooner or later pass through the same stages of one and the same trajectory of development. This common trajectory is none other than the one that Europe (including America) has traversed. It will be recalled from Chap. 2 that Needham completely shared this view. According to this view, having completed traversing this trajectory, Europe stands, if not exactly at 'the end of history', in Francis Fukuyama's (1999) famous (or, for that matter, infamous) phrase, then, at least at the hitherto highest stage. Further, this view implies the belief that being placed in this privileged position, Europe enjoys the right to comment on and judge other societies still stuck at lower stages of the trajectory. The demand of Euro-American gender pluralists that other societies also adopt their position is a stark illustration of the strength of this deep-seated, centuries-old view and the mentality that flows from it even among those who claim, or believe themselves, to be multi-cultural pluralists. One wonders whether or not such a mentality could be behind the disposition among Eurocentric scholars to jump to conclusions about other civilizations without first making sure that one has acquired a sufficient understanding of the latter.

Bibliography

Bala, Arun. 2006. *The Dialogue of Civilizations in the Birth of Modern Science*. Palgrave Macmillan.

Bala, Arun. 2024 (in press). 'Explaining the rise of modern science: A dialogical perspective'. Chapter 16 of Arun Bala, Raymond W.K. Lau and Mei Jianjun (eds.), *Multicivilizational Exchanges in the Making of Modern Science. Needham's Dialogical Vision*. Palgrave Macmillan.

Bodde, Derk. 1991. *Chinese Thought, Society, and Science: The Intellectual and Social Background of Science and Technology in Pre-Modern China*. Honolulu: University of Hawaii Press.

Chen Cheng-I. 1996. *Early Chinese Work in Natural Science*. Hong Kong: University of Hong Kong Press.

Chen Meidong. 2008. *Ancient Chinese Astronomical Thoughts*. 陈美东,《中国古代天文学思想》(北京:中国科学技术出版社, 2008).

Fukuyama, Francis. 1999. *The End of History and the Last Man*. Free Press.

Goody, Jack. 2006. *The Theft of History*. Polity.

Huff, Toby. 2003. *The Rise of Early Modern Science: Islam, China and the West*, 2nd ed. New York: Cambridge University Press.

Joseph, George G. 2011. *The Crest of the Peacock Non-European Roots of Mathematics*, 3rd edition. Princeton: Princeton University Press.

Li Yan and Du Shiran. 1987. *Chinese Mathematics A Concise History*, translated by John N. Crossley and Anthony W.-C. Lun. Oxford: Clarendon.

Lloyd, G.E.R. and Nathan Sivin. 2002. *The Way and the Word: Science and Medicine in Early China and Greece*. New Haven: Yale University Press.

Martzloff, Jean-Cluade. 1997. *A History of Chinese Mathematics*. Translated from the French original of 1987 by Stephen S. Wilson. Berlin: Springer.

Needham, Joseph, and Kenneth Robinson. 1962. *Science and Civilization in China, Volume 4. Physics and Physical Technology, Part 1. Physics*. Cambridge: Cambridge University Press.

Needham, Joseph, with the collaboration of Kenneth Girdwood Robinson and Ray Huang (Huang Jen-yu), and including contributions by them, Introduction by Mark Elvin, edited by Kenneth G. Robinson. 2001. *Science and Civilisation in China, Volume 7, Part II: General Conclusions and Reflections*. Cambridge University Press.

CHAPTER 11

Appendix - Deconstructing Cultural Essentialism at Root: The 'Maritime vs Continental Civilizations' Narrative

At the end of Chap. 9, it was mentioned that there exists a 'maritime versus continental civilizations' (MvC) narrative in relation to the NQ. This narrative lies at the root of most cultural-essentialist arguments of both the conventional and neo varieties of Eurocentrism. It contrasts Europe with China as non-Europe, in which the former is said to constitute a 'maritime civilization', while the latter a 'continental civilization'.[1] This narrative was spawned by the 'voyages of discovery' narrative, which constitutes the 'M' component of the MvC narrative—hereinafter, the 'M' and 'C'

See note 3 of Chap. 4 on the non-Derridean sense of the word 'deconstruct' used by us.

[1] There are, in fact, two types of claims made from the perspective of 'maritime civilization' versus 'continental civilization'. The first type is a form of geographical determinism, arguing that in a 'maritime civilization' such as Greece, there was a need to conquer nature, hence leading to the emergence of rational, logical thinking (e.g. Ji Shuli 1988, p. 71; Gao Chenyang 1994, pp. 20–1). In Chinese literature, this view was extremely popular in the 1980s and 1990s after the translation (from a Russian translation of the early 1930s) of Lucien Levy-Bruhl's *How Natives Think* (1985, French original published in 1910) into Chinese in the early 1980s, introducing his much-critiqued notion of different peoples possessing different 'mentalities' ('mental functioning' in Levy-Bruhl's own words). This type of 'maritime versus continental civilizations' view does not concern us here. The second type concerns claims about cultural values, drives, and so on, and it is this second type that forms the subject matter of the present appendix.

R. W. K. Lau, *Rethinking the Needham Question*,
https://doi.org/10.1007/978-981-97-9472-0_11

components refer to that of the MvC narrative. This spawning has occurred because the 'voyages of discovery' narrative pertains only to the positive NQ, hence, under the circumstances that, like Needham, Eurocentric, and neo-Eurocentric scholars invariably raise the NQ in dual form, a complementary argument to address the negative NQ that matches the 'voyages of discovery' narrative in form is required. This is what the MvC narrative provides, with its 'M' component that speaks of 'maritime civilization' addressing the positive NQ, and its 'C' component that speaks of 'continental civilization', as the exact reverse of 'maritime civilization', addressing the negative NQ. The 'voyages of discovery' narrative has been adequately critiqued in Chaps. 8 and 9, hence the 'M' component need not be addressed again. However, in critiquing Needham's various claims about China constituting a unified empire since −221 and so on in Chap. 9, we have addressed only some of the core claims made under the 'C' component. In view of the special prominence of cultural essentialism today, a more comprehensive critique of the 'C' component is clearly required.

At the end of Chap. 9, it was remarked that to our knowledge, the 'C' component has never been subject to a systematic and comprehensive critique up to today. Actually, this is not exactly correct. In the 1950s, the Chinese-American historian Lo Jung-Pang drafted a manuscript titled *China as a Seapower, 1127–1368*. Its argument constitutes an indirect critique of the 'C' component. Lo's draft, however, was never published. It was eventually dug out from the archives by Bruce A. Elleman of the US Naval War College and published in 2012, edited by Elleman himself who also wrote a forward to it. In his work, Lo argued that the historical record shows that not only did dynastic China possess a maritime tradition, when it was unified and strong, that tradition enabled China to project itself as a formidable seapower: 'The history of China has seen many occasions when the Chinese pushed out to sea…when China embarked on maritime enterprises…the height was reached when China was strong and unified' (Lo 2012, pp. 353–4). With China's rise now being seen by America (followed by its allies) as constituting a 'threat', it is due to this argument of Lo's that his five-decades-old manuscript has finally seen the light of day. Elleman voices such Euro-American worries openly: 'China is now seeking to fill the vacuum left by the USSR's collapse and build an ocean-going blue water navy. The question of the hour is "can China succeed in becoming a sea power?"' (p. 11). The argument of the present appendix had been arrived at by us before we read Lo's book, but not only does that book confirm many of our own arguments, it has also provided valuable data to support them, though, of course, some of Lo's data obviously require updating today. Besides Lo's book, some Chinese scholars have also recently begun to subject the 'C' component to critical scrutiny. For instance, in 2015, Xi Longfei published an authoritative work in Chinese on the history of shipbuilding and the shipping industry of ancient China; in 2021, Xi and his collaborator Shi Hequn wrote a summary of Xi's earlier work, which has been translated into and published in English (Xi and Shi 2021). This latter work opens with the following statement:

> For a long time, there has been a prejudice that ancient Chinese civilization was basically the product of farming economy and had nothing to do with marine culture. They described enterprising and adventurous marine culture as a symbol of western civilization...A large number of historical documents and unearthed cultural relics irrefutably prove that the ancient Chinese were not only used to sailing but also better at sailing. Ancient China had a brilliant marine culture. (Xi and Shi 2021, p. 348)

Thus, while much of the overall argument of the present appendix is original, we are not completely alone in subjecting the 'C' component to critique. Just as Needham's cultural-essentialist claims about China mentioned in Chap. 9 are at odds with the simplest historical facts about China that have been overlooked, ignored, or otherwise forgotten by those who should have known better, the same applies to the 'C' component. That this should have happened was due, to repeat, to the cultural dominance of the 'voyages of discovery' narrative. There is a historical event that advocates of the 'C' component invariably refer to as 'proof' of their view, namely, the banning of private maritime trade for two centuries (from 1370 to 1567) by the Ming Dynasty. According to them, this ban was symptomatic of China's allegedly age-old 'inward orientation' resulting from its being a 'continental civilization'. To be able to interpret this event properly, a pre-condition is needed, namely, the delineation of the development of dynastic China's, in Xi's and Shi's words, 'brilliant marine culture' from the Han up to and including the Ming dynasties. This delineation requires a broad analysis, involving an examination of China's territorial extension, its ecological and economic diversity, the resulting role of commerce in the national economy, and the shift of the country's centre of gravity (economic and otherwise) over time from the interior middle course of the Yellow River, where dynastic China initially hailed, to the eastern and southeastern coastal regions.

To re-state the MvC narrative, according to it, Europe constituted a 'maritime civilization', 'lured' by the sea and 'fearless' of its dangers, whereas China constituted a 'continental civilization', 'bound', in 'self-sufficiency', to the land, 'shying away' from the hazards of the sea. Among Chinese supporters of this narrative, the Yellow River is often used, up to the present day, as a reference point for China's purported 'inward-looking', 'self-enclosing', and 'unadventurous' mentality resulting from its 'self-sufficient land-based characteristic'. The Yellow River is used for this purpose because dynastic China initially hailed from the middle course of the Yellow River in a region known as the Central Plains (today's Henan province and its surrounding regions). For instance, in the latter 1980s, some pro-Euro-American Chinese cultural intellectuals wrote a six-part television series on the MvC theme titled *River Elegy* (《河殇》)—the river in the title being, of course, Yellow River—that was all the rage at the time, as barely a decade into China's 'opening up' policy initiated by paramount leader Deng Xiaoping, most Chinese fell prey to the 'voyages of discovery' narrative.

From the time of the Zhou Dynasty (−1046 to −256), China's territory extended far beyond the Central Plains area, right up to the coast in the eastern direction, with the states, counting from the northeast bordering today's Korea down south to the region subsequently known as Jiangnan (江南, literally south of the [lower course of the Yangtze] River), of Yan (燕), Qi (齐), Wu (吴) and Yue (越) all being coastal states.[2] These four states were not peripheral states, but *key* ones: Of the hundreds of fiefdoms-turned-states (see note 2) that initially existed as the central authority of the Zhou court waned, only seven remained by −476, among which were Yan and Qi, while Wu and Yue had been amalgamated into the state of Chu (楚); initially, Qi and Chu, one after another, overshadowed the state of Qin (秦) in terms of power and military strength, before the latter finally conquered the whole of China and established the Qin Dynasty. Under the Qin (−221 to −206) and Han (−206 to 220) dynasties, China's territory extended further to the entire southeastern coastal region from the just-mentioned Jiangnan region, through today's coastal

[2] These states were originally fiefdoms of the Zhou court, but as the latter's central authority progressively waned, they increasingly became de facto states, pledging only nominal allegiance to the Zhou throne.

provinces of Fujian, Guangdong, and Guangxi, all the way to the northern half of today's Vietnam, extending China's continental coast line alone to well over 180,000 kilometres.

China's expansive territorial extension entailed great ecological and economic diversity between different parts of the country. This, in turn, provided fertile ground for commerce to thrive. Not counting grain, the most important commodities in ancient China for domestic consumption were iron and salt, and, for export and the upper class, silk and, later, porcelain as well. Iron production and silk and porcelain manufacture were geographically dispersed, salt was produced along the coast and around salt-water lakes. Hence, all these commodities were heavily traded across the country. During the Han Dynasty, when the Central Plains constituted the overall centre of gravity of the country (in political, economic, and other terms), there were five national commercial cities, namely, Luoyang (洛阳 in today's Henan province), Wan (宛 also in Henan province), Handan (邯郸 in today's southern Hebei province), Linzi (临淄 in the coastal Shandong province where the above-mentioned state of Qi was once located), and Chengdu (成都 a centre of silk manufacture in today's Sichuan province in the country's southwest) (Fu Zhufu 1981, p. 419); of these, only the first two were in the core Central Plains region, the third was on its northeastern periphery, while the last two were in other parts of the country.

Heavy trading in these commodities led to the rapid rise of a merchant class as early as during the Warring States period (−475 to −221). As Marx often stressed, commerce was an efficient and destructive solvent of agriculture-based socio-economic relations. Which was why Warring-States ruling elites and Confucian scholars were keen to put a check on the merchant class. In the Confucianism-based value system, individuals were grouped into four categories, namely, in declining order of social status and honour, scholars, farmers, artisans/labourers, and merchants, denigrating merchants to the bottom. This social status-and-honour ladder became institutionalized and enforced throughout the centuries even during periods when rulers depended considerably on merchants in financial terms. Merchants' children were disqualified from taking part in the civil service examination system that selected scholar-officials, which was clearly intended to prevent the merchant class from acquiring political power. China's greatest poet Li Bai (701–762) failed to achieve an official career precisely because he was born to (Silk Road) merchant parents.

Occasionally, some officials did speak in support of the merchant class. During the Northern Song period, for instance, chancellor Wang Anshi (王安石1021–1086) criticized opponents to his economic reforms in the following way: 'half of the book *Zhou Li* [《周礼》a document on the socio-political institutions of the Zhou Dynasty] is concerned with the management of finances. Could the Duke of Zhou [the inventor of the legitimation myth justifying the Zhou's overthrowing of the Shang mentioned in Chap. 3] be said to be greedy of profit?'; indeed, during this period, 'Government officials and members of the court started commercial and industrial enterprises, bought shares in trading companies, and sent ships out to trade' (Lo 2012, pp. 112–3). Nonetheless, such occasions were infrequent. The socio-political discrimination against merchants shows precisely the strength and vitality of commerce in society and the existence of a strong commercial ethos among significant numbers of the population. As testimony to the vigorous development of commerce despite official discrimination, the world's first-ever bank draft, known, aptly, as 'flying money' (*Feiqian*飞钱), appeared during the Tang Dynasty in the early ninth century; it remained in use until the late 10th to early 11th centuries when it was replaced by the invention of the world's first-ever paper currency known as the *Jiaozi* (交子). Initially an invention of the private economy, the *Jiaozi* subsequently became a state-issued fiduciary currency introduced on a national scale in 1005; its use was officially terminated in the early twelfth century as a result of inflationary pressures due to over-issuance by the state to cover mounting budget deficits (Lo 2012, p. 200). The development of the money economy, as an indicator of thriving commerce, reached its peak during the Song Dynasty. Even during the Tang period, government revenue collected in money form still constituted only 3.9% of the total in the year 749; but by 1021 (during the Northern Song period), it had climbed to 17.6%, then further to 30.9% in 1049, and reaching an astounding (for an agrarian society) all-time high of 51.6% in 1065. This rapid rise was due not to shrinking taxes collected in the form of grain, which in volume terms remained steady, but to the huge increase in taxes paid by merchants as well as manufacturers (*ibid.*, pp. 197–8).

Unsurprisingly, China's coastal regions began developing a maritime tradition from very early on, both military and civilian. The use of sails in open sea sailing appeared during the Warring States period at the latest (Lin Huadong 1986). The first documented sea battle in China's history occurred in the year −485, when the above-mentioned state of Wu sent a naval expedition up north to attack the state of Qi.[3] According to an unofficial historical chronicle[4] titled the *Spring and Autumn Annals of Wu and Yue* (《吴越春秋》), during the Warring States period, there were altogether more than twenty sea battles between Wu and Yue (Xi and Shi 2021, p. 357). The people of Wu and Yue, which covered today's Jiangsu and Zhejiang provinces, developed an especially strong maritime including naval tradition. Some one-and-a-half millennia after the states of Wu and Yue of the Warring States period had ceased to exist, when the Southern Song Dynasty (1127–1279) considered strengthening its navy, an official document spoke of 'to fight on ships like the men of Wu and Yue', and in a memorial to the throne of 1136 on the matter, it was stated that 'The navy is the strong weapon of Wu and Yue' (quoted in Lo 2012, p. 153). It is not known when the stern rudder was first invented in China, but the earliest extant archaeological evidence dates from the Han Dynasty (Xi and Shi 2021, p. 365). After the fall of the Han in 220, China became fractured and entered the period of the Three Kingdoms. The kingdom that occupied the eastern and southeastern coastal regions down to today's northern part of Vietnam is known as Wu (or Eastern Wu). As mentioned in Chap. 9, it was Wu which upkept diplomatic relations with states in SE Asia and today's Sri Lanka, first established by the Marshall Emperor of Han through the sea route.

There is an unofficial and unconfirmed account that the First Emperor (reigned as First Emperor from −221 to −210) sent an expedition led by a 'man of the secret arts' (*fangshi* 方士) called Xu Fu (徐福) to the eastern seas in search of the elixir of immortality. Xu's expedition departed from today's Shandong peninsula (where the previous Qi state was located) and is said to have landed in Japan, with its members numbering several thousand settling there; while this account is unconfirmed by Chinese sources, it is taken to be true in Japan, where there is a park in his memory called Jofuku Park in Shinomiya, Wakayama prefecture, the area where Xu Fu reputedly landed. If this is true, then, it means that by the late -third century, Chinese sailors from the coastal regions already possessed the navigational means and skills to sail some 1500 kilometres into the open ocean.

[3] *Zuo's Commentary*, 10th year of Duke Ai. *Zuo's Commentary* is an authoritative year-by-year chronicle of events occurring from −722 to −468, arranged in accordance with the reigns of the various ducal heads of the state of Lu (鲁国). The 10th year of Duke Ai corresponds to −485.

[4] On the distinction between official and unofficial histories, see note 16 of Chap. 9.

In any case, by the time of the Marshall Emperor of Han (reigned −141 to −87), coastal sailing all the way from the northeast down to the northern part of today's Vietnam (then a part of Han) had been well established (Xi and Shi 2021, p. 362).

If Xu Fu's alleged trip to Japan remains unconfirmed, by the early Three Kingdoms period, China's ability to sail into the open ocean is well documented. In 230–31, the king of the state of Wu mentioned in the previous but one paragraph sent a naval expedition to the eastern seas in search of two islands, one of which is likely to be today's Taiwan, that were reputed to exist there. With regard to this expedition, Lo (2012, p. 52, our emphasis) remarked: 'the expedition of 230-31 was sent to "seek" (…) the two islands, so it was a voyage of discovery'. Lo wrote in the late 1950s (see the beginning of this appendix), hence it is unsurprising that he spoke in language borrowed from the 'voyages of discovery' narrative; the point of our quoting him is not to endorse this, nor to comment on its appropriateness with regard to the expedition of 230–31, but to note the fact of the early development of a maritime tradition all along China's coastal regions. For the same purpose, let us quote Lo again, where he commented on the development of a maritime tradition along China's coastal regions during the period down to the fifth century: '*The urge of the Chinese to the sea* and the activities of the Chinese navy in the period down to the fifth century had, to this point, included tales of magical islands, which convinced Chinese to embark upon *voyages of exploration and discovery*' (p. 57, our emphases).

Two inventions made in the early fifth century greatly enhanced the navigational capabilities of Chinese ships (Xi and Shi 2021, pp. 366–8). The first is the watertight bulkhead, which divided the hull of the ship into a number of individual watertight cabins by means of cross-sectional boards. The watertight bulkhead is able to serve two purposes at the same time; firstly, any leak can be restricted to the cabin concerned without causing the whole ship to be inundated by water, thereby much minimizing the risk of sinking due to water leakage; secondly, by fixing mast poles to the cross-sectional boards of the watertight cabins, the use of multiple sails was greatly enhanced. The second invention was the paddle wheel as a propelling tool, which greatly enhanced the manoeuvrability of the ship in shallow waters.

The ability of Chinese sailors to sail into the open ocean during the period from the third to the 5th centuries is illustrated by another incident. From 386 to 557, a large part of northern China was under the rule of the state of Beiwei (北魏), established by an invading nomad. During the reign of Beiwei's Emperor Taiwu (太武帝), the king engaged in a campaign to eradicate Buddhism that lasted from 446 to 452, whereupon a Buddhist monk named Hui Shen (慧深) left and sailed towards the east reaching a place that he called the state of Fu Sang (扶桑国), *Fu Sang* being the name he gave to a tree found there. There exist fierce debates among scholars, both Chinese and Euro-American, as to where this place exactly is, with some speculating, on the basis of Hui Shen's descriptions of various things he saw there, that it might be Japan, Sakhalin, or even as far as Mexico. In any case, what is certain is that it is somewhere to the east or northeast of China a considerable distance into the open ocean. The point about Hui Shen's trip is that the ability of Chinese mariners to sail deep into the open ocean was something that he could take for granted.

As mentioned in Chap. 9 and reiterated in the previous but three paragraphs, the Marshall Emperor of Han sent envoys to the south reaching today's SE Asia and Sri Lanka; from our sources, we are unable to ascertain whether or not his envoys travelled on Chinese ships, but it would seem that given the already well-established maritime network between SE Asia and the south of India and Sri Lanka and Chinese mariners' unfamiliarity with the region, it is most probable that for at least part of the journey, Han's envoys travelled on board SE Asian merchant ships. Be that as it may, it also seems clear that, in view of Wu's above-mentioned expedition of 230–231, by the early third century, it would have been technically feasible for Wu's envoys to SE Asia and Sri Lanka to have travelled on Chinese ships all the way.

Thus, maritime connections between China's southeastern coast all the way to Sri Lanka had been established as early as the time of the Marshall Emperor of Han. Unsurprisingly, trade along this route had also been established by that time. There is archaeological evidence to confirm this. Dispatched by the Marshall Emperor, a Chinese military commander established a kingdom known as Nanyue (南越国) in today's southern Guangdong, Guangxi, and northern Vietnam area. This kingdom was nominally part of the Han empire, though its rulers enjoyed de facto autonomy. Among the funerary objects in the tomb of its second king (reigned −137 to −122—note that this was during the reign of the Marshall Emperor) is a silver box of Persian origins, which is now on display at the museum built at the site of the tomb of the king in today's city of Guangzhou. By the time of the Eastern Han Dynasty (25 to 220), as seen (see note 19 of Chap. 9), trade between Rome and China through the mid-way station of India had become firmly established. After the Han, it will be recalled from Chap. 9 that the Buddhist monk Fa Xian travelled to India in 399 on a trip that lasted thirteen years, due to his old age, he decided to return to China by sea, for which purpose he went to Sri Lanka to pick up a trading ship destined for China. This indicates that by that time, trading along the maritime route had become highly regularized. In his travelogue, Fa Xian mentioned meeting a Chinese merchant in Sri Lanka, indicating that Chinese maritime merchants took full part in this trading network. In view of the existence of a strong commercial ethos among significant numbers of the Chinese population, that the people in the coastal regions were quick to engage in overseas maritime trade was altogether natural and to be expected.

Before continuing with our account of the maritime tradition of China's coastal regions, a brief historical detour is necessary. As mentioned, after the fall of the Han in 220, China was fractured into three separate and rival kingdoms, eventually, one of the three kingdoms managed to defeat the other two in 263 and 280 respectively, establishing a unified dynasty known as Western Jin (西晋), which, however, was short-lived, lasting only until 316. During the Western Jin period, from 304 onwards, various nomadic tribes invaded from the north, until the Jin was completely driven out of the part of the country north of the Yangtze River in 317—hereinafter, unless specified otherwise, whenever we speak of northern China or the north, and southern China or the south, we mean north of the Yangtze River and south of the Yangtze River respectively. From then until 420, the Jin, which occupied less than half the country in the south, is known as the Eastern Jin Dynasty. In the north, the invading tribes established various states, the strongest of which was the state of Beiwei (北魏), founded by the *Xianbei* (鲜卑) tribes in 386. In 439, Beiwei managed to unite the entire north. The period from 304 (when nomadic tribes began invading the Western Jin) to 439 (year of Beiwei's unification of the north) is known as the period of Five Foreign Peoples and Sixteen States (五胡十六国), with the five foreign peoples referring to the various invading nomads, and the sixteen states referring to the various states that they established until Beiwei's unification of the north. The period from 386 (the year Beiwei was founded) to 581 (the year when the reunification of the whole of China began—see below) is known, with reference to the north, as the Northern Dynasty (北朝). In the south, the Eastern Jin fell in 420, succeeded, in succession, by four different Han-Chinese states until 598 (the year when the process of the reunification of China begun in 581 was completed). This period, with reference to the south, from 420 to 589, is known as the Southern Dynasty (南朝). To sum up, from 317 to 581, for 274 years, the whole of northern China was under the rule of invading nomads. The importance of this will be explained presently. The reunification of the whole of China was accomplished by Yang Jian (杨坚), a Han-Chinese general serving under one of the states founded by nomadic tribes in the north. On unifying the north, Yang declared the founding of the Sui Dynasty (隋朝) in 581 and became known as Emperor Wen (文帝). His unification of the south was accomplished in 589. Emperor Wen died in 604, succeeded by his second son who became Emperor Yang (杨帝). In 618, Yang was killed in a military rebellion, bringing an end to the Sui Dynasty.

The Sui Dynasty is important for our discussion for two reasons, the first to be discussed here, the second in a later context. In 598, Emperor Wen sent a combined land and naval military expedition to invade the state of Koguryo in today's Korean Peninsula; in 605, he sent a naval military expedition against the state of Linyi (林邑) in today's central Vietnam.[5] Like his father, Emperor Yang also adopted an outwardly oriented foreign policy: 'Emperor Yang was fascinated by tales of lands beyond the Eastern Sea, and aspired to bring them under his political sway' (Lo 2012, p. 59). In 605, he sent a naval expedition out to the east which reached today's Okinawa, though without actually attacking it militarily. Emperors Wen's

[5] Koguryo was the most powerful of the states formed by indigenous Koreans on the Korean peninsula. China's military conflicts with the indigenous Korean states and states in today's Vietnam, which occurred every now and then over the centuries, need to be put in context. During the Zhou Dynasty (−1046 to −256), èmigrès from China's state of Yan (燕), occupying today's Hubei and Liaoning provinces, with the latter bordering the Korean peninsula, had established a state in the northern part of the Korean peninsula, which probably predated the formal rise of indigenous Korean states on the peninsula. This led to conflicts between Chinese èmigrès (as well as the Chinese of Liaoning) and indigenous Koreans which occurred every now and then. This was the background to the sometimes-difficult relations between China and the indigenous Korean states. As to Vietnam, the indigenous population of much of the northern part of today's Vietnam belonged to the Yue (越) tribes, the same tribes that occupied China's southeastern corner. Hence, when that area came under the authority of the Han Dynasty and became part of it, this included much of the northern part of today's Vietnam's. On the other hand, non-Yue peoples hailing from the southern half of today's Vietnam were also keen to extend into the areas in the north. This was the background of the periodic conflicts between China and Vietnamese states over the centuries.

and Yang's foreign policy has sometimes been described as being vainglorious. However, whether or not 'vainglorious', this needs to be situated in the context of a particular background, namely, that for 274 years before then, the whole of northern China was under foreign rule. Wen's and Yang's 'vaingloriousness' represented an effort to re-assert the honour and power of China after such a humiliating experience.[6] As we shall shortly see, the famous voyages to the west made by Zheng He in the early fifteenth century occurred under very similar circumstances. Wen's and Yang's naval expeditions greatly enhanced the navigational capabilities in open-ocean sailing of Chinese sailors (Lo 2012, p. 60). This paved the way for the great flourishing of China's maritime ventures in the ensuing Tang, Song, Yuan, and Ming dynasties.

So far, we have discussed the expansiveness of China's territory; the resulting ecological and economic diversity, which provided fertile soil for the vigorous growth of commerce, and hence the development of a strong commercial ethos among significant numbers of the population; the development of a maritime culture and tradition along the coastal regions, and the early growth of maritime trade along the maritime trade route. What we are going to discuss next is this: While dynastic China initially hailed from the Central Plains area, over the centuries, the country's centre of gravity, economic as well as in other respects, gradually shifted east to the coastal regions, as a result of which, in flat contradiction to the MvC narrative, maritime trade, and maritime relations in general, became increasingly important to the country.

A most graphic illustration of this shift in the country's centre of gravity is the changes of the location of the capital city over the centuries. As can be expected, China's capital city was originally located in the Central Plains region, either in Chang'an (today's Xi'an) or, to its east, the above-mentioned commercial hub of Luoyang. The capital city of the Western Zhou Dynasty (−1046 to −771) was in Chang'an, it moved to Luoyang during the Eastern Zhou period (−771 to −256) as Chang'an was overran by invading nomads; the Qin Dynasty (−221 to −206) built its capital in

[6] In Chap. 9, we critiqued Needham's claim that after −221, China had been a unified (implying and connoting stable) country as being in flat contradiction to the historical facts, where we remarked that 'those who fail to realize' that 'China was, to borrow Needham's expression, "uneasy within her boundaries" ***throughout the ages***' 'are ill-equipped to understand China's historical development'. Situating Emperor Wen's and Yang's 'vaingloriousness' as occurring in the context of 274 years of humiliation at the hands of invading nomads is an illustration of this; other illustrations can be found throughout the present appendix.

the outskirts of Chang'an (in an area known as Xian Yang 咸阳 where today's Xi'an's international airport is located); the Western Han Dynasty (−202 to 9) and the Eastern Han Dynasty (25 to 220) located their capitals in Chang'an and Luoyang respectively. After the post-Han Period of the Three Kingdoms (220–280), China was briefly re-unified by the short-lived Western Jin Dynasty (266/280 to 316), whose capital city was in Luoyang. As noted, from 317 to 581, the whole of northern China was under alien rule; the Eastern Jin Dynasty (317–420) was limited to the south, and established its capital in today's Nanjing, on the fringe of the Yangtze River delta area. Following Eastern Jin, the four Han-Chinese dynasties reigning from 420 to 589 that ruled over the south all established their capitals in Nanjing. By this time, the region around the lower course of the Yangtze River centring upon the above-mentioned Jiangnan (south of the Yangtze River) area, not to mention further south, had surpassed the Central Plains, not only in manufacture (which had for some time been the case), but even in agriculture:

> As early as the Han period, observers [as recorded in the official historical chronicles *Shiji* 《史记》 and *Han Shu* 《汉书》] had commented on the potential wealth and resources of the Lower Yangzi [Yangtze River] region [potential only because during the Han Dynasty, the area was as yet undeveloped]. The development of the region, begun by the state of Wu [of the Three Kingdoms period]...was vastly stimulated by the transfer of the nation's capital to Nanjing [during the Eastern Jin Dynasty] in the fourth century. It was at this time that the transition of the economic base of China from the Yellow River to the Yangzi took place. (Lo 2012, p. 74)

When the Sui Dynasty re-unified the country between 581 and 589, it relocated its capital initially back to Chang'an (called Daxing 大兴 at that time) and then to Luoyang. Why? There were two reasons, firstly, it did so certainly because Chang'an possessed institutionalized legitimacy implications, while the Sui Dynasty had been established after China had gone through more than 360 years (220 to 581/589) of fracturing after the fall of Han. Secondly, besides legitimacy implications, relocating back to Chang'an also served strategic purposes, for, as previously pointed out, Sui was founded after 274 years of nomadic occupation of northern China, hence, the need to remain on guard against the nomadic tribes to the north. However, despite relocating back to Chang'an and then Luoyang, and, indeed, precisely because of this, in view of the economic shift to the coastal regions, Emperor Yang built a system of canals, which remains in busy use up to the present day, to link the economically rich Yangtze River lower-course area to Luoyang as well as to the north up to today's Beijing so as to safeguard economic supplies to the latter regions. This is the second great contribution of the Sui Dynasty to China's historical development in addition to the first one previously mentioned, though, ironically, the enormous resources that had to be invested in the building of the canal system directly contributed to Emperor Yang's unpopularity and eventual demise of the Sui. As can be expected, the building of this canal system much enhanced the weight of commerce in the national economy.

The powerful Tang Dynasty that followed the Sui re-located the capital back to Chang'an, undoubtedly for the same reasons as the Sui, with Luoyang serving as the accompanying 'eastern capital' (东都, while Chang'an was called the 'western capital' 西都). But Chang'an during Tang times was no longer the Chang'an of old, as, firstly, its great prosperity was built, not on the basis of self-sufficient Central Plains agriculture, but on that of flourishing Silk Road trade which reached its peak development during Tang times; and, secondly, as the starting point and centre of Silk Road trade, it was an immensely cosmopolitan and outwardly oriented city, with sizable foreign communities of merchants, students, emissaries, artiste troupes, monks of various religions and so on residing in the city, some of whose descendants still live there today. The immense development of commerce during the Tang led, as already noted, to the appearance of the world's first-ever bank draft 'flying money'. However, despite the renewed, commerce-based prosperity of Chang'an, the Tang scholar-statesman Han Yu (韩愈) remarked: 'nine-tenths of the nation's revenue comes from Jiangnan' (quoted in Lo 2012, p. 74), the area south of the lower course and delta area of the Yangtze River. In terms of agriculture, the shift towards the coastal regions is well illustrated by the percentage of water control works undertaken in those regions out of the national total. During Han times, it stood at 18%, rising to 45% during the Tang period, then further to 71.5% during the Song period; during the Yuan (1271–1368) and Ming (1368–1644) periods, as, for reasons to be explained, northern China (though not the Central Plains area) regained considerable prominence, the figure still averaged 64% (*ibid.*).[7] In terms of sources of government revenue, the coastal to national proportion during the Northern Song period reached 60% in 1077 and an all-time high of

[7]This almost halved to 36% during the Qing period; though we do not know the reasons for this drastic drop, it does not seem to be accidental. We return to this later on.

82% in 1116; even during the Yuan Dynasty established by the invading Mongols, it stood at 74% in 1324; for the entire Ming period (1368–1644), it averaged at the mid-50s level.

After the fall of the Tang in 907, China again became fractured with invading nomads occupying much of the north, which led to the closure of Silk Road trade. Of the five successive Han-Chinese dynasties that ruled over part of northern China after the fall of Tang until 960, one located its capital in Luoyang, while the other four located their capital cities in today's Kaifeng city. While Kaifeng is located in the Central Plains area, it is further to the east than Luoyang, as well as being on the above-mentioned canal system, with commerce playing a core role in its economy, as we shall presently see. In 960, China became re-unified, though still with much of the country's northern part under nomadic rule, under the Song Dynasty, which was divided into two periods, the Northern Song period (960 to 1127) and the Southern Song period (1127 to 1279). Under the Northern Song, the just-mentioned Kaifeng, which the dynasty retained as the capital, became probably the most highly developed commercial city in the world up until then, with a population of around a million and whose commercial prosperity is immortalized in the much-studied mural painting titled 'Along the river at Qingming Festival' (清明上河图) completed during the Northern Song period.

As invading nomads progressively drove the Song out of the north, the Southern Song Dynasty located its capital city in today's Hangzhou in the Jiangnan area. Upon overthrowing the Song, the Mongolian Yuan Dynasty (1271–1368) moved the capital to today's Beijing, which is unsurprising since the Mongols were from the north—they would certainly feel vulnerable were they to establish their capital in the Jiangnan area. It should be noted that Beijing was far away from both the Central Plains and the Yellow River. On overthrowing the Yuan in 1368, the Ming Dynasty initially located the capital back to the rich coastal region in Nanjing. However, defeating the Yuan Dynasty did not mean the end of the threat of the Mongols, for while the Mongols had been driven back to the north, they were always preparing for another invasion of China, which eventually occurred in 1449. As mentioned in Chap. 9, on this occasion, they managed to hold the Ming emperor captive for four years, almost causing the collapse of the Ming. This was why in 1403, Emperor Yongle, Ming's third emperor, with decades of experience battling the nomads, re-located the capital from Nanjing to Beijing. This was *not a retreat to the inwardly oriented agrarian self-sufficiency of the Central*

Plains. Firstly, Yongle was precisely the emperor who was to send Zheng He on his famous sea voyages beginning just two years later in 1405. Secondly, as noted, Beijing was far away from both the Central Plains and the Yellow River. As also pointed out by Lo (2012, p. 82), this relocation was a move entirely driven by strategic military considerations. Indeed, despite this relocation to Beijing, it was widely acknowledged during the Ming that the centre of gravity of the country, in all except strategic military terms, was located in the south. As one Ming scholar remarked: 'The trend of the world (*tianxia* [literally all under heaven]) is from the west to the east and from the north to the south', while another observed: 'Although the capital Yen [Beijing] is located in the north, it faces the south. With the solution of the problem of food and transportation,[8] we can extend our power abroad and maintain peace at home...we have the riches of the southeast to support the military strength of the northwest' (both quotes taken from *ibid.*). The Ming was later overthrown by the Qing founded by the invading Manchus who were from the northeast; unsurprisingly, the Manchus retained Beijing, which was close to the northeast, as their capital. Like the Mongols, the Manchus would have felt vulnerable were they to locate their capital in the Jiangnan area or anywhere else in the area around the lower course of the Yangtze River.

To sum up, the changes in the location of the capital over the centuries provide a graphic indication of the shift of China's centre of gravity from the Central Plains area east to the coastal regions, and especially the regions around the lower course of the Yangtze River as well as further in the southeast. In the above, this shift has already been shown in economic terms, but it went far beyond economic terms, as there are several other dimensions of this shift as well.

There is, firstly, a shift in the distribution of the country's population. At the beginning of the first millennium (during the interregnum between the Western and Eastern Han dynasties), the population of the southeast was only one-tenth of that of the whole country. The population of the eastern and southeastern provinces as a proportion of the national total then began to grow over the centuries. For instance, from the Han to the Tang dynasties (−206 to 907), the population of the six coastal provinces in the east to the national total ranged from 34% to 38%; that proportion rose to 53% to 57% during the Song period (960–1279), continuing to grow under the

[8] The strategic importance of the canal system built by Emperor Yang of the Sui Dynasty is clear from this.

Yuan Dynasty, reaching a high of 73% in 1330; it then began to fall, but remained at over half of the national total on average during the entire Yuan and Ming period (1279–1644); under the Qing, however, it began to fall considerably, reaching a low of 37% in 1785 (Lo 2012, pp. 83–5). This substantial relative depopulation of the coastal provinces under the Qing went hand in hand with the halving of the percentage of water control works undertaken in the coastal provinces under the Qing in comparison to the previous period (see note 7). The probable reason for this double reversal under the Qing will be explained in due course.

At this point, let us address the cause of the shift of the centre of gravity in economic and population terms from the Central Plains to the east and southeast over the centuries, disregarding the partial reversal of this long-term trend under the Qing. There were both push and pull factors involved. Firstly, concerning push factors, as repeatedly stressed previously, China was frequently invaded by nomads from the north; as noted, from 316 to 581, the entire northern half of the country was ruled by invading nomads; a similar pattern recurred after the fall of Tang in 907, when the country again became seriously fractured. For twenty-eight years from 923 to 951, the north was ruled by invading Turkic peoples. While the Song Dynasty was able to unite the northern and southern parts of the country from 960 to 979, the threat of invasions from the north never ceased; in fact, it was such invasions that finally drove the Song to the south of the Yangtze River in the Southern Song period (1127–1279), with the north completely under the rule several kingdoms established by nomadic invaders. Southern Song was, of course, eventually overthrown by the Mongols, who founded the Yuan Dynasty ruling over the whole of China. It was due to the incessant invasions from the north and to being subject much of the time to the rule of invading nomads that the north suffered immense economic decline and relative depopulation over the centuries. Lo (2012, p. 73) observes with reference to manufacture: 'Factories making porcelain, lacquerware, textiles, and paper moved from North China to South-east China...The great ceramics centers in Zhejiang, Fujian, and in Jiangxi [all located in the south], were established at this time [during the Song]'. With regard to pull factors, the Silk Road reached its peak development during the Tang Dynasty, when due to Tang's strength and prestige, nomadic states all the way to west Asia voluntarily submitted to it. However, the Silk Road went into decline when Tang's power began to wane, leading to the rise in the relative weight of the maritime trade route. As an indication, by the Northern Song period, 65% of foreign missions arrived by sea; by the time of the Southern Song, all such missions came by sea as the north was entirely under alien rule (Lo 2012, p. 105). Prospering maritime trade at the expense of the Silk Road also attracted manufacturers to relocate to the south, in addition to the push factors mentioned above.

This large-scale migration of population and manufacture to the south produced a most intriguing side effect or, in the terminology of our theoretical-methodological framework, *unforeseen consequence*. Lo (2012, pp. 91–2) quotes a remark made by Ulu of the Jurchen (女真) Jin Dynasty (金朝1115–1234) that was established as one of the alien states in northern China during the Southern Song period, Ulu said: '[Han-Chinese in the north were] by habit sly and submissive', whereas, in contrast, the southerners were 'bold, upright, and dare to speak'. In other words, it was the relatively more resourceful (such as manufacturers, artisans, the more educated, and the like) and those who were less willing to submit to alien rule who migrated to the south. This relative shift in the distribution of quality human resources between the north and the south resulted in a shift of the country's cultural centre of gravity to the latter region. This shift was recognized by a Ming Dynasty writer writing in 1623:

> When the Jin court crossed [the Yangtze River] to the East,[9] it brought the robes and hats [i.e. culture], and the rites and music of several tens of generations, which it continued to foster and develop so that what used to the called *Zhongyuan* [the Central Plains] became a land torn by war and overgrown with brambles, so that there was a reversal [of cultural development between the north and the southeast].[10] Later, the Song court crossed [the Yangtze River] to the south[11] and brought with it the imperial charts and the sacred books to preserve national culture of millions [of people], so what used to be called the Two Capitals [Chang'an and Luoyang] became a field of black felt tents [of the invading nomads]. [The cultural difference between the north and the south] became even more pronounced. (quoted in Lo 2012, p. 94)

Shortly earlier, we noted the halving of the proportion of water control works undertaken in the coastal provinces and the relative depopulation of

[9] This refers to the driving of the Jin Dynasty (晋) by invading nomads to the south in 317, resulting in the entire north coming under alien rule for 274 years mentioned previously.

[10] During Han times, the southeast was referred to as the 'Hundred Yue peoples' (*Baiyue* 百越) with the connotation that they constituted a cultural backwater.

[11] Referring to the driving of the Song south by invading nomads and the beginning of the Southern Song period in 1127.

those provinces under the Qing, in view of the migration of the more resourceful and those less willing to submit to alien rule to the coastal and southern regions, the following query arises. Could it be that the Qing had a deliberate discriminatory policy against the coastal regions, which would have been beneficial to assuring the rule of the Manchus? This is worth investigating into, if there are not already studies concerned that we are unaware of. We mention this because it relates to a point that we made in Sect. 7.3 of Chap. 7, namely, that it is most unfortunate and misleading to draw conclusions about China by making reference to what happened under the Qing. We stress this not out of any xenophobia against the Manchus, but because it is of great importance and yet, in the literature on the NQ, it is unrecognized.

Lo (2012, pp. 94–6) has provided important data to document the shift of the country's cultural centre of gravity. Comparing the number of scholars and literary men (in ancient times, these were overwhelmingly men) between four northern provinces (Henan, Shanxi, Shaanxi, and Gansu, with Henan, southern Shanxi, and eastern Shaanxi comprising the Central Plains), and four eastern and southeastern provinces (Jiangsu, Zhejiang, Fujian, and Jiangxi, the first two constituting the Jiangnan area), Lo found the following (Table 11.1):

Again, the considerable drop in the proportion accounted for by the four eastern and southeastern coastal provinces under the Qing is notable. The number of 'notable personalities' of the six coastal provinces as a percentage of the national total averaged 38.09% during the Han Dynasty;

Table 11.1 Percentage of national total of scholars and literary men, Song to Qing dynasties

	Four northwest provinces	*Four eastern and southeastern provinces*
Northern Song period (Confucian scholars)	31.38%	58.08%
Southern Song period (Confucian scholars)	4.81%	85.46%
Yuan Dynasty (literary men)	10.11%	87.36%
Ming Dynasty (*Jinshi*进士 –those who passed the civil service examination at the national level)	15.43%	46.80%
Ming Dynasty (First among the *Jinshi*'s)	6.08%	80.50%
Qing Dynasty (scholars)	11.28%	57.25%

Note: The comparatively low figure of 46.8% for *Jinshi* under the Ming was possibly a consequence of its moving of the capital to Beijing in 1403, but the figure for first among the *Jinshi*'s is significant the figure stood at 34.77% during the Tang (which, for reasons previously explained, chose to relocate its capital to Chang'an in today's eastern Shaanxi); during the entire Song period, it averaged 46.67%; during the Ming, despite relocating its capital from Nanjing to Beijing in 1403, the figure remained at a high 47.60%.[12]

We have now established that whereas dynastic China initially hailed from the interior Central Plains, over the centuries, the country's centre of gravity, in terms of economic weight and quality human resources, shifted decisively to the coastal eastern and southern regions[13]; these regions had long developed a strong maritime tradition with an outward orientation, and in the context of the existence of a strong commercial ethos among significant numbers of the population, they were quick to establish overseas maritime trading relations. From the Han to the Sui dynasties, shipbuilding technology and the navigational capabilities of Chinese mariners in open-ocean sailing steadily improved. All these developments paved the way for the flourishing of an outwardly oriented maritime projection of China to the rest of the world for many centuries beginning with the Tang Dynasty.

As noted, the Silk Road reached its peak development under the Tang due to its might and prestige. But, at the same time, maritime trade also developed rapidly. The Tang established three ports for long-distance maritime trade, namely, Hangzhou, Mingzhou (today's Ningbo), both in the Jiangnan area, and Guangzhou in the southeastern Guangdong province, with the first two largely focusing on trade with Korea and Japan, while the last was devoted to trade through SE Asia all the way to west Asia. In 714, a Maritime Shipping Superintendent was established to supervise maritime trade and the shipping industry including the licensing of Chinese ships (Heng 2017, p. 143). Whereas maritime trade in the Indian Ocean was dominated by Arabic maritime merchants from the mid-700s to the mid-800s under the strong Abbasid Dynasty (Heng 2017, p. 148), Chinese merchant ships also sailed to the Persian Gulf regularly during this period (Li Jinming 2015, p. 12; Lo 2012, p. 126). In fact, 'In the Tang Dynasty, Chinese seagoing ships were famous in the

[12] Figures for the Yuan and Qing dynasties are not very meaningful since Mongols and Manchus, respectively, dominated the top governmental echelons.

[13] To highlight this for one last time, those who believe that since –221, China had been a unified and stable state are simply not in a position to understand such monumental changes, the occurrence of which had a lot to do with incessant invasions by nomadic tribes.

Pacific and Indian Oceans for their large hull, wide volume, strong structure, great ability to resist wind and waves, and skillful navigation skills of Chinese sailors' (Xi and Shi 2021, p. 374). In the year 751, Tang suffered a military defeat in Central Asia,[14] whereupon the Silk Road was blocked, giving the maritime trade route a great boost—as previously noted, the rise of trade in porcelain, which was unsuitable for land transport, was another key factor in the increasing importance of maritime trade.

The Silk Road remained blocked during the Song Dynasty; Quanzhou (泉州) in Fujian province (located south of the Jiangnan area and north of Guangdong) became a new port city with its own Maritime Shipping Superintendent, which subsequently became China's greatest sea port of the dynastic period, inspiring awe in Marco Polo, among other foreigners, when he visited the port. Maritime trade became so prosperous during Song times that, in one exceptional year (1128), customs duties accounted for as much as an astounding twenty percent of total government cash revenue; though the proportion subsequently fell back to more 'normal' levels such as that of five percent in 1159, these were still very substantial (Lo 2012, p. 202). By that time (the early 1100s), Chinese mariners had begun using another pioneering innovation in navigational technology, namely, the mariners' compass (Xi and Shi 2021, p. 376). With the waning of the might of the Abbasids, during the Southern Song period, 'the Chinese were able to wrest control of the sea-borne commerce and trade routes in the China Seas and the Indian Ocean from the Arabs' (Lo 2012, pp. 126, 202). As Marco Polo, among others, was later to testify (see shortly below), China had by then become undisputed masters of the maritime trade route in both technological (shipbuilding and navigation) and commercial terms.

Upon completely defeating the Southern Song in 1279, the Mongolian Yuan Dynasty initially made maritime trade a state monopoly, but it was never enforced. With the formal lifting of the monopoly in 1294, 'Private merchants, government officials, members of the court, even the empress of one of the khans, sent ships to trade or bought shares in joint commercial ventures', and maritime trade remained as prosperous as previously, 'Many of the merchants became multi-millionaires' (Lo 2012, p. 322). This was the case despite the reopening of the Silk Road after the

[14] The mighty Tang was in the process of rapid decline at that time, with the then reigning emperor Xuanzong (玄宗), once a great leader, totally immersed in a legendary affair with his well-known concubine Yang Guifei (杨贵妃).

founding of the Yuan Dynasty. During the Yuan period, Marco Polo, the famed Arabic mariner Ibn Batuta, among others, visited China by sailing on Chinese ships. All of them stated that Chinese ships were better built, more seaworthy, safer, larger, and more comfortable than the rest plying the maritime trade route. Indeed, so many foreigners preferred to travel on Chinese ships that the Yuan government had to draw up regulations governing the conduct of foreign passengers (see Lo 2012, pp. 126–33 for details of the praise heaped upon Chinese ships and the Quanzhou port by Polo, Ibn Batuta, and others).[15]

Zheng He's seven voyages to the 'western ocean' from 1405 to 1433 are so famous that there is no need to go into any details concerning them here, except to note a few things. Zheng was born a Muslim, captured by Ming soldiers when he was around ten, made a eunuch, and sent to serve in the court of Zhu Di (朱棣), the fourth son of the founder of the Ming Dynasty, and became a trusted aide of the latter. It was Zhu, reigning as Emperor Yongle (永乐), who sent Zheng on his first six voyages, while the seventh voyage was ordered by Emperor Xuande (宣德), the grandson of Zhu. When Ming's founder died, he passed the throne to his grandson (whose father, the Crown Prince, had died), instead of to his other sons, who reigned as Emperor Jianwen (建文). Zhu Di, a great warrior of exceptional abilities, was unhappy and overthrew Jianwen in a civil war that lasted from 1399 to 1402. Jianwen apparently managed to escape when the coup forces stormed his palace, as he was not found among the corpses in the palace compound after the coup. This posed a great legitimacy threat to Zhu's reign, since the surviving Jianwen might serve as a rallying point for a possible counter-coup against Zhu. Rumours were rife that Jianwen had escaped overseas, and this has traditionally been regarded as one key reason why Zhu sent Zheng He on his overseas voyages, in an effort to locate Jianwen.

While the above interpretation should not be lightly dismissed in the context of Chinese dynastic politics, there is another crucial background that we believe is of much greater importance in explaining the motives of Emperor Yongle's sending of Zheng He on his voyages. After the fall of

[15] Incidentally, there was one 'anomaly' in this pattern of the superiority achieved by Chinese shipbuilding and navigation during Song and Yuan times: Chinese mariners did not adopt the astrolabe from the Arabs, though they must certainly be well aware of it. One modern mariner has suggested that, for reasons that we skip for simplicity, this could be because 'The Chinese navigator has never felt the lack of means for determining longitude' (Lo 2012, p. 125).

Tang in 907, as noted, the whole of China fell into disarray, with the north being ruled by invading nomads from 923 to 951. The founder of the Song Dynasty managed to reunify the north in 960, and eventually the whole country in 979. Despite that, several nomadic tribes to the north continued to threaten to invade China, and the Song had to buy peace with them on crippling financial terms; even worse, the Song throne was also forced to accept that Song was a vassal state of the nomads. Despite that, the nomads eventually invaded again, completely driving the Song to the south in 1127. Southern Song was, of course, eventually overthrown by the Mongols who established the Yuan Dynasty. Under the Yuan, society was divided into four categories, namely, in descending order of social status, the Mongols, the *Semu* (色目 all other non-Han-Chinese living in China), Han-Chinese in the north, and Han-Chinese in the south, the last of which, officially called 'southerners' (*nanren* 南人), were also referred to as *manzi* (蛮子 semi-barbarians)—the southerners were discriminated against more severely than Han-Chinese in the north because they were the least willing to submit to the Mongols. The laws of 1337 forbade Han-Chinese from learning foreign languages, denied them employment in government, prohibited them from carrying arms, and refused them the right to retaliate when attacked by Mongols or *Semu*'s (Lo 2012: 99–100). In sum, from the fall of Tang in 907 to the Ming's overthrow of the Yuan in 1368, Han-Chinese had suffered 461 years of utter humiliation at the hands of nomadic invaders. For a civilization that regarded itself to be 'superior' to its neighbours[16] including the various nomadic tribes, this was an especially painful experience. As previously mentioned, on an earlier occasion, the Sui Dynasty's 'vainglorious' foreign policy had to be understood in the context of China's humiliation by invading nomads for 274 years preceding Sui's unification of the country. China's humiliation on this occasion was even greater. By the time of Emperor Jianwen's overthrow and Zhu Di's reign as Emperor Yongle in 1402, the Ming, founded in 1368, had recovered sufficiently in economic and other terms. Yongle's sending of Zheng He on his voyages beginning in 1405 needs to be understood in this context, as an attempt by Yongle to reassert China's status and honour among other states after those humiliating 461 years.

What was the nature of this projection of power overseas by China? How did it compare to the case of the Europeans who came to Asian

[16] The only exception was India, but this was confined to the Buddhists.

waters almost a century after Zheng He's first voyage? To answer this question, a brief note on China's historical territorial expansion will be helpful. The Zhou Dynasty (−1046 to −256) extended, very roughly, from today's Shaanxi province and eastern Gansu province to the west; Shanxi, Hebei, and part of Liaoning to the north and northeast; all the way to the coast to the east; the Yangtze River delta area to the southeast; along the middle course of the Yangtze River to the south and southwest. As an agrarian society with relatively plentiful natural resources, China had no resource-driven incentive to make territorial conquests to its northeast (today's Jilin and Heilongjiang provinces, from where the Manchus hailed, and the Korean peninsula), north (today's Inner Mongolia and Mongolia), west (today's Xinjiang), southwest (today's Tibet),[17] and south (today's Yunnan province and its surrounding areas), all of which areas were, for agriculturalists, resource-poor. On the contrary, as stressed repeatedly, it was the other way round: China was under perennial threat and invasion by nomadic tribes from these areas, except the south.

Empire building occurred mainly under the Marshall Emperor of Han. He made four major conquests or attempts at conquest. Firstly, he conquered the rest of today's Gansu province and Xinjiang province. This was not resource-driven but done completely out of defence considerations against the invading Xiongnu; a symbolic indication of this is the name of a city in southeast Gansu that still exists today, namely, Dingxi (定西), which literally means to stabilize the west. Secondly, he tried to invade the south, but unsuccessfully.[18] This was also done out of defence considerations, for he was looking for an alternative route to central Asia in his search for allies against the Xiongnu. Thirdly, he conquered part of the northern part of the Korean peninsula. The Marshall Emperor originally had no resource-driven imperial design on Korea, this conquest was the outcome a military intervention in the conflict between an indigenous Korean state and the state founded by Chinese èmigrès that we have previously mentioned (see note 5 above). The only resource-driven conquest engaged in by the Marshall Emperor, which was successful, was to conquer the southeast, today's Fujian, Guangdong, and Guanxi provinces,

[17] No Han-Chinese dynasties harboured any interest in conquering the physically formidable Qinghai-Tibetan plateau region, this area was first conquered and incorporated into China by the invading Mongols who founded the Yuan Dynasty.

[18] Generally, no post-Marshall Emperor Han-Chinese dynasties harboured any imperial interest in this inhospitable area, which was eventually conquered and incorporated into China, again, by the invading Mongols.

down to the middle of today's Vietnam—the northern part of Vietnam was populated by Yue tribes, the same tribes populating Guangdong and Guangxi. As already noted, the southeast, along with the Jiangnan area and its surroundings, subsequently became the economic centre of gravity of the country.

After the Marshall Emperor, empire building basically stopped. Conflicts with indigenous Korean states and with Vietnamese states occasionally occurred, as a historical legacy of the circumstances mentioned in note 5 above, instead of being an ongoing imperial enterprise. Incidentally, in addition to what is mentioned in notes 17 and 18, it is worth mentioning that no Han-Chinese dynasty ever harboured any imperial design on Japan; the only Chinese dynasty which attempted to invade Japan was the Mongolian Yuan, to be mentioned later on. In the post-Marshall Emperor era, the only significant territorial conquests made by a Chinese dynasty were, firstly, the conquest of Taiwan by the Manchurian Qing in 1684. This conquest was very unique. The regime established on Taiwan island had been established by Fujianese migrants under the leadership of a famous Fujianese maritime merchant who pledged allegiance to the Ming that the Qing had overthrown. Thus, the Qing's conquest of the Ming Zheng (明郑) regime in Taiwan, as the regime was called ('Ming' refers to the Ming Dynasty, Zheng was the family name of the maritime merchant), was not at all a resource-driven imperial enterprise. The second was the conquest of Tibet and the neighbouring south by the Mongolian Yuan Dynasty (see notes 17 and 18). Both conquests were carried out by invading nomads who had conquered the whole of China. In pointing out the above, our intention is not to advance any essentialist argument that Han-Chinese did not possess, so to speak, 'imperialistic[19] genes'—the conquest of the southeast was an imperial enterprise in the full sense, notwithstanding the population's subsequent cultural assimilation and identification of themselves as Han-Chinese. Rather, it is to point out that except for the Marshall emperor's conquest of the southeast, the absence of resource-driven imperial ambitions on the part of Han-Chinese dynasties was the

[19] As is the case in Chap. 8 and the present work generally, we use the term 'imperial' and its cognates in the sense of empire building, not in the Marxist sense, which applies only to a much later era, namely, *The Age of Empire: 1875–1914* analysed by the Marxist historian Hobsbawm (1989[1987]).

outcome of the geographically determined comparative endowment of resources between China and its neighbours in the pre-industrial age.[20]

The absence of such ambitions, however, did not mean that China was, as cultural-essentialists allege, inwardly oriented, self-enclosing, incurious about the outside world, and so on. As is obvious from our above account and the account in Sect. 9.4 of Chap. 9, whereas the Confucian ethic did nurture the Chinese people to be restrained in manners and disposition on an individual level, on a national level, spearheaded especially by the coastal regions, China developed a strong outward orientation and intense curiosity about the outside world.[21] The fact was simply that what China produced was in great demand overseas, hence, just as the circumstances analysed in the previous paragraph led to the formation of a non-imperialistic orientation towards the outside world in political-military terms, in economic terms, China was able to trade in accordance with strictly commercial principles to acquire what it wanted but did not produce itself from its overseas trading partners. In a word, both sets of circumstances entailed that China developed over the centuries a ***non-predatory political culture in foreign relations, whether political-military or economic.***

Like Greece and Rome, China regarded itself to be culturally 'superior', but this sense of 'superiority' is different from a predatory orientation. It was precisely in order to restore this sense of 'superiority' that, after it had suffered extreme humiliation by nomadic invaders for 461 years, Emperor Yongle sent Zheng He on his voyages. Though these voyages were in order to re-assert China's status and honour, because China's political culture in foreign relations was as just analysed, they were of a non-predatory nature. In other words, the objective was not to subjugate other states in order to plunder them economically, but to make them submit in the sense of recognizing China's 'superior' status, whether or not this submission took the form of a formal suzerain-vassal relationship. As is well-known, the tributes received by China in this relationship were far exceeded in economic value by the gifts that China gave back in return.

[20] We specify in the pre-industrial age because some areas that were resource-poor in that age can, on the contrary, be resource-rich today, given today's level of industrial and technological developments. Incidentally, it can be seen that there is nothing wrong with making use of geographical factors as an explanatory variable, the MvC narrative is wrong because, without any analysis of historical developments in concrete terms with regard to both Europe and China (either out of ignorance or some other factors), it draws simple-minded deterministic conclusions that cannot stand up to even the most preliminary empirical scrutiny.

[21] This curiosity also existed independently of the coastal regions, as exemplified by the Buddhist pilgrims, envoys sent overseas who wrote travelogues, and so on.

Zheng He's voyages were generally peaceful, many states readily submitted, for instance, the kings of states in the East Indies and southern India received investiture from the Ming emperor, and as many as four kings went personally to the Chinese court to pay homage; Korea, the regime on Liuqiu (today's Okinawa), and Japan acknowledged the suzerainty of China; and so on. But on a couple of occasions, force was used: The Singhalese in Sri Lanka attacked the Chinese on their visit, Zheng He violently suppressed them and brought their king back as prisoner; the pro-Chinese king in northern Sumatra was overthrown by the Battaks, Zheng He defeated the latter and replaced their leaders with his own pro-China appointees (Lo 2012, pp. 347–9).

It is from this angle that we can understand the nature of China's use of naval power up to the time of the Ming. For simplicity, let us only consider the period beginning with the Song. One of the nomadic tribes which threatened and eventually invaded the Song was the Jurchen (女真) from the northeast, who established the Jin state, referred to as the Jin Dynasty (金朝) by historians, occupying the entire northeastern part of China between 1115 and 1234. Because the northeastern part of China encircled the Bohai (part of the Yellow Sea), the Jurchen had substantial naval experience, and were often engaged in sea battles with the Korean states. On three occasions, the Southern Song strengthened its naval force.

> Each time it had been primarily their fear of a sea-borne invasion from the north that stimulated and inspired their building and expansion of naval forces and establishment of naval bases to guard the rich provinces of the lower Yangzi [Yangtze River], the national capital at Hangzhou, and the commercial ports and sea lanes of the south. (Lo 2012, p. 195)

In other words, it was purely for defence. The Ming possessed a strong navy, what was it used for?

> The Ming fleet not only patrolled the East and South China Seas to guard against the depredation of Japanese sea-rovers[22] but also protected the sea lanes against pirates in the Malay Archipelago and the waters of Southern India. The destruction of the fleet of Chen Zuyi [a Chinese pirate operating in Indonesia] at Palembang and the capture of the king of Ceylon were punitive actions, according to the *Ming Shi* [official history of the Ming], against the acts of piracy committed by these chieftains. Both Majapahit and Siam were warned against interfering with the passage of merchants and

[22] These were predatory sea gangs that engaged in pillage along the coastal regions of Korea and China. We discuss them shortly.

> envoys through their territorial waters. To safeguard the peace in the Orient, a responsibility he assumed, [Emperor] Yongle sent the Ming navy to intervene in disputes between foreign states. The first instance of this function of the Ming navy was in 1403 when it responded to the appeal of the king of Champa [a Vietnamese state otherwise often engaged in conflict with China], Jaya Sinhavarman, whose capital was besieged by the forces of the Annamese usurper, Ho Han-th'o'ng. At the appearance of nine Chinese warships, the Annamese fleet withdrew without giving battle. (Lo 2012, pp. 349–50)

In contrast to the Song and the Ming, the Mongols used naval power for a different purpose:

> In 1266, Qubilai [Kublai Khan] sent a special envoy to instruct the Korean king [then a Mongolian vassal] to prepare military supplies and to build ships for the invasion of Song and Japan[23]...The ship-building program [for invading Japan] began in the winter months of 1273-4. (Ibid., pp. 257–8)

Altogether, the Mongols sent out five naval expeditions of conquest, in 1274, 1281 (against Japan on both occasions), 1282 (against southern Vietnam), 1287 (against Annam in northern Vietnam), and 1292 (against Java) (Xi and Shi 2021: 381). Lo (2012, pp. 350–1, our emphases) sums up the contrast between the Song and the Yuan as follows:

> the naval ascendency of China actually began in the Southern Song period when the Chinese first organized a national, permanent, sea-going navy[24]...Driven out of North China by the invasion of the Jurchens, the Song court established itself in the southeastern coast of China. There, in a region endowed by nature with the topography and resources for maritime activities, with a people who had...a naval tradition and a nautical spirit, facilitated by the technological advances of the time, and stimulated by the

[23] Unlike the Jurchen, the Mongols possessed no sea-borne experience, hence, they had to rely on the Koreans (on their first expedition against Japan), and later, the Chinese (on their second expedition against Japan as well as other expeditions) to build ships and do the sailing for them. It has been argued by some historians that Chinese shipbuilders deliberately sabotaged Kublai Khan's second expedition against Japan in 1281 by building under-par ships, which were unable to withstand the storm that destroyed the fleet off the coast of Japan; indeed, Chinese (as well as Korean) poets wrote songs to rejoice on learning of the Mongolian debacle (Lo 2012, pp. 285, 293).

[24] During the Northern Song period, naval forces were attached to the provincial armies. In 1132, the Southern Song Dynasty put them under a centralized national command.

> exigencies of war, the Chinese built a strong navy. But it was *a weapon for defense* in their life-and-death struggle against the northern invaders...Remnants of the Song navy were incorporated into the Yuan navy, which became *an instrument of aggression*. Yuan fleets carried out overseas campaigns against Japan, Formosa, Annam, Champa, and Java...efforts made [by the Yuan] in achieving political imperialism by military means were the attempts to build an overseas economic empire.

As seen, the Ming navy was also used for defensive purposes (against the Japanese sea-rovers) and the safeguarding of maritime trade against threats from pirates and hostile foreign states and rulers. Up to and including the fifteenth century, China was the most powerful naval power in the world. In Zheng He's voyages, China's naval might was, in addition to the purposes of defence and safeguarding maritime trade, also used to re-assert China's status and honour as a 'superior' civilization, but even then, China's naval might was never used to pursue predatory objectives. 'In strictly technical and quantitative terms, China was much better placed to exploit the wealth of the Eastern seas than the European powers that in the end did so. In the language of a later era of navalism, China had the ships, had the men, and had the money, too' (Dreyer 2007, p. 9). But China did not do so. The reasons should be clear from China's historically developed external political culture analysed above.

The 'C' component of the MvC narrative would have us believe that having hailed initially from the interior, landlocked middle course of the Yellow River, China had remained a 'continental civilization' ever since, inwardly oriented, self-enclosing, incurious about the outside world, apprehensive of the hazards of the sea, and so on. This is such a far cry from China's actual historical development that it is amazing how it could have been ceaselessly repeated—even by those who should have known better about China.

There was one event that, taken out of context, seems to lend credence to the 'C' component's view, and hence has been seized upon by one Eurocentric scholar after another who are unfamiliar with China's history. This is the banning of private maritime trade (hereinafter referred to as sea ban) for almost two centuries by the Ming. The Eurocentric Harvard scholar Landes (1998, pp. 96, 98) remarks:

> The abandonment of the program of the great voyages [referring to Zheng He's voyages] was part of a larger policy of closure, of retreat from the haz-

ards and temptations of the sea…Isolationism became China. Round, complete, apparently serene, ineffably harmonious, the Celestial Empire purred along for hundreds of years more, impervious and imperturbable.

Never mind Landes' deliberate use of derogatory and offensive language, as is obvious, 'hazards and temptations of the sea' comes straight from the 'voyages of discovery' narrative; more relevant to our present concern is that the caricatural cliches previously repeated by Needham (see Chap. 9) are again repeated by Landes, in a mutually reinforcing vicious cycle feeding on unfamiliarity with China's history: 'Round, complete' reminds us of Needham's assertion of China's supposed 'self-sufficiency', 'serene, ineffably harmonious' of Needham's claim that since −221, China had been a 'unified' implying and connoting stable country—in relation to which and to 'impervious and imperturbable', was Landes aware that the defeated Mongols invaded China again in 1449, taking the Ming emperor captive for four years? That for more than five decades, the Ming state was engaged in a long, bitter struggle against Chinese maritime merchants-cum-pirates rebelling against the ban on private maritime trade that will be discussed shortly? That the Manchus, then organized under the name of Great Jin (*Dajin* 大金, known in history as Later Jin [*Houjin* 后金]),[25] began their invasion of China in 1618? With a little more familiarity with China's history, one would not have spoken of post-Zheng He Ming in Landes' terms at all. We shall have more to say about Landes in due course.

The 'larger policy of closure, of retreat from the hazards and temptations of the sea' spoken of by Landes refers to Ming's sea ban. Why did the Ming ban private maritime trade for such a long time? Firstly, the sea ban was not unique to China or the Ming, it was practised at various times by Korea and Japan as well during the historical period concerned. With regard to China, it pre-dated the Ming for it was the Mongolian Yuan Dynasty which first implemented it on four different occasions, albeit for a total of only eleven years; further, after the Ming, it was also adopted by the Manchurian Qing Dynasty. During the Ming itself, it began right from the beginning of the dynasty, and not, as many unfamiliar with China's history mistakenly believe, only after the ending of Zheng He's voyages.

[25] The Manchus were formed from a reconstitution of Jurchen tribes; these tribes ruled over much of northeast China under the name of the previously mentioned Jin Dynasty (金朝) from 1115 to 1234; this was why the Manchus referred to their state of that time as Great Jin; historians call Great Jin Later Jin because it came after the just-mentioned Jin Dynasty.

That the ban was adopted by the Koreans, the Japanese, the Mongolians and the Manchurians in addition to the Han-Chinese Ming suggests that there must have been some common factor(s) that applied to most, if not all, of these peoples in that particular historical period. In other words, to jump to essentialist conclusions about China on the basis of Ming's sea ban betrays an a priori position rather than a substantiated argument.

The common factor that applied to the above cases, except for that of the Qing (whose case is to be discussed in due course), was that Japan fell into civil war during the period from 1336 to 1392, as a result of which many bankrupted warriors became sea-rovers or *wokou* (倭寇), turning to piracy as well as raiding and pillaging of the entire coastline from Korea to southeast China (Han Qing 2011, p. 87). As an illustration, the problem became so serious that by the 1350s, *wokou* raids on Korea became an annual occurrence, and one raid in 1363 involved as many as 263 ships (Lo 2012, pp. 334–5). The sea ban was a direct reaction to the *wokou* threat. With reference to the Ming, during the entire reign of the dynasty's founding emperor (1368–1398), *wokou* raids on China were recorded almost every year; that was why sea-ban decrees were repeatedly issued during his reign in 1370, 1374, 1381, 1390, 1394, and 1397 (Han Qing, *ibid.*, pp. 87–88). On acceding to the throne in 1402, Emperor Yongle continued with his father's sea-ban policy with a decree in 1404, the year before he sent Zheng He on his first voyage.

Besides the *wokou* raids, there was another crucial reason why the Ming imposed the sea-ban policy right from the beginning. Ming's founder, Zhu Yuanzhang (朱元璋) was not the only one who rose up against the Mongolian Yuan; of the other rebel forces, one was led by Fang Guozhen (方国珍1319–1374), hailing from the coastal province of Fujian, another by Zhang Shicheng (张士诚1321–1367), hailing from the coastal province of Jiangsu; although Fang himself surrendered to Zhu Yuanzhang in 1359 after initially refusing to accept Zhu's leadership, while Zhang died one year before the founding of the Ming, their followers refused to submit to Zhu, and set up bases on off-shore islands as holdouts against Zhu; and in their anti-Zhu activities, they often collaborated with the Japanese *wokou* (Li Rui 2022, p. 129). Thus, Zhu's sea-ban policy was aimed as much against Fang's and Zhang's followers as against the *wokou*.

In 1419, the same year as Zheng He completed his fifth voyage, Emperor Yongle, in alliance with the Koreans, inflicted a heavy blow upon the *wokou*, after which, for the rest of the fifteenth century, the scale of the *wokou* dropped significantly (Han Qing 2011, p. 87). But neither Yongle, who died several years later in 1424, nor his successors lifted or, at least, relaxed the sea ban. Given the reduction of the *wokou* threat, and that the resistance of the followers of the above-mentioned Zhang Shicheng and Fang Guozhen had died down, why was the sea-ban policy not changed? To answer this question, it has to be noted that the sea ban did not mean the cessation of China's maritime trade, it was simply replaced by a state monopoly. As Zheng He's voyages succeeded in establishing a broad network of relations between China and its trading partners, as vassals or otherwise, along the maritime route, maritime trade took the form of tributary trade (*chaogong maoyi* 朝贡贸易).[26] Tributary missions became a cover for trading missions: 'All cargoes in excess of the official tribute were landed with it and labelled "ballast on board tribute ships"…if the foreign ship needed to leave, it had to take on ballast in order to assure safe passage. It therefore bought Chinese goods for ballast on its voyage home' (Curtin 1984, p. 169). The most valuable incoming merchandise was reserved for the state monopoly, only the leftovers were made available to private merchants. 'In many cases foreign merchants, especially those from Central Asia, presented themselves as the bearers of fictitious tribute from imaginary states solely for the purpose of conducting trade' (Rodzinski 1979, p. 197). As it was China's policy to return gifts to tribute missions far exceeding the tributes in economic value, hence the Ming imposed strict limits on the frequency of such tribute missions to prevent them from becoming a financial burden on the state. As Kong Xiaoting (2015) points out from the perspective of neo-institutionalist economics, the Ming state acquired a vested, institutional interest in maintaining the sea ban to protect the state monopoly in maritime trade. This interest extended beyond financial terms, for at the same time, the state monopoly built on the sea ban served to maintain China's position in its diplomatic relations with other states. It is from this angle that one can appreciate why the sea ban was maintained after 1419.

The sea ban forced private maritime merchants to trade illegally and engage in piracy; by the early 1500s, as many as 130 illegal maritime merchant-cum-pirate groups had come into existence (Han Qing 2011,

[26] Technically, there is a distinction between the tributes of vassals, and the exchange of gifts between China and its non-vassal partners; for simplicity, we bypass this detail. Incidentally, whereas the Tang Dynasty also had various vassals who regularly brought tributes, tributary trade, the nature of which is to be explained presently, started only under the Ming and was due to its sea-ban policy (Chen and Sun 2021).

p. 90). By then, the Japanese *wokou* had become their junior partners in raiding the coastal regions. Officially, their activities were lumped under the *wokou* category, but they were fake *wokou*; in terms of numbers, the fake-to-Japanese ratio was around eighty percent to twenty percent, a situation that some Ming officials were aware of (Nie Zuoping 2020, p. 49). To illustrate the scale of the fake *wokou*, the most famous of them was one Wang Zhi (汪直 who changed his family name to 王 so as to protect his family and clan from being implicated by his activities),[27] who used Goto and Hirado, both under the Nagasaki prefecture, in southern Japan as his base. Wang commanded a force of over 100,000 (some 2000 of whom followed him to live in Japan), and was called the King of Hui (徽王), since he hailed Huizhou (徽州) in Anhui province situated next to the Jiangnan area. In 1552, he appealed to the Ming to lift the sea ban, but was refused. In 1557, the Ming general Hu Zongxian (胡宗宪), responsible for dealing with the *wokou* (both fake and real) problem, successfully persuaded Wang to surrender, promising him leniency. But after intense debate, the Ming court failed to honour Hu's promise and executed Wang in 1559. There were other well-known fake *wokou* groups hailing, like Wang Zhi, from Huizhou, such as one operated by several brothers with the family name Xu (许), who collaborated with the Portuguese to trade illegally (Fan Shuzhi 2000, p. 43; Han Qing 2011, p. 90).

The execution of Wang Zhi in 1559 did not, of course, solve the problem. It was clear to many Ming officials, especially those in the coastal provinces, that the problem of the fake *wokou* was the direct result of the sea ban. Kong Xiaoting (2015) observes, from the perspective of neo-institutionalist economics, that by the second quarter of the 1500s, the cost of maintaining the sea ban (the military outlays as well as the lost customs duties on the increasingly thriving illegal trading) had far outweighed the benefits from the state monopoly on maritime trade. At the same time, coastal officials often turned a blind eye to the illegal trade in return for bribes, creating huge corrupting influences (Lin Chunhong 2019, p. 70). Calls for lifting the sea ban within officialdom were increasingly heard. Eventually, in 1567, upon the death of Emperor Shizong (世宗) in January, the ban was lifted in the same year by the new emperor Muzong (穆宗).

Private maritime trade exploded immediately after the lifting of the ban. For instance, from 1570 to 1642 (two years before the fall of the

[27] On Wang, see Nie Zuoping (2020); Fan Shuzhi (2000).

Ming), on average, a total of twenty-five ships sailed to Manila every year (Deng 1997, p. 108). Chinese maritime exports grew so rapidly that by the seventeenth century, the last four decades or so of the Ming, much of the world's production of silver was sucked into China to pay for China's huge export surplus, which well served China's need to increase its currency supply to facilitate increasingly vigorous commerce (Chuan 1969). China's sucking in of world silver is well-known among those who pay even cursory attention to the history of that period, yet, Landes could still remark that since the sea ban 'the Celestial Empire purred along…impervious…'.

To sum up, Ming's sea ban was imposed not out of a flourishing of an allegedly age-old inward orientation, but under very specific circumstances; the ban was maintained despite the waning of the Japanese *wokou* and the ending of the resistance of the followers of Zhang Shicheng and Fang Guozhen because of the benefits (both economic and diplomatic-political) brought to the Ming by the state monopoly on maritime trade; the sea ban forced private Chinese mariners to engage in smuggling, piracy and pillage, and by the early 1500s, they, as fake *wokou*, had grown to alarming scale, greatly threatening Ming's political stability; the lifting of the sea ban in 1567 in order to overcome the fake *wokou* problem can satisfactorily be explained from the perspective of neo-institutionalist economics. Maritime trade resumed with a vengeance afterwards, leading to the sucking into China of the greater part of the world's production of silver. Only those unfamiliar with China's history are able to seize upon Ming's sea ban as an argument to support the 'C' component of the MvC narrative.

To continue with the story of the sea ban a little longer, upon overthrowing the Ming, the Qing restored the ban in 1665. Even though the Qing was an exceptionally conservative regime, this also had nothing to do with any inward orientation at all—the Manchus were, after all, a nomadic tribe with centuries of an outwardly oriented conquering tradition:

> After the establishment of the Qing dynasty, in order to prevent the residents of the southeast coast and the officials of the late Ming dynasty from taking overseas bases as the bridgeheads to attack and endanger the safety of the Dynasty, in…(1665), the Qing court…renewed the ban [on private sea trade]. In [1661], after Zheng Chenggong [郑成功] occupied Taiwan, the Qing court issued another order 'to move far away from sea', forcing the

> coastal residents of Fujian, Guangdong, Jiangsu, and Zhejiang to move into the inland 30 li (500 m), and people who crosses the boundary will be beheaded instantly. (Xi and Shi 2021, pp. 388–9)

Zheng Chenggong mentioned in the quote was a maritime merchant from the coastal Fujian province. During Ming times, under the leadership of his maritime-merchant father Zheng Zilong (郑芝龙), there occurred mass migrations of Fujianese to Taiwan. Upon Ming's overthrow, Zheng Zilong pledged allegiance to the Manchus, but Zheng Chenggong rebelled against his father and continued to pledge allegiance to the Ming. He succeeded in driving away the Dutch who had meanwhile colonized Taiwan and established the previously mentioned Ming Zheng (明郑) regime on the island, vowing to restore the Ming by overthrowing the Qing. This was why the Manchus restored the sea ban as mentioned in the quote. In 1684, Emperor Kangxi (康熙) eventually succeeded in subjugating the Ming Zheng regime and put Taiwan under the administration of Fujian province. As a result, the next year, the sea ban was lifted.

Finally, while Ming's imposition of the sea ban had nothing to do with any alleged age-old inward orientation, it did entail fateful *unforeseen consequences* for China. Though maritime trade did not stop but took the form of a state monopoly, private Chinese maritime merchants no longer sailed long-distance to the Indian Ocean, with illegal smugglers-cum-pirates operating only in the East and South China Seas. The development over the centuries of the long-distance shipbuilding and navigational technologies of an outwardly oriented maritime tradition, which enabled Zheng He to make his voyages, was stifled and came to a stop. As can be imagined, Zheng He's voyages were extremely costly; since they had served their political-diplomatic purposes as analysed above, they were stopped after Zheng's last voyage in 1431–33 (Zheng died in 1433 on this voyage while in Calicut on the way back), as the continuing threat of the Mongols from the north (who were, indeed, to invade again a mere fifteen years later) required concentrating resources on the northern front. But the occurrence of these unforeseen consequences provides no ground to make any cultural-essentialist claims about China in the way done by Needham, Cohen, Landes, and so many others who were ideologically dominated by the 'voyages of discovery' narrative. To end on a note to show how such cultural essentializing in accordance with the MvC narrative can sometimes reach farcical proportions, the Great Wall, like Rome's Hadrian Wall, was built in order to keep invaders *out*, however, up to the present day, cultural essentialists under the complete spell of the MvC narrative, such as the authors of *River Elegy* mentioned at the beginning of this appendix, claim that it was built because China wanted to self-enclose itself *within*.

Bibliography

Chen Kun, and Sun Xiubing. 2021. 'Maritime trade under Ming's tributary system'. 陈昆、孙秀冰, '明代朝贡体系下的海上贸易', 《安阳师范学院学报》, 2021年4期, 页79-100.

Chuan, Han-sheng. 1969. 'The inflow of American silver into China from the late Ming to the mid-Ch'ing period'. *Journal of the Institute of Chinese Studies of the Chinese University of Hong Kong* 2, pp 61-75.

Curtin, Philip D. 1984. *Cross-Cultural Trade in World History*. Cambridge: Cambridge University Press.

Deng, Gang. 1997. *Chinese Maritime Activities and Socioeconomic Development, c. 2100 BC-1900 AD*. London: Greenwood Press.

Dreyer, Edward L. 2007. *Zheng He. China and the Oceans in the Early Ming Dynasty, 1405-1433*. Pearson Longman.

Fan Shuzhi. 2000. 'A review of "Japanese pirates"'. 樊树志，'"倭寇"新论'，《复旦学报（社会科学版）》，2000年1期，页37-46.

Fu Zhufu. 1981. *Economic History of Chinese Feudal Society, Volume 1*. 傅筑夫，《中国封建社会经济史》，卷一（北京：人民出版社）。

Gao Chenyang. 1994. *The Chinese Traditional Mode of Thinking*. 高晨阳，《中国传统思维方式研究》(济南:山东大学出版社，1994).

Han Qing. 2011. 'Probing into the reasons of Ming's sea-ban policy'. 韩庆，'明朝实行海禁政策的原因探究'，《大连海事大学学报(社会科学版)》，2011年10月(卷10，期5)，页87-91.

Heng, Derek. 2017. 'The Tang shipwreck and the nature of China's maritime trade during the late Tang period'. Pp 142-59 in Alan Chong and Stephen A. Murphy (eds.), *The Tang Shipwreck: Art and Exchange in the Ninth Century*. Singapore: Singapore National Heritage Board.

Hobsbawm, Eric. 1989[1987]. *The Age of Empire: 1975-1914*. Vintage reprint edition.

Ji Shuli. 1988. 'Holistic thinking and formal thinking'. 纪树立，〈整体思维与形式思维〉，收入李志林、陈卫平等，《中西哲学比较面面观》(上海:华东师范大学出版社，1988)，页56-99。

Kong Xiaoting. 2015. 'On the changes of Ming's overseas trading system—an analysis from the perspective of neo-institutionalism'. 孔晓婷，'论明朝海外贸易制度变迁—基于新制度经济学的分析'，《经济研究参考》，2015年28期，页54-63。

Landes, David S. 1998. *The Wealth and Poverty of Nations.* London: Little, Brown.
Levy-Bruhl, Lucien. 1985. *How Natives Think,* authorized translation by Lilian A. Clarke. Princeton: Princeton University Press.
Li Jinming. 2015. 'The development and changes in China's ancient maritime silk road'. 李金明，'中国古代海上丝绸之路的发展与变迁'，《新东方》，2015年第一期，页10-15.
Li Rui. 2022. 'Analyzing the *wokou* problem of the mid-Ming'. 李睿，〈明代中叶倭寇问题探析〉，《西部学刊》，期165(2022年6月)，页129-133.
Lin Huadong. 1986. 'The exploration of Chinese sailing'. *Study on the History of Marine Transportation*, No. 2, pp 85-88.
Lin Chunhong. 2019. 'Ming's sea-ban policy and the commercial thinking of Fujian's scholar-officials'. 林春虹，'明朝海禁政策与闽海士大夫的商贸思想'，《福建商学院学报》，2019年6期，页66-71.
Lo, Jung-Pang. 2012. *China as a Seapower, 1127-1368.* National University of Singapore Press.
Nie Zuoping. 2020. 'Wang Zhi: From maritime merchant to *wokou*'. 聂作平，'汪直:从海商到倭寇'，《人物春秋》，2020年5期，页47-50.
Rodzinski, Witold. 1979. *A History of China.* Oxford: Pergamon Press.
Xi, Longfei and Shi Hequn. 2021. 'The shipbuilding and shipping industry in ancient China'. Pp 347-398 in X. Jiang (ed.), *The Origins of Sciences in China, History of Science and Technology in China.* Singapore: Springer Nature Singapore.

Bibliography

Aminzade, Ronald. 1992. 'Historical sociology and time'. *Sociological Methods and Research* 20(4), pp 456-80.

Bajaj, Jatinder K. 1990. 'Francis Bacon, the first philosopher of modern science: a non-Western view'. Pp 24-67 in Ashis Nandy (ed.) *Science, Hegemony and Violence: A Requiem for Modernity* (1990[1988]). Oxford University Press.

Bala, Arun. 2006. *The Dialogue of Civilizations in the Birth of Modern Science.* Palgrave Macmillan.

Bala, Arun. 2024 (in press). 'Explaining the rise of modern science: A dialogical perspective'. Chapter 16 of Arun Bala, Raymond W.K. Lau and Mei Jianjun (eds.), *Multicivilizational Exchanges in the Making of Modern Science. Needham's Dialogical Vision*. Palgrave Macmillan.

Bassin, M. 1992. 'Russia between Europe and Asia: The ideological construction of geographical space'. *Slavic Review* 50(1), pp 1-17.

Bassin, M. 1998. 'Asia'. In N. Rzhevsky (ed.), *The Cambridge Companion to Modern Russian Culture.* Cambridge University Press.

Bauman, Zygmunt. 1991. *Modernity and Ambivalence.* London: Polity.

Berger, Peter, and T. Luckmann. 1967. *The Construction of Social Reality.* New York: Doubleday Anchor.

Bernal, Martin. 1987. *Black Athena: The Afroasiatic Roots of Classical Civilization, Volume 1 The Fabrication of Ancient Greece 1785-1985.* New Brunswick: Rutgers University Press.

Bernal, Martin. 1991. *Black Athena: The Afroasiatic Roots of Classical Civilization Volume 2 The Archaeological and Documentary Evidence.* New Brunswick: Rutgers University Press.

R. W. K. Lau, *Rethinking the Needham Question*,
https://doi.org/10.1007/978-981-97-9472-0

Bernal, Martin. 2001. *Black Athena Writes Back: Martin Bernal Responds to His Critics.* Durham: Duke University Press.

Bernal, Martin. 2006. *Black Athena: The Afroasiatic Roots of Classical Civilization Volume 3 The Linguistic Evidence.* New Brunswick: Rutgers University Press.

Blaut, James. 1993. *The Colonizer's Model of the World.* London: Guilford Press.

Bodde, Derk. 1991. *Chinese Thought, Society, and Science: The Intellectual and Social Background of Science and Technology in Pre-Modern China.* Honolulu: University of Hawaii Press.

Bonnett, Alastair. 2003. 'From White to Western: "racial decline" and the idea of the West in Britain, 1890-1930'. *Journal of Historical Sociology* 16(3), pp 320-348.

Bonnett, Alastair. 2004. *The Idea of the West: Culture, Politics and History.* Palgrave Macmillan.

Bose, D.M., S.N. Sen, and B.V. Subbarayappa. 1971. *A Concise History of Science in India.* New Delhi: Indian National Science Academy.

Brinton, Mary, and Victor. Nee (eds). 1998. *The New Institutionalism in Sociology.* New York: Russell Sage Foundation.

Brook, Timonty. 2001. 'Xu Guangqi in his context'. Chapter 2 of Jami, Catherine, Peter Engelfriet, and Gregory Blue (eds.). *Statecraft and Intellectual Renewal in Late Ming China: The Cross-Cultural Synthesis of Xu Guangqi (1562-1633).* Leiden: Brill.

Brook, Timothy and Gregory Blue. 2002. *China and Historical Capitalism: Genealogies of Sinological Knowledge.* Cambridge University Press.

Brotton, Jerry. 2000. *The Renaissance Bazaar: From the Silk Road to Michelangelo.* Oxford: Oxford University Press.

Browning, R. 2000. 'Education in the Roman Empire'. *The Cambridge Ancient History*, vol. XIV. Cambridge: Cambridge University Press.

Burkett, Walter. 1992. *The Orientalizing Revolution: Near Eastern Influence on Greek Culture in the Early Archaic Age.* Cambridge, Mass.: Cambridge University Press.

Campbell, John, and Philip Sherrard. 1965. 'The Greeks and the West'. Pp. 69-86 in Raghavan Iyer (ed.), *The Glass Curtain Between Asia and Europe.* London: Oxford University Press.

Chalmers, Alan. 2013. *What is This Thing Called Science?*, 4th edition. University of Queensland Press and Open University Press.

Chemla, Karine. 2018. 'Abstraction as a value in the historiography of mathematics in ancient Greece and China: a historical approach to comparative history of mathematics'. Pp. 290-325 in G.E.R. Lloyd and Jenny Jingyi Zhao (eds.), *Ancient Greece and China Compared.* Cambridge: Cambridge University Press.

Chen Cheng-I. 1987. 'A comparative study of early Chinese and Greek work on the concept of limit'. Pp. 3-52 in Idem. (ed.), *Science and Technology in Chinese Civilization.* Singapore: World Scientific.

Chen Cheng-I. 1996. *Early Chinese Work in Natural Science.* Hong Kong: University of Hong Kong Press.

Chen Fangzheng. 2002. 'On the development of Chinese mathematics and its relationships with the rise and fall of dynasties and external influences'. 陈方正，'论中国数学发展与皇朝盛衰和外来影响的关系'，收入氏著，《在美妙的新世纪的门上》（香港：中文大学中国文化研究中心，2002），页479-92。

Chen Fangzheng. 2009. *Inheritance and Rebellion: Why Modern Science Emerged in the West.* 陳方正，《继承与叛逆:现代科学为何出现于西方》(北京:三联，2009)。

Chen Guying. 2012a. *The* Zhuangzi*: Contemporary Annotations and Interpretation.* 陈鼓应，《庄子今注今译》，修订重校本(香港:中华书局，2012)。

Chen Guying. 2012b. *The* Laozi*: Annotations and Interpretation with Introduction and Commentary*, revised and expanded edition. 陳鼓應，《老子注譯及評介》，修訂增補本(香港:中華書局, 2012).

Chen Kun, and Sun Xiubing. 2021. 'Maritime trade under Ming's tributary system'. 陈昆、孙秀冰, '明代朝贡体系下的海上贸易', 《安阳师范学院学报》, 2021年4期, 页79-100.

Chen Meidong. 2008. *Ancient Chinese Astronomical Thoughts.* 陈美东, 《中国古代天文学思想》(北京:中国科学技术出版社, 2008).

Choi, Mihwa. 2012. 'Extreme asceticism: Confucian practice and Riesebrodt's religious virtuoso', *Journal of the Scientific Study of Religion* 51(3), pp 456-67.

Chuan, Han-sheng. 1969. 'The inflow of American silver into China from the late Ming to the mid-Ch'ing period'. *Journal of the Institute of Chinese Studies of the Chinese University of Hong Kong* 2, pp 61-75.

Cohen, H. Floris. 1994. *The Scientific Revolution: A Historiographical Inquiry.* University of Chicago Press.

Cohen, H. Floris. 2011. *How Modern Science Came into the World: Four Civilizations, One 17th-Century Breakthrough.* Amsterdam University Press.

Cohen, H. Floris. 2012. 'From west to east, from east to west? Early science between civilizations', book review of Arun Bala, *The Dialogue of Civilizations in the Birth of Modern Science*, Avner Ben-Zaken, *Cross-Cultural Scientific Exchanges in the Eastern Mediterranean*, Sylvain Gouguenheim, *Aristote au Mont-Saint-Michel. Les racines grrcques de l'Europe Chretienne*, Toby Huff, *Intellectual Curiosity and the Scientific Revolution. A Global Perspective*, and Jonathan Lyons, *The House of Wisdom. How the Arabs Transformed Western Civilization*'. *Early Science and Medicine* 17, pp 339-350.

Cohen, H. Floris. 2015. *The Rise of Modern Science Explained. A Comparative History.* Cambridge: Cambridge University Press.

Cohen, H. Floris. 2017. 'Review of *Imagined Civilizations: China, the West, and Their First Encounter* by Roger Hart'. *East Asian Science, Technology, and Medicine* 2017, pp 225-31.

Cohen, H. Floris. 2024. (in press) 'Dialogue and comparison compared'. Chapter 18 of Arun Bala, Raymond W.K. Lau, and Mei Jianjun (eds.) *Multicivilizational Exchanges in the Making of Modern Science: Needham's Dialogical Vision.* Palgrave Macmillan.

Collins, Randall. 1985. *Max Weber*. London: Sage.

Creel, Herrlee. 1970. *The Origins of Statecraft in China, Volume 1, The Western Zhou Empire.* Chicago: University of Chicago Press.

Crombie, A.C. 1959. 'The significance of medieval discussion of scientific method for the scientific revolution'. Pp 79-101 in Marshall Clagett (ed.) *Critical Problems in the History of Science.* University of Wisconsin Press.

Cullen, Christopher. 2004. 'Book review of G.E.R. Lloyd and Nathan Sivin, *The Way and The Word*'. (See Lloyd and Sivin 2002). *Studies in History and Philosophy of Science* 35, pp 357-62.

Curtin, Philip D. 1984. *Cross-Cultural Trade in World History.* Cambridge: Cambridge University Press.

David, Paul A. 1985. 'Clio and the economics of QWERTY'. *American Economic Review* 75(2), pp 332-7.

de Santillan, Georgio. 1963. 'On forgotten sources in the history of science'. Pp 813-28 in A.C. Crombie (ed.), *Scientific Change: Historical Studies in the Intellectual, Social and Technical Conditions for Scientific Discovery and Technical Invention from Antiquity to the Present. Symposium on the History of Science Held at the University of Oxford, 9-15 July 1961.* London: Heinemann.

Dear, Peter. 1998. *Mersenne and the Learning of the Schools.* Ithaca, NY: Cornell University Press.

Delanty, Gerard. 1995. *Inventing Europe: Idea, Identity, Reality.* Basingstoke: Macmillan.

Deng, Gang. 1997. *Chinese Maritime Activities and Socioeconomic Development, c. 2100 BC-1900 AD.* London: Greenwood Press.

Ding Yuanzhi. 1998. *The Guodian* Laozi*: Explication, Analysis and Research.* 丁原植，《郭店竹簡老子釋析與研究》(臺北:萬卷樓圖書有限公司，1998)。

Dionisotti, Carlo. 1971. *Europe in the Sixteenth-Century Italian Literature.* Oxford: Clarendon.

Dobbs, B.J. Teeter. 2008[1975]. *The Foundations of Newton's Alchemy.* Cambridge University Press.

Dreyer, Edward L. 2007. *Zheng He. China and the Oceans in the Early Ming Dynasty, 1405-1433.* Pearson Longman.

Duara, Pransenjit. 2024 (in press). 'The oceanic paradigm of historical flows'. Chapter 2 of Arun Bala, Raymond W.K. Lau and Mei Jianjun (eds.), *Multicivilizational Exchanges in the Making of Modern Science. Needham's Dialogical Vision.* Palgrave Macmillan.

Duchesne, Ricardo. 2006. 'Asia First?'. *The Journal of the Historical Society* 6(1), pp 69-91.

Elman, Benjamin A. 2000. *A Cultural History of Civil Examinations in Late Imperial China.* Berkeley: University of California Press.

Elvin, Mark. 2001. 'Vale atque ave'. Pp xxiv-xliii in Needham, Joseph, with the collaboration of Kenneth Girdwood Robinson and Ray Huang (Huang Jen-yu), and including contributions by them, Introduction by Mark Elvin, edited by Kenneth G. Robinson. 2001. *Science and Civilisation in China, Volume 7, Part II: General Conclusions and Reflections*. Cambridge University Press.

Etymology of Words. 1987. Combined edition compiled by the Editorial Department of the Commercial Press, China. 商务印书馆编辑部，《辞源(缩印合订本)》(香港:商务印书馆香港分馆，1987)。

Fan Shuzhi. 2000. 'A review of "Japanese pirates"'. 樊树志，'"倭寇"新论'，《复旦大学（社会科学版）》，2000年1期，页37-46.

Fang Meimeng. 2011. 'The phenomenon of "the Manchurization of the Hans" under the Eight Banner System'. 方妹孟，'八旗制度下的"汉人满化"现象'，《历史研究》，2011， 页139-140.

Feng Shi. 2007. *New Theses on Ancient Writing and Ancient History*. 冯时，《古文字与古史新论》(台北:台湾书房，2009).

Feng Youlan. 1993. *History of Chinese Philosophy*, Volume 1, expanded edition. 馮友蘭，《中國哲學史》, 增訂本, 上冊(臺北:商務，1993).

Finley, Moses. 1970. *Early Greece: The Bronze and Archaic Ages*. London: Chatto & Windus.

Fischer, Steven Roger. 1999. *A History of Language*. London: Reaktion.

Fischer, Steven Roger. 2001. *A History of Writing*. London: Reaktion.

Fu Zhufu. 1981. *Economic History of Chinese Feudal Society, Volume 1*. 傅筑夫，《中国封建社会经济史》，卷一（北京：人民出版社）。

Fukuyama, Francis. 1999. *The End of History and the Last Man*. Free Press.

Gao Chenyang. 1994. *The Chinese Traditional Mode of Thinking*. 高晨阳，《中国传统思维方式研究》(济南:山东大学出版社，1994).

Ge Yinghui. 2007. 'Deciphering the name of King Yao, advancing investigation of the sources of civilization'. 葛英会，'破译帝尧名号，推进文明探源'，北京大学震旦古文明研究中心编，《古文明研究通讯》，总32期(2007)，页1-6.

Geary, Patrick. 2002. *The Myth of Nations: The Medieval Origins of Europe*. Princeton: Princeton University Press.

Gernet, Jacques. 1985. *China and the Christian Impact: A Conflict of Cultures*, trans. Janet Lloyd. New York: Cambridge University Press

Giddens, Anthony. 1984. *The Constitution of Society*. Berkeley: University of California Press.

Glover, Ian C. 1989. *Early Trade Relations Between India and South-East Asia*. University of Hull Centre for South-East Asian Studies occasional paper.

GoGwilt, Christopher. 1995. *The Invention of the West: Joseph Conrad and the Double-Mapping of Europe and Empire*. Stanford.

Goldstone, Jack A. 1998. 'Initial conditions, general laws, path dependence, and explanation in historical sociology'. *American Journal of Sociology* 104(3), pp 829-45.

Goldway, David. 1967. 'Appearance and reality in Marx's *Capital*'. *Science and Society*, 31(4), pp 428-47.

Goody, Jack. 2004. *Islam in Europe*. Polity.

Goody, Jack. 2006. *The Theft of History*. Polity.

Gould, Steven Jay. 1989. *Wonderful Life: The Burgess Shale and the Nature of History*. New York: W.W. Norton.

Gould, Steven Jay. 2002. *The Structure of Evolutionary Theory*. Cambridge, MA: Belknap.

Gould, Steven Jay. 2006. *The Richness of Life: The Essential Stephen Jay Gould*, edited by Paul McGarr and Steven Rose, with an Introduction by Steven Rose and a foreword by Oliver Sacks. New York: W.W. Norton.

Graham, Angus C. 1973. 'China, Europe, and the origins of modern science: Needham's The Grand Titration'. Pp 45-69 in Shigeru Nakayama and Nathan Sivin (eds.), *Chinese Science*. Cambridge, Mass: MIT Press.

Graham, Angus C. 1989. *Disputers of the Tao*. Chicago: Open Court.

Grant, Edward. 1996. *The Foundations of Modern Science in the Middle Ages*. New York: Cambridge University Press.

Grant, Edward. 2004. *Science and Religion 400B –AD 1550*. Baltimore: John Hopkins University Press.

Griffen, Larry. 1992. 'Temporality, events, and explanation in historical sociology'. *Sociological Methods and Research* 20(4), pp 403-27.

Guo Shuchun. 1984. 'Investigation into the origins of the thoughts of Liu Hui'. 郭书春，〈刘徽思想探源〉，《中国哲学史研究》，1984年2期，页57-62.

Gutas, Dimitri. *1998. Greek Thought, Arabic Culture. The Graeco-Arabic Translation Movement in Baghdad and Early Abbasid Society (2nd-4th/8th-10th centuries)*. Routledge.

Guthrie, W.K.C. 1962. *A History of Greek Philosophy, Vol. 1, The Early Presocratics and the Phythagoreans*. Cambridge: Cambridge University Press.

Guthrie, W.K.C. 1965. *A History of Greek Philosophy, Vol. 2, The Presocratic Tradition from Parmenides to Democritus*. Cambridge University Press.

Guthrie, W.K.C. 1967[1950]. *The Greek Philosophers: From Thales to Aristotle*. Routledge.

Guthrie, W.K.C. 1969. *A History of Greek Philosophy, Vol. 3, The Fifth Century Enlightenment*. Cambridge University Press.

Guthrie, W.K.C. 1978. *A History of Greek Philosophy, Vol. 5, The Later Plato and the Academy*. Cambridge University Press.

Gutting, Gary. 1989. *Michel Foucault's Archaeology of Scientific Reason*. Cambridge: Cambridge University Press.

Hacking, Ian. 1983. *Representing and Intervening*. New York: Cambridge University Press.

Hall, A. Rupert. 1962. 'General Introduction'. In Marie Boas, *The Scientific Renaissance 1450-1630*. London: Collins.

Hall, Edith. 1989. *Inventing the Barbarians: Greek Self-Definition through Tragedy*. Oxford: Clarendon.

Halsall, Guy. 2007. *Barbarian Migrations and the Roman West, 376-568*. Cambridge: Cambridge University Press.

Han Qing. 2011. 'Probing into the reasons of Ming's sea-ban policy'. 韩庆，'明朝实行海禁政策的原因探究'，《大连海事大学学报(社会科学版)》，2011年10月(卷10，期5)，页87-91.

Han Wei. 2013. 'The textual features and scholarly value of the Western Han bamboo-strip *Laozi*'. 韓巍，〈西漢竹書《老子》的文本特徵和學術價值〉，收入陳鼓應(編)，《道家文化研究》，第二十七輯(北京:三聯，2013).

Harbsmeier, Christoph. 1998. *Science and Civilisation in China, Volume 7 Part I: Language and* Logic. Cambridge: Cambridge University Press.

Harley, J.B. and David Woodward (eds.). 1987. *Cartography in Prehistoric, Ancient, and medieval Europe and the Mediterranean*. University of Chicago Press.

Harley, J.B. and David Woodward (eds.). 1992. *Cartography in the Traditional Islamic and South Asian Societies*. University of Chicago Press.

Harley, J.B. and David Woodward (eds.). 1994. *Cartography in the Traditional East and Southeast Asian Societies*. University of Chicago Press.

Harley, J.B. and David Woodward (eds.). 1998. *Cartography in the Traditional African, American, Artic, Australian and Pacific Societies*. University of Chicago Press.

Hart, Roger. 2011. *The Chinese Roots of Linear Algebra*. Baltimore: John Hopkins University Press.

Hart, Roger. 2013. *Imagined Civilizations: China, the West, and Their First Encounter*. Baltimore: John Hopkins University Press.

Hart, Roger. 2024 (in press). 'Ancient Chinese origins of modern western science; or, the early history of linear algebra'. Chapter 8 of Arun Bala, Raymond W.K. Lau and Mei Jianjun (eds.), *Multicivilizational Exchanges in the Making of Modern Science. Needham's Dialogical Vision*. Palgrave Macmillan.

Hartt, Frederick and David Wilkins. 2010. *A History of Italian Renaissance Art*, 7th edition. Pearson.

Hay, Denys. 1968. *Europe: Emergence of an Idea*, 2nd ed. Edinburgh: Edinburgh University Press.

Hay, Denys. 1980. 'Europe revisited: 1979'. *History of European Ideas* 1(1), pp 1-6.

Hegel, Georg Wilhelm Friedrich. 1975. *Lectures on the Philosophy of World History: Introduction: Reason in History*, trans. by H.B. Nisbet, edited by Duncan Forbes. Cambridge University Press.

Heller, Peggy. 2010. 'The Russian Dawn: How Russia Contributed to the Emergence of "the West" as a Concept.' Pp 33-52 in Christopher Browning and Marko Lehti (eds.), *The Struggle for the West: A Divided and Contested Legacy*. Routledge.

Heng, Derek. 2017. 'The Tang shipwreck and the nature of China's maritime trade during the late Tang period'. Pp 142-59 in Alan Chong and Stephen A. Murphy (eds.), *The Tang Shipwreck: Art and Exchange in the Ninth Century.* Singapore: Singapore National Heritage Board.

Hess, David. 1995. *Science and Technology in a Multicultural World.* Columbia University Press.

Hobsbawm, Eric. 1989[1987]. *The Age of Empire: 1975-1914.* Vintage reprint edition.

Hobson, John M. 2006. *The Eastern Origins of Western Civilisation.* 5th printing with corrections. Cambridge: Cambridge University Press.

Hobson, John. 2012. 'Global dialogical history and the challenge of neo-Eurocentrism'. Pp 13-33 in Arun Bala (ed.), *Asia, Europe and the Emergence of Modern Science: Knowledge Crossing Boundaries.* New York: Palgrave Macmillan.

Hu Jiacong. 1995. *New Investigations into the* Guanzi. 胡家聪，《管子新探》(北京:中国社会科学出版社，1995)。

Huff, Toby. 2003. *The Rise of Early Modern Science: Islam, China and the West*, 2nd ed. New York: Cambridge University Press.

Huff, Toby. 2010. *Intellectual Curiosity and the Scientific Revolution: A Global Perspective.* Cambridge University Press.

Ibn Warraq. 2007. *Defending the West.* New York: Prometheus Books.

Ingram, Paul, and K. Clay. 2000. 'The choice-within-constraints new institutionalism and implications for sociology'. *Annual Review of Sociology* 26, pp 525-46.

Jami, Catherine, Peter Engelfriet, and Gregory Blue (eds.). 2001. *Statecraft and Intellectual Renewal in Late Ming China: The Cross-Cultural Synthesis of Xu Guangqi* (1562-1633). Leiden: Brill.

Ji Shuli. 1988. 'Holistic thinking and formal thinking'. 纪树立，〈整体思维与形式思维〉，收入李志林、陈卫平等，《中西哲学比较面面观》(上海:华东师范大学出版社，1988)，页56-99。

Joseph, George G. 2011. *The Crest of the Peacock Non-European Roots of Mathematics*, 3rd edition. Princeton: Princeton University Press.

Kabbani, Rana. 1986. *Europe's Myth of the Orient.* Indiana University Press.

Kahn, Charles H. 2001. *Pythagoras and the Pythagoreans A Brief History.* Indianapolis: Hackett.

Katz, Victor. 1993. *A History of Mathematics: An Introduction.* New York: HarperCollins.

Katz, Victor. (ed.). 2007. *The Mathematics of Egypt, Mesopotamia, China, India, and Islam: A Sourcebook.* Princeton, NJ and Oxford: Princeton University Press.

Katz, Victor and B. Lumpkin. 1995. 'The relationship of Egyptian to Greek mathematics'. *History and Pedagogy of Mathematics Newsletter* 35, pp 10-3.

Keay, John. 2005. *The Spice Route.* London: John Murray Publishers.

Keay, John. 2011. *India: A History*, revised and expanded edition. Grove Press.

Kim Yung Sik. 2020. 'A philosophy of machines and mechanics in seventeenth-century China: Wang Zheng's characterization and justification of the study of machines and mechanics in the Qiqi tushuo'. *East Asian Science, Technology, and Medicine* 31, pp 64-95.

Kirk, G.S., J.E. Raven, and M. Schofield. 1983. *The Presocratic Philosophers*, 2nd edition. Cambridge: Cambridge University Press.

Kong Xiaoting. 2015. 'On the changes of Ming's overseas trading system—an analysis from the perspective of neo-institutionalism'. 孔晓婷，'论明朝海外贸易制度变迁—基于新制度经济学的分析'，《经济研究参考》，2015年28期，页54-63。

Kuhn, Thomas. 1957. *The Copernican Revolution*. Harvard University Press.

Kuhn, Thomas. 1996[1962]. *The Structure of Scientific Revolutions*, 3rd edition. Chicago: The University of Chicago Press.

Kvasz, Ladislav. 2015. 'Mathematical language as a bridge between conceptualization of motion and experimental practice'. Pp. 229-48 in Raffaele Pisano (ed.) *A Bridge Between Conceptual Frameworks. Sciences, Society and Technology Studies.* Dordrecht: Springer.

Landes, David S. 1998. *The Wealth and Poverty of Nations.* London: Little, Brown.

Lao Siguang. 1981a. *New Edition History of Chinese Philosophy*, Volume 1. 勞思光, 《新編中國哲學史(一)》(台北: 三民書局, 1981a).

Lao Siguang. 1981b. *New Edition History of Chinese Philosophy*, Volume 2. 勞思光, 《新編中國哲學史(二)》(臺北: 三民書局, 1981b).

Lao Siguang. 1983. *New Edition History of Chinese Philosophy*, Volume 3(1). 勞思光, 《新編中國哲學史(三【上】)》(臺北: 三民書局, 1983).

Lau, Raymond W.K. 2013. 'On the achievements of Xunzi's theory of *Xing* from the social science perspective'. 劉榮錦，〈從社會科學角度看荀子性惡論的成就〉，《新亞學報》第三十一卷[上] 2013年6月，頁65-109.

Lau, Raymond W.K. 2015. 'Examining Laozi's concept of Dao from the perspective of modern scientific realism'. 劉榮錦〈以現代科學實在論為參照看老子之道〉，收入《新亞學報》第三十二卷，2015年7月，頁231-96.

Lau, Raymond W.K. 2020. *Intellectual Developments in Greece and China: Contingency, Institutionalization and Path-Dependency.* Newcastle upon Tyne: Cambridge Scholars Publishing.

Lau, Raymond W.K. 2021. 'Re-visiting the role of craft in Zhuangzi's philosophy'. *Asian Philosophy*, 31(4), pp 368-384.

Lefkowitz, Mary and Guy Rogers (eds.). 1996. *Black Athena Revisited*, 2nd edition. The University of North Carolina Press.

Levy-Bruhl, Lucien. 1985. *How Natives Think,* authorized translation by Lilian A. Clarke. Princeton: Princeton University Press.

Leyser, Karl. 1992. 'Concepts of Europe in the Early and High Middle Ages'. *Past and Present* no. 137 (Nov), pp 25-47.

Li Jinming. 2015. 'The development and changes in China's ancient maritime silk road'. 李金明，'中国古代海上丝绸之路的发展与变迁'，《新东方》，2015年第一期，页10-15.

Li Rongxi. 1996. *The Great Tang Dynasty Record of the Western Regions*. Berkeley: Numata Center for Buddhist Translation and Research.

Li Rui. 2022. 'Analyzing the *wokou* problem of the mid-Ming'. 李睿，〈明代中叶倭寇问题探析〉，《西部学刊》，期165(2022年6 月)，页129-133.

Li Xiaolin. 1981. 'Searching for the origins of our country's double-entry bookkeeping'. 李孝林，〈我国复式记账法溯源〉，中国商务部《财会参考资料》, 1981(26).

Li Xiaolin. 1988. 'Exploring our country's indigenous double-entry bookkeeping'. 李孝林，〈我国固有复式簿记的探讨〉，《北京商学院学报》1988年2期.

Li Yan and Du Shiran. 1987. *Chinese Mathematics A Concise History*, translated by John N. Crossley and Anthony W.-C. Lun. Oxford: Clarendon.

Lin Huadong. 1986. 'The exploration of Chinese sailing'. *Study on the History of Marine Transportation*, No. 2, pp 85-88.

Lin Chunhong. 2019. 'Ming's sea-ban policy and the commercial thinking of Fujian's scholar-officials'. 林春虹，'明朝海禁政策与闽海士大夫的商贸思想'，《福建商学院学报》，2019年6期，页66-71.

Lin, Z. Jun. 1992. 'Chinese double-entry bookkeeping before the nineteenth century'. *The Accounting Historians Journal* 19(2), pp 103-22.

Lindberg, David. 1978. 'The transmission of Greek and Arabic learning to the West'. Pp 52-90 in Idem. (ed.) *Science in the Middle Ages*. Chicago: Chicago University Press.

Lindberg, David. 2007. *The Beginnings of Western Science*, 2nd edition. Chicago: University of Chicago Press.

Lloyd, G.E.R. 1973. *Greek Science After Aristotle*. London: W.W. Norton.

Lloyd, G.E.R. 1990. *Demystifying Mentalities*. Cambridge: Cambridge University Press.

Lloyd, G.E.R. 1996. *Adversaries and Authorities: Investigations into Ancient Greek and Chinese Science*. Cambridge: Cambridge University Press.

Lloyd, G.E.R. 2015. *Analogical Investigations: Historical and Cross-Cultural Perspectives on Human Reasoning*. Cambridge: Cambridge University Press.

Lloyd, G.E.R. 2018. 'Introduction: methods, problems and prospects'. Pp 1-29 in G.E.R. Lloyd and Jenny Jingyi Zhao (eds.) *Ancient Greece and China Compared*. Cambridge University Press.

Lloyd, G.E.R. 2022. *Expanding Horizons in the History of Science. The Comparative Approach*. Cambridge University Press.

Lloyd, G.E.R. and Nathan Sivin. 2002. *The Way and the Word: Science and Medicine in Early China and Greece*. New Haven: Yale University Press.

Lo, Jung-Pang. 2012. *China as a Seapower, 1127-1368*. National University of Singapore Press.

Lumpkin, B. 2002. 'The mathematical legacy of ancient Egypt: A response to Robert Palter'. In Martin Bernal (ed.) *Debating Black Athena*. Durham, NC: Duke University Press.

Machamer, P. and D.M. Miller. 2021. 'Galileo Galilei'. *Stanford Encyclopedia of Philosophy*, available online, accessed 1 Jun 2022.

Mahoney, James. 2000. 'Path dependence in historical sociology'. *Theory and Society* 29(4), pp 507-48.

Malik, Kenan. 2008. *Strange Fruit: Why Both Sides Are Wrong in the Race Debate*. Oxford: Oneworld.

March, James and H. Simon. 1985. *Organizations*. New York: John Wiley.

Martzloff, Jean-Cluade. 1997. *A History of Chinese Mathematics*. Translated from the French original of 1987 by Stephen S. Wilson. Berlin: Springer.

Merton, Robert. 1936. 'The unanticipated consequences of purposive social action'. *American Journal of Sociology* 1(6), pp 894-904.

Nee, Victor. 1998. 'Sources of the new institutionalism'. Pp. 1-16 in Mary Brinton and Victor Nee (eds.), *The New Institutionalism in Sociology*. New York: Russell Sage Foundation.

Needham, Joseph. 1954. *Science and Civilization in China, Vol. 1: Introductory Orientations*. Cambridge: Cambridge University Press.

Needham, Joseph, with the collaboration of Wang Ling. 1959. *Science and Civilization in China, Vol. 3: Mathematics and the Sciences of the Heavens and the Earth*. Cambridge: Cambridge University Press.

Needham, Joseph, and Kenneth Robinson. 1962. *Science and Civilization in China, Volume 4. Physics and Physical Technology, Part 1. Physics*. Cambridge: Cambridge University Press.

Needham, Joseph. 1969a. *The Grant Titration*. Taylor and Francis.

Needham, Joseph. 1969b. *Within the Four Seas: The Dialogue of East and West*. London: George Allen and Unwin.

Needham, Joseph. 1981. *Science in Traditional China*. Hong Kong: The Chinese University Press.

Needham, Joseph, with the collaboration of Kenneth Girdwood Robinson and Ray Huang (Huang Jen-yu), and including contributions by them, Introduction by Mark Elvin, edited by Kenneth G. Robinson. 2001. *Science and Civilisation in China, Volume 7, Part II: General Conclusions and Reflections*. Cambridge University Press.

Neumann, I. 1996. *Russia and the Idea of Europe: A Study in Identity and International Relations*. Routledge.

Newman, William and Lawrence Principe. 2002. *Alchemy Tried in the Fire: Starkey, Boyle, and the Fate of Helmontian Chymistry*. Chicago: University of Chicago Press.

Newman, William. 2004. *Promethean Ambitions*. Chicago and London: University of Chicago Press.

Newman, William. 2006. *Atoms and Alchemy. Chymistry and the Experimental Origins of the Scientific Revolution*. Chicago and London: University of Chicago Press.

Nie Zuoping. 2020. 'Wang Zhi: From maritime merchant to *wokou*'. 聂作平，'汪直:从海商到倭寇'，《人物春秋》，2020年5期，页47-50.

Noble, Thomas (ed.). 2006. *From Roman Provinces to Medieval Kingdoms*. London: Routledge.

O'Connor, Michael. 1996. 'Epigraphic Semitic scripts'. Pp. 88-107 in P. Daniels and W. Bright (eds.), *The World's Writing Systems*. Oxford: Oxford University Press.

Palter, Robert. 1993. 'Black Athena, Afro-centrism, and the history of science'. *History of Science* 31, pp 227-87.

Parsons, Talcott. 1951. *The Social System*. New York: Free Press.

Pinyin *Chinese-English Dictionary*. 1979. Compiled by the English Department of the Beijing Foreign Languages College. 北京 外国语学院英语系《汉英词典》编写组(编), 《汉英词典》(香港:商务印书馆, 1979)。

Poskett, James. 2023[2022]. *Horizons: A Global History of Science*. UK: Penguin Books.

Pomeranz, Kenneth. 2000. *The Great Divergence: China, Europe and the Making of the Modern World Economy*. Princeton: Princeton University Press.

Pomeranz, Kenneth, and Steven Topik. 2006. *The World that Trade Created*, 2nd edition. Armond: M.E. Sharpe.

Poo Mu-chou. 1998. *In Search of Personal Welfare A View of Ancient Chinese Religion*. Albany: State University of New York Press.

Powell, Walter, and P. DiMaggio. 1991. 'Introduction'. Pp. 1-38 in Idem. (eds.). *The New Institutionalism in Organizational Analysis*. Chicago: The University of Chicago Press.

Raphals, L. 1998. 'Book review of G.E.R. Lloyd, *Adversaries and Authorities*' (see Lloyd 1996). *Journal of Asian Studies* 57(4), pp 1132-3.

Reding, Jean-Paul. 2004. *Comparative Essays in Early Greek and Chinese Rational Thinking*. Aldershot: Ashgate.

Rees, Graham. 1975. 'Francis Bacon's semi-Paracelsian cosmology'. *Ambix* 22, pp 81-101.

Renn, J. and M. Valleriani. 2001. 'Galileo and the challenge of the Arsenal', lecture delivered in Florence on 21 March 2001, available online, accessed 23 May 2022.

Rodzinski, Witold. 1979. *A History of China*. Oxford: Pergamon Press.

Rogers, Henry. 2005. *Writing systems A Linguistic Approach*. London: Blackwell.

Ronan, Colin. 1983. *The Cambridge Illustrated History of the World's Science*. Cambridge: Cambridge University Press.

Ruelle, David. 1991. *Chance and Chaos*. Princeton: Princeton University Press.

Saliba, George. 1994. *A History of Arabic Astronomy: Planetary Theories during the Golden Age of Islam*. New York and London: New York University Press.

Saliba, George. 2007. *Islamic Science and the Making of the European Renaissance*. Cambridge, Mass: MIT Press.

Schwarz, E.H.L. 1936. 'The Chinese connection with Africa'. *Journal of the Royal Asiatic Society Bengal Branch*, No. 4 of 1936.

Scott, W. Richard. 2001. *Institutions and Organizations*, 2nd edition. Thousand Oaks: Sage.

Selznick, Philip. 1949. *TVA and the Grass Roots*. Berkeley: University of California Press.

Sen, Tansen. 2006. 'The travel records of Chinese pilgrims Faxian, Xuanzang, and Yijing'. *Education about Asia* 11(3), pp 24-33.

Sewell, Jr., W.H. 1996. 'Three temporalities: Toward an eventful sociology'. Pp. 262-4 in Terence J. McDonald (ed.), *The Historic Turn in the Human Sciences*. Ann Arbor: University of Michigan Press.

Seyyed Hossein Nasr. 2003. *Science and Civilization in Islam*, 2nd edition. Islamic Texts Society.

Shuowen jiezi (Explaining Graphs and Analyzing Characters)) (n.d.) Compiled in the early 100s by Xu Shen (许慎). Accessible online. 许慎(东汉), 《说文解字》。

Sivin, Nathan. 1995. *Medicine, Philosophy and Religion in Ancient China Researches and Reflections*. Variorum.

Sivin, Nathan. 2018. 'Why some comparisons make more difference than others'. Pp. 33-9 in G.E.R. Lloyd and Jenny Jingyi Zhao (eds.), *Ancient Greece and China Compared*. Cambridge University Press.

Southern, R.W. 1962. *Western Views of Islam in the Middle Ages*. Cambridge, MA: Harvard University Press.

Stinchcombe, Arthur. 1987. *Constructing Social Theories*. Chicago: University of Chicago Press.

Stock, Brian. 1978. 'Science, technology, and economic progress in the early Middle Ages'. Pp. 1-51 in David Lindberg (ed.), *Science in the Middle Ages*. Chicago: Chicago University Press.

Sun Hong'an. 2008. *Ancient Chinese Mathematical Thoughts*. 孙宏安, 《中国古代数学思想》(大连:大连理工大学出版社, 2008).

Sun Kaitai. 1995. 'The relationship between Zou Yan and the Daoists'. 孫開泰，〈鄒衍與道家的關係〉，收入陳鼓應(編)，《道家文化研究》，第八輯(上海:上海古籍出版社，1995)，頁213-225。

Sun Xue. 2016. 'A brief analysis of the changes in the relationship between the Manchus and the Hans with reference to the Qing Dynasty's policies'. 孙雪，'从清代政策简析满汉关系变化'，《边疆经济与文化》，2017年2期，页49-51.

Swiggers, Pierre. 1996. 'Transmission of the Phoenician script to the West'. Pp. 261-70 in P. Daniels and W. Bright (eds.), *The World's Writing Systems*. Oxford: Oxford University Press.

Tan Jiajian, and Sun Zhongyuan. 2009. *The* Mozi*: Contemporary Annotations and Interpretation.* 谭家健、孙中原(注译)，《墨子今注今译》(北京:商务，2009).

Todd, Malcolm. 1992. *The Early Germans*. Oxford: Blackwell.

Ullmann, Walter. 1969. *The Carolingian Renaissance and the Idea of Kingship*. London: Methuen.

Varouxakis, Georgios. 2019. 'The godfather of "occidentality": Auguste Comte and the idea of "the West"'. *Modern Intellectual History* 16(2), pp 411-441.

Varouxakis, Georgios. 2020. 'When did Britain join the Occident? On the origins of the idea of "the West" in English'. *History of European Ideas* 46(5), pp 563-581.

Vogel, Klaus and Alisha Rankin. 2016. 'Cosmography'. Pp. 469-96 in *The Cambridge History of Science, vol 3 Early Modern Science*. Cambridge: Cambridge University Press.

Wang Bi (Wei Dynasty, annotator) and Lou Yulie (proof and explication). 2008. *The Canon of Dao and De of Laozi: Annotations, Proof and Explications.* 〔魏〕王弼注·楼宇烈校释，《老子道德经注校释》(北京:中华书局，2008).

Wang Chong. 1974. *Lunheng* (meaning assessing the relative merits of different doctrines and views). 王充，《论衡》(上海:上海人民出版社，1974).

Wang Shanbo. 1996. *In Search of the Scientific Spirit: Comparing Chinese and Western Science from a Philosophical Perspective*. 王善博，《追求科学精神:中西科学比较与融通的哲学透视》(南宁:广西人民出版社，1996).

Weber, Max. 1992. 'Author's introduction'. Pp xxviii-xlii in Max Weber *The Protestant Ethic and the Spirit of Capitalism,* trans. by T. Parsons, Introduction by A. Giddens. Routledge.

Webster, Charles. 1980. *From Paracelsus to Newton, Magic and the Making of Modern Science*. Cambridge University Press.

West, Martin. 1997. *The East Face of Helicon: West Asiatic Elements in Greek Poetry and Myth*. Oxford: Clarendon.

Whimster, Sam and S. Lash. 1987. 'Introduction'. Pp. 1-34 in Idem. (eds.), *Max Weber: Rationality and Modernity*. London: Allen & Unwin.

Williams, Raymond. 1976. 'Civilization'. Pp. 48-50 in Idem., *Keywords: A Vocabulary of Culture and Society*. London: Fontana.

Winchester, Simon. 2009. *The Man Who Loved China*. New York: Harper Perennial.

Wolf, Eric R. 1997[1982]. *Europe and the People Without History*. Berkeley: University of California Press.

Woolf, Stuart. 1989. 'French civilization and ethnicity in the Napoleonic Empire'. *Past and Present* 124(1), pp 96-120.

Woolhouse, R.S. 1988. *The Empiricists.* Oxford and London: Oxford University Press.Xi Longfei. 2015. *History of Shipbuilding in Ancient China.* 席龙飞，《中國古代造船史》(武汉:武汉大学出版社, 2015).

Xi, Longfei and Shi Hequn. 2021. 'The shipbuilding and shipping industry in ancient China'. Pp 347-398 in X. Jiang (ed.), *The Origins of Sciences in China, History of Science and Technology in China.* Singapore: Springer Nature Singapore.

Xu Fuguan. 1975. *History of Chinese Doctrines of Human Nature: The Pre-Qin Period.* 徐复观,《中國人性論史先秦篇》(台北:商務, 1975).

Xu Hong. 2021. *The Earliest China: The Rise of the* Erlitou *Civilization.* 许宏，《最早的中国:二里头文明的崛起》，(北京:三联书店，2021).

Yates, Frances. 1967. 'The Hermetic tradition in Renaissance science'. Pp. 255-74 in Charles S. Singleton (ed.), *Art, Science, and History in the Renaissance.* Baltimore: John Hopkins University Press.

Yates, Frances. 1972. *The Rosicrucian Enlightenment.* London: Routledge and Kegan Paul.

Yates, Frances. 1979[1957]. *Giordano Bruno and the Hermetic Tradition.* University of Chicago Press.

Zuo Qiuming, Hu Zhihui and Chen Kejiong. 1996. *Zuo's Commentary.* 左丘明、胡志辉及陈克炯,《左传》(长沙:湖南人民出版社, 1996).

Index[1]

[1] Note: Page numbers followed by 'n' refer to notes.

R. W. K. Lau, *Rethinking the Needham Question*,
https://doi.org/10.1007/978-981-97-9472-0

D

T

SPRINGER NATURE

GPSR Compliance

The European Union's (EU) General Product Safety Regulation (GPSR) is a set of rules that requires consumer products to be safe and our obligations to ensure this.

If you have any concerns about our products, you can contact us on ProductSafety@springernature.com

In case Publisher is established outside the EU, the EU authorized representative is:

Springer Nature Customer Service Center GmbH
Europaplatz 3
69115 Heidelberg, Germany

The manufacturer's authorised representative in the EU is Springer Nature Customer Service Centre GmbH, Europaplatz 3, 69115 Heidelberg, Germany. If you have any concerns regarding our products, please contact ProductSafety@springernature.com

Printed and bound by CPI Group (UK) Ltd, Croydon, CR0 4YY

01/10/2026

02225912-0015